Regular and Chaotic Motions in Dynamic Systems

NATO ASI Series

Advanced Science Institutes Series

A series presenting the results of activities sponsored by the NATO Science Committee, which aims at the dissemination of advanced scientific and technological knowledge, with a view to strengthening links between scientific communities.

The series is published by an international board of publishers in conjunction with the NATO Scientific Affairs Division

A	**Life Sciences**	Plenum Publishing Corporation
B	**Physics**	New York and London
C	**Mathematical and Physical Sciences**	D. Reidel Publishing Company Dordrecht, Boston, and Lancaster
D	**Behavioral and Social Sciences**	Martinus Nijhoff Publishers
E	**Engineering and Materials Sciences**	The Hague, Boston, and Lancaster
F	**Computer and Systems Sciences**	Springer-Verlag
G	**Ecological Sciences**	Berlin, Heidelberg, New York, and Tokyo

Recent Volumes in this Series

Volume 114—Energy Transfer Processes in Condensed Matter
edited by Baldassare Di Bartolo

Volume 115—Progress in Gauge Field Theory
edited by G.'t Hooft, A. Jaffe, H. Lehmann, P. K. Mitter, I. M. Singer, and R. Stora

Volume 116—Nonequilibrium Cooperative Phenomena in Physics and Related Fields
edited by Manuel G. Velarde

Volume 117—Moment Formation in Solids
edited by W. J. L. Buyers

Volume 118—Regular and Chaotic Motions in Dynamic Systems
edited by G. Velo and A. S. Wightman

Volume 119—Analytical Laser Spectroscopy
edited by S. Martellucci and A. N. Chester

Volume 120—Chaotic Behavior in Quantum Systems: Theory and Applications
edited by Giulio Casati

Series B: Physics

Regular and Chaotic Motions in Dynamic Systems

Edited by

G. Velo
Institute of Physics
University of Bologna
Bologna, Italy

and

A. S. Wightman
Princeton University
Princeton, New Jersey

Plenum Press
New York and London
Published in cooperation with NATO Scientific Affairs Division

Proceedings of the Fifth International School
of Mathematical Physics and NATO Advanced Study Institute on
Regular and Chaotic Motions in Dynamic Systems,
held July 2-14, 1983,
at the Ettore Majorana Center for Scientific Culture,
Erice, Sicily, Italy

Library of Congress Cataloging in Publication Data

International School of Mathematical Physics (5th: 1983: Ettore Majorana International Centre for Scientific Culture)
Regular and chaotic motions in dynamic systems.

(NATO ASI series. Series B, Physics; v 118)
"Proceedinqs of the Fifth International School of Mathematical Physics and NATO Advanced Study Institute on Regular and Chaotic Motions in Dynamic Systems held July 2-14, 1983, at the Ettore Majorana Center for Scientific Culture, Erice, Sicily, Italy".
"Published in cooperation with NATO Scientific Affairs Division."
Bibliography: p.
Includes index.
1. Dynamics—Congresses. 2. Chaotic behavior in systems—Congresses. 3. Mathematical physics—Congresses. I. Velo, G. II. Wightman, A. S. III. NATO Advanced Study Institute on Regular and Chaotic Motions in Dynamic Systems (1983: Ettore Majorana International Centre for Scientific Culture). IV. Title. V. Series.
QC133.I58 1983 531′.11 84-26369
ISBN-13: 978-1-4684-1223-9 e-ISBN-13: 978-1-4684-1221-5
DOI: 10.1007/978-1-4684-1221-5

Softcover reprint of the hardcover 1st edition 1985

A Division of Plenum Publishing Corporation
233 Spring Street, New York, N.Y. 10013

PREFACE

The fifth International School of Mathematical Physics was held at the Ettore Majorana Centro della Culture Scientifica, Erice, Sicily, 2 to 14 July 1983. The present volume collects lecture notes on the session which was devoted to Regular and Chaotic Motions in Dynamical Systems.

The School was a NATO Advanced Study Institute sponsored by the Italian Ministry of Public Education, the Italian Ministry of Scientific and Technological Research and the Regional Sicilian Government.

Many of the fundamental problems of this subject go back to Poincaré and have been recognized in recent years as being of basic importance in a variety of physical contexts: stability of orbits in accelerators, and in plasma and galactic dynamics, occurrence of chaotic motions in the excitations of solids, etc. This period of intense interest on the part of physicists followed nearly a half a century of neglect in which research in the subject was almost entirely carried out by mathematicians. It is an indication of the difficulty of some of the problems involved that even after a century we do not have anything like a satisfactory solution.

The lectures at the school offered a survey of the present state of the theory of dynamical systems with emphasis on the fundamental mathematical problems involved. We hope that the present volume of proceedings will be useful to a wide circle of readers who may wish to study the fundamentals and go on to research in the subject. With this in mind we have included a selected bibliography of books and reviews which the participants found helpful as well as a brief bibliography for four seminars which were held in addition to the main lecture series.

There were sixty-one participants from sixteen countries.

G. Velo and
A.S. Wightman
Directors of the School

CONTENTS

Introduction to the Problems 1
A.S. Wightman

Applications of Scaling Ideas to Dynamics
L.P. Kadanoff
Lecture I. Roads to Chaos: Complex Behavior from Simple Systems . . 27
II. From Periodic Motion to Unbounded Chaos: Investigations of the Simple Pendulum . . 45
III. The Mechanics of the Renormalization Group 60
IV. Escape Rates and Strange Repellors 63

Introduction to Hyperbolic Sets 73
O.E. Lanford III

Topics in Conservative Dynamics 103
S. Newhouse

Classical Mechanics and Renormalization Group 185
G. Gallavotti

Measures Invariant Under Mappings of the Unit Interval. . . 233
P. Collet and J.-P. Eckmann

Integrable Dynamical Systems 267
E. Trubowitz

Appendix (Seminars)
Iteration of Polynomials of Degree 2, Iterations of Polynomial-like Mappings 293
A. Douady

Boundary of the Stability Domain around the Origin for Chirikov's Standard Mapping 295
G. Dôme

Incommensurate Structures in Solid State Physics and Their Connection with Twist Mappings 296
S. Aubry

Julia Sets - Orthogonal Polynomials Physical Interpretations and Applications. 300
D. Bessis

Scaling Laws in Turbulence 303
J.-D. Fournier

Index 309

REGULAR AND CHAOTIC MOTIONS IN DYNAMICAL SYSTEMS

INTRODUCTION TO THE PROBLEMS

A.S. Wightman

Departments of Mathematics and Physics
Princeton University
Princeton, N.J. 08544 USA

The purpose of this introduction is twofold; first, to sketch the origin of some of the problems that will be discussed in detail later, and, second, to introduce some of the concepts which will be used. In a subject like analytical mechanics, with such a long history and such hard problems, a little sense of history is both enlightening and consoling.

A dynamical system is loosely specified as a system with a state at time t given by a point x(t) lying in a phase space, M, and a law of evolution given by an ordinary differential equation (= ODE)

$$\frac{dx}{dt}(t) = v(x(t)) \tag{1}$$

Here v is a vector field on the phase space M. M is customarily assumed to be a differentiable manifold such as an open set in n-dimensional Euclidean space. Alternatively, one can consider the dynamical system specified by its set of possible histories, the set of mappings, $t \mapsto x(t)$, of some time interval $a < t < b$ into the phase space satisfying the ODE (1). When $-\infty < t < \infty$, the solutions are said to define a flow; when $0 \leq t < \infty$ a semi-flow.

A discrete dynamical system is one in which the time takes integer values. Then the dynamics is given by the iterates of a mapping of the phase space M into itself. If t runs over all the integers, $\mathbb{Z}$, one sometimes speaks of a cascade; if over the positive integers, $\mathbb{Z}_+$, of a semi-cascade. Although the extension of these definitions to infinite dimensional M is of obvious physical interest (fluid dynamics!), in what follows, for lack of time, attention will be mainly confined to the finite dimensional case.

Poincaré's Bequest

The analysis of dynamical systems (= analytical mechanics = classical mechanics = rational mechanics) is one of the oldest parts of physics, but, in a sense, the modern period begins with Poincaré. It is notorious that the physicists of most of the twentieth century had little appreciation of Poincaré's work. Nevertheless, it is his outlook which dominates the field today. To appreciate this, it helps to have been brought up, as I was, on a really old-fashioned version of the subject, say that in E.T. Whittaker's *A Treatise on the Analytical Dynamics of Particles and Rigid Bodies*. That is a remarkable book, which has some coverage of Poincare's technical results but scarcely a word about his general point of view. Nearly a hundred years later, we find our thinking completely dominated by Poincare's geometric attitude, whether we prefer it in the super-Smalean version of R. Abraham and J. Marsden's *Foundations of Mechanics* or the proletarian version of V. Arnold's *Classical Mechanics*.

What then did Poincaré do to exert all this influence? Here is a little list - far from complete.

1) Qualitative Dynamics
   Generic behavior of flows as a whole, the classification of phase portraits.

2) Ergodic Theory
   Probabilistic notions, recurrence theorem.

3) Existence of Periodic Orbits; Detailed Analysis of the Structure of a Flow Near a Periodic Orbit.

4) Bifurcation Theory
   General ideas for systematic theory; detailed study of rotating fluid with gravitational attraction.

First, I will comment briefly on 2). It sounds somewhat anachronistic to call Poincaré a pioneer of ergodic theory but there is a sense in which it is true. In that sense, the first theorem of ergodic theory was the invariance of the Liouville measure while the second was Poincaré's Recurrence Theorem. By the invariance of the Liouville measure. I refer to the fact that

$$dq_1 \ldots dq_n \quad dp_1 \ldots dp_n$$

defines a measure on 2n-dimensional phase space invariant under the flow defined by a Hamiltonian system of differential equations

$$\frac{dq_i}{dt} = \frac{\partial H}{\partial p_i}, \qquad \frac{dp_i}{dt} = -\frac{\partial H}{\partial q_i}, \quad i = 1,\ldots n$$

In modern language, the recurrence theorem can be stated as follows

Theorem

Let T be a mapping of a phase space M into itself which preserves a measure μ on M:

$$\mu(X) = \mu(T^{-1}X) \quad \text{for any measurable subset X of M}$$

Suppose μ is finite i.e.

$$\mu(M) < \infty$$

Then, if A is any measurable subset of M, almost every point x of A returns to A infinitely often i.e. for an infinite set of positive integers, n, $T^n x \varepsilon A$.

Poincaré emphasized that his proof required only the finiteness and invariance of his measure, although the argument used the language of the theory of incompressible fluids. He had already gone far in the direction of generality in these matters by introducing the general notion of integral invariants. These are invariant integrals of differential forms over subsets of M.

Incidentally, for those who may wish to read the original, I should note that Poincaré did not call this result a recurrence theorem; he referred to it as stabilité à la Poisson. You can find it, along with a magistral exposition of his theory of integral invariants in his Prize Memoir which won (21 January 1889) the Prize offered by King Oscar II of Sweden. It is published in Acta Math 13 (1890) 1-270.

It is interesting to compare this stunningly general result with what was going on in physics at that time. Maxwell and Boltzmann had constructed statistical models of gases leading to quantitative predictions of thermodynamic phenomena, and Boltzmann had published a proof of the so-called H-Theorem giving a mechanical interpretation of the increase of entropy in accord with the Second Law of Thermodynamics. Boltzmann's proof was greeted with skepticism because of the Recurrence Theorem and the invariance under time inversion of the usual Hamiltonian models. Both Maxwell and Boltzmann made independent efforts to justify statistical procedures on the basis of what Boltzmann called the Ergodic Hypothesis:

the trajectory of a Hamiltonian system in phase space passes through every point of its surface of constant energy. Poincaré thought it very unlikely that a single trajectory could fill a whole surface of constant energy.[1] (A theorem to this effect was proved much later by A. Rosenthal and M. Plancherel.[2]) He immediately replaced it by the more plausible assumption that every orbit is dense, a property later called the Quasi-Ergodic Hypothesis by the Ehrenfests, in their well-known article in the Mathematical Encyclopedia.[3] Even in the 1890's Poincaré knew too much about the behavior of orbits in concrete dynamical systems to believe in the general validity of the Ouasi-Ergodic Hypothesis. He pointed out that in the restricted problem of three bodies (interacting with gravitational attractions) there are orbits not dense on the surfaces of constant values of the integrals of motion. His general attitude was summarized[1]

> "Il est possible et même vraisemblable que le postulat de Maxwell est vrai pour certains systèmes et faux pour d'autres, sans qu'on ait aucun moyen certain de discerner les uns des autres."

As will be discussed in the following, we now have some means of distinguishing ergodic from non-ergodic systems and the first part of the sentence has turned out to be exactly right.

After the Ehrenfests most theoretical physicists stayed away from the problem. The only exception I know was Enrico Fermi.[4] In 1923, he extended a theorem of Poincaré to show that for Hamiltonian systems of n degrees of freedom with $n > 2$ satisfying certain conditions of genericity, there could exist no smooth hypersurface of dimension $2n-1$ invariant under the flow except for the surfaces of constant energy. He then applied his result to prove that such systems would have to be ergodic because, if there existed an open subset of the phase space invariant under the flow, its boundary would be a hypersurface to which the preceding theorem would apply. This argument assumes that the only subsets which have to be considered are those with smooth boundaries, an assumption which is now known to fail in general as a result of the KAM theorem. (See Giovanni Gallavotti's lectures.) Fermi's argument would have meant that generically there could be no nontrivial invariant decomposition of the energy surface. (We now know there are large interesting classes of Hamiltonian systems for which the orbits are dense on the energy surface and large interesting classes for which they are not. In the latter case, the boundary of an invariant subset is rough, violating Fermi's assumption.)

Now I turn to 3), the analysis of a flow near a periodic orbit. It was here that Poincaré uncovered many of the problems that have evolved into the main subjects of the lectures that follow.

When Poincaré began his work, the result which we now know as the rectification theorem was standard knowledge In a somewhat modernized form, it is as follows.

Theorem

If v is a continuously differentiable vector field defined in a neighborhood of a point x_0, where $v(x_0) \neq 0$, then in some sub-neighborhood of x_0, there exists a continuously differentiable change of coordinates $x \to y = f(x)$ such that the differential equation

$$\frac{dx}{dt} = v(x)$$

is reduced to the form

$$\frac{dy_1}{dt} = 1, \quad \frac{dy_2}{dt} = \frac{dy_3}{dt} = \dots \frac{dy_n}{dt} = 0$$

In pictures

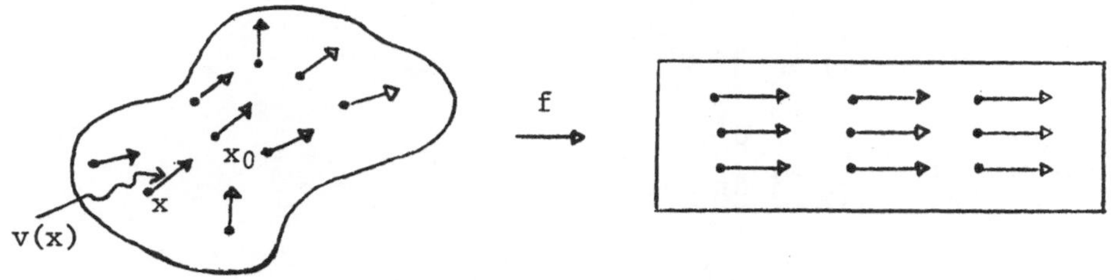

Figure 1. The vector field v in a neighborhood of x is rectified by the mapping f.

In the coordinates y, y_1 increases proportionally to t, and $y_2, \dots y_n$ are integrals of motion. The Rectification Theorem asserts that a smooth flow is very simple in a neighborhood of a non-singular point i.e. a point where the vector field is non-vanishing. If a flow is this simple locally, it is natural to ask why one cannot use the n-1 local integrals of motion to obtain a simple global labeling of orbits. The answer is complicated in general. Sometimes, one can and then one has an integrable system. Sometimes, one can do it over a large region but runs into difficulties when trying to extend to the whole phase space. The extension may be impossible because the vector field v has a singular point at some x_0 i.e. $v(x_0) = 0$. Alternatively, it may be impossible because, as one approaches some subset , the neighborhoods that the rectification theorem provides get smaller and smaller. For any finite time the local rectifications could be patched together but in the limit of infinite time, it would be impossible.

However, the labeling of orbits by $y_1 \dots y_n$ may fail for a simpler reason based on the phenomenon of recurrence, which we know to be very general from Poincaré's Recurrence Theorem. If the orbit of a particle goes away but returns to a neighborhood of its starting point, it may not be possible to patch together the coordinate systems so that they give the orbit the same values of $y_2 \dots y_n$ as when it started out. This phenomenon does not necessarily involve any lack of smoothness or singularity of the vector field; it is a matter of global geometry of phase space. (You will hear this eloquently expounded by Trubowitz in his lectures on integrable systems.)

In the context of the Newtonian N-body problem (N bodies of masses $m_1 \dots m_N$ interacting by gravitational attraction) the question of the existence of global integrals of motion in addition to the ten well-known integrals

Energy $$E = \sum_{j=1}^{N} \frac{1}{2} m_j \left(\frac{d\vec{x}_j}{dt}\right)^2 + \sum_{j<k} \frac{Gm_j m_k}{|\vec{x}_m - \vec{x}_k|}$$

Momentum $$\vec{P} = \sum_{j=1}^{N} m_j \frac{d\vec{x}_j}{dt}$$

Angular Momentum $$\vec{J} = \sum_{j=1}^{N} [\vec{x}_j \times m_j \frac{d\vec{x}_j}{dt}]$$

Center of Mass $$M\vec{X} - \vec{P}t \quad \text{where} \quad \vec{X} = \frac{1}{M} \sum_{j=1}^{N} m_j \vec{x}_j, \quad M = \sum_{j=1}^{N} m_j$$

was a famous nineteenth century problem. It was a result of Bruns that no additional independent integrals exist depending algebraically on the coordinates momenta and time. Poincaré generalized this result; he showed that algebraic could be replaed by holomorphic - this was another result of his Prize Memoir of 1889, the result which Fermi extended in the work referred to above.

Clearly, to go farther one has to understand the nature of a flow near a singular point. (This is also the simplest case of a periodic orbit: all positive real numbers are periods.)

Poincaré's first results in this direction were contained in his thesis (1879) although the main emphasis there is on other matters.[5] The thesis, in fact, is mainly about first order partial differential equations (= PDEs), and Poincaré was principally concerned with completing the Cauchy-Kowalewski theory for those equations, so that it covered certain cases which previous authors had not treated. However, in the course of this investigation, he discussed the ODEs for a bicharacteristic strip in a neighborhood of an equilibrium point.

Recall that if the PDE is

$$F(x_1 \ldots x_n, u, p_1 \ldots p_n) = 0$$

where u is the unknown function of $x_1 \ldots x_n$ and $p_1 = \partial u/\partial x_1 \ldots p_n = \partial u/\partial x_n$, then associated with it is a system of ODEs

$$\frac{dx_j}{dt} = \frac{\partial F}{\partial p_j}(x_1 \ldots x_n, u, p_1 \ldots p_n) \qquad j = 1, \ldots n$$

$$\frac{dp_j}{dt} = -\left[\frac{\partial F}{\partial x_j}(x_1 \ldots x_n, u, p_1 \ldots p_n) + p_j \frac{\partial F}{\partial u}\right] \qquad j = 1, \ldots n$$

$$\frac{du}{dt} = \sum_{j=1}^{n} p_j \frac{\partial F}{\partial p_j} \qquad (2)$$

in $\mathbb{R}^{2n+1}$, the ODEs for a bicharacteristic strip.[6] When the right-hand side of (2) vanishes at the initial point, it is called an equilibrium point. More generally, for the ODE (1), one calls x_0 a <u>singular point</u> of the vector field v if $v(x_0) = 0$. For the general purposes of the present discussion, the special features of the system (2) for the bicharacteristics are not significant (that was also the case in Poincaré's thesis) so I will continue the discussion in terms of the general equation.

For convenience, assume that the singular point x_0 is at the origin of coordinates. There is then a special case in which the qualitative behavior of the solutions is determined by the eigenvalues of A. For a general differentiable vector field with a singular point at 0, the Taylor expansion of v provides a linear transformation with $A = (\nabla v)/_0$ and nonlinear correction terms. Poincaré posed the problem: Under what conditions will a change of coordinates $x \mapsto \phi(x)$ leaving the origin fixed, $\phi(0) = 0$, reduce a vector field to its linear part in a neighborhood of zero:

$$v(\phi(x)) = \phi(Ax)$$

Poincaré assumed v analytic, solved the problem when ϕ is regarded as a formal power series, and then gave sufficient conditions that the formal power series converge. The key to the first step is the idea of resonance:

Definition Let $\lambda_1 \ldots \lambda_n$ be the eigenvalues of the matrix A, and $x^m = \sum_{j=1}^{n} x_j^{m_j}$ be a monomial occurring in the Taylor expansion of v. Then x^m is resonant if for some s, $1 \leq s \leq n$

$$\lambda_s = \sum_{j=1}^{n} \lambda_j m_j \quad \text{with} \quad \sum_{j=1}^{n} m_j \geq 2$$

In the recursive procedure which Poincaré found for the determination of the coefficients of the power series for ϕ the quantities

$$\lambda_s - \sum_{j=1}^{n} \lambda_j m_j$$

occur in denominators. For a resonant term the procedure fails. Even if there are no resonant terms there may be trouble at the next stage if the denominators get small - that is the famous problem of small divisors - and Poincaré's sufficient condition for the convergence of the power series for ϕ yields uniform boundedness away from zero of the denominators. Later on, Dulac modified Poincare's procedure so as to give a solution of the normal form problem: find a formal power series which transforms v(x) to the linear term Ax plus resonant terms.[7] All this leaves open what happens to the flow in the presence of resonances or where there are no resonances but Poincaré's sufficient condition for convergence is not satisfied. Poincaré's thesis uncovered the hard nut of problems which are still with us. For further details of the Poincaré-Dulac theorems, see Chapter V of Arnold's book Chapitres Supplémentaires... .[8]

The previous analysis treats orbits in a neighborhood of a singular point by finding a coordinate system in which the vector field on the right-hand side of the ODE is reduced to normal form, which will be linear in favorable cases. When the procedure goes through, the linear approximation describes the exact behavior in the new coordinate system. This analysis is not directly applicable to periodic orbits of strictly positive period, $T > 0$, but Poincaré developed a different approach which makes an analogous analysis possible. Given a periodic orbit, one picks a point on it and chooses an n-1 dimensional manifold passing through x_0 and not tangent to the periodic orbit i.e. it is transversal to the orbit. It is sometimes called a Poincaré section. See Figure 2. Then through a point x on the section near x there passes an orbit which will come back and hit the section near x, but not necessarily at x because not every orbit near x_0 need be periodic.

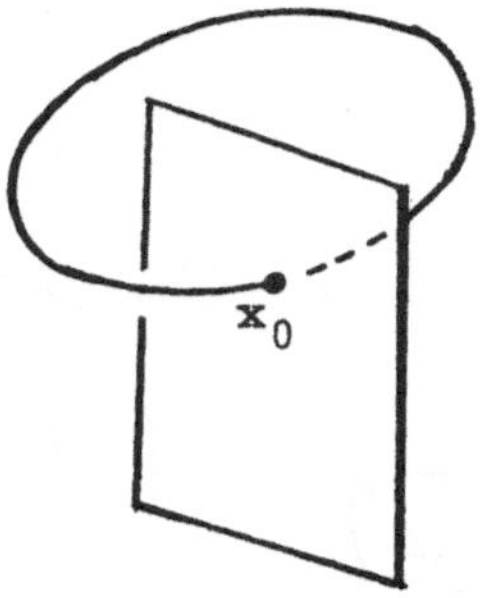

Figure 2 The Poincaré Section is transversal to the periodic orbit at x_0. The Poincaré mapping ϕ carries the section into itself.

Thus the flow defines a mapping, ϕ, of a neighborhood of x_0 on the section into itself, the "once around" mapping, usually called the Poincaré mapping associated with the periodic orbit and x_0 and the Poincaré section. The point x_0 is a fixed point of the Poincaré mapping

$$\phi(x_0) = x_0$$

An expansion of $\phi(x)$ in Taylor series about x_0 has no constant term and the linear term $[(\nabla\phi)(x_0)](x-x_0)$ can be regarded as an analogue of the linear approximation Ax to a vector field near an equilibrium point. One can study the asymptotic behavior of orbits near the periodic orbit by studying the iterates $\phi^n x$ under the Poincaré mapping. Notice that the first derivative of ϕ^n evaluated at x_0 is just the nth power of the matrix $(\nabla\phi)(x_0)$ so the asymptotics in linear approximation can be read off from this matrix. Afterwards, one will put the nonlinear terms back in and see what qualitative features survive.

Up to this point, everything that has been said applies to a general ODE. Now, I turn to the more special results which hold if the ODE is Hamiltonian. Here we have the remarkable fact that the Poincaré mapping ϕ is always symplectic i.e. on the surface of section it preserves the symplectic structure which that surface of section inherits from the general symplectic structure of the phase space. This has the consequence that the eigenvalues of $(\nabla\phi)(x_0)$ always appear in quadruples $\lambda, \bar{\lambda}, \lambda^{-1}, \bar{\lambda}^{-1}$. If λ is on the unit circle, the traditional terminology calls it elliptic while if $|\lambda| \neq 1$ it is called hyperbolic. The associated patterns of flow are strikingly different, as one sees from the two dimensional example illustrated in Figure 3. This two-dimensional example becomes relevant in a Hamiltonian system of two degrees of freedom where the energy surface, $E(q_1 q_2 p_1 p_2) = \text{const}$, is three-dimensional and a Poincaré section is two-dimensional. The invariant curves

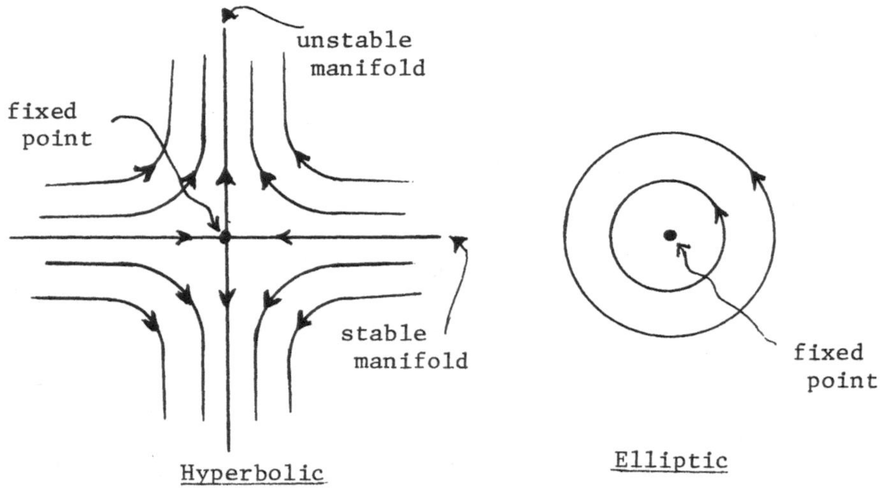

Figure 3 Qualitative behavior of points in the plane of the Poincaré section under the linear approximation to the Poincaré mapping. Successive iterates lie on the indicated curves.

of the elliptic case correspond to the invariant two-dimensional tori nesting around the periodic orbit in the energy surface. In the hyperbolic case there is an unstable manifold along which the successive iterates of the linear approximation to the Poincaré mapping applied to a point run away from the fixed point and the stable manifold where they run into the fixed point.

These striking pictures beg a question: to what extent is the linear approximation a good guide to the behavior of the full Poincaré mapping in a Hamiltonian system? The answer for the hyperbolic case is that the picture persists: under the full non-linear time development, points run away along an unstable and approach along a stable manifold. Poincaré foresaw that associated with this behavior in the hyperbolic case and the analogous elliptic case there would be problems of great difficulty which could lead to geometric complexity both in the large and in the small. In his great treatise Les Méthòdes Nouvelles de la Mécanique Celeste (Vol I 1892, Vol II 1893, Vol III 1899) he studied these questions in the context of the Newtonian 3-body problem. Un the famous last Chapter XXXIII of the third and last volume, after a seemingly interminable sequence of changes of variables used to select an appropriate Poincaré section, he examines the possibility that the stable and unstable manifolds of hyperbolic fixed points may cross. He calls such a point homoclinic (from the Greek "inclined to the same") if the fixed points belong to the same periodic orbit and heteroclinic "inclined to the different") if they come from

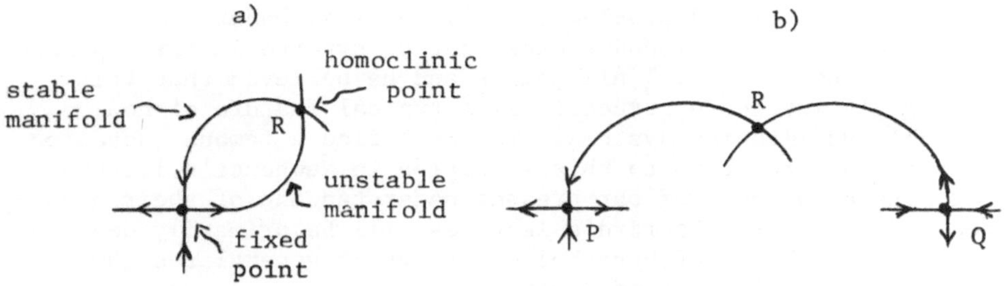

Figure 4 a) The stable and unstable manifolds of a fixed point intersect in a homoclinic point. b) If P and Q are points of a periodic orbit, R is by definition homoclinic; if P and Q. do not belong to the same periodic orbit, R is heteroclinic.

different periodic orbits. (If ϕ is the Poincaré mapping and $\phi^n(x) = x$ but $x, \phi(x), \phi^2(x), \ldots \phi^{n-1}(x)$ are all distinct, they constitute a <u>periodic orbit of period</u> n. Each of the points $\phi^k(x)$ is then a fixed point of ϕ^n.) Now, as Poincaré explained, a homoclinic point x has successors $\phi(x), \phi^2(x), \ldots$ lying on the unstable manifold and antecedents $\phi^{-1}(x), \phi^{-2}(x), \ldots$ lying on the stable manifold, but because the Poincaré mapping is single-valued the first sequence must also lie in the stable manifold and the second in the unstable. Furthermore, the stable and unstable manifolds cannot intersect themselves. Thus the existence of a single homoclinic point implies the existence of an infinity of others. Some faint idea of the thrashing about which the stable and unstable manifolds manifest is indicated in Figure 5.

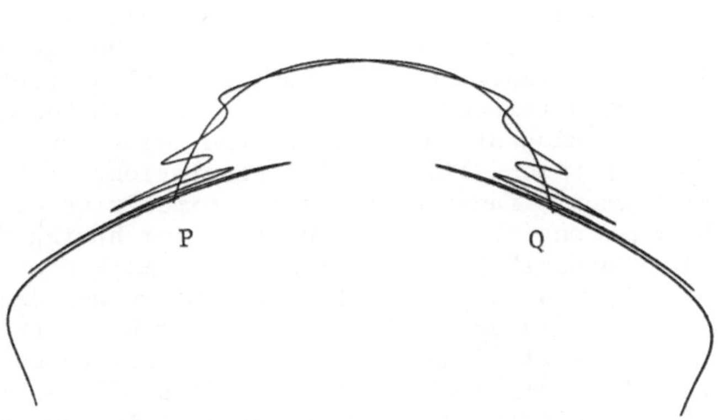

Figure 5 P and Q are hyperbolic points on a periodic orbit of the celebrated Hénon map $\{x,y\} \to \{y+1-ax^2, bx\}$.[9] The unstable manifold of P wiggles wildly as it approaches Q, and the same for the stable manifold of Q as it approaches P.

In the restricted problem of three bodies, under restrictions Poincaré found a dense set of hyperbolic fixed points and a dense set of homoclinic points and he believed that this astoundingly intricate structure is a typical feature of non-integrable Hamiltonian systems. You will find a famous quotation concerning his reaction to this situation in Newhouse's lectures, along with an account of our present understanding of their occurrence and behavior. Lanford's lectures will be primarily devoted to the modern theory of hyperbolic sets which generalizes the theory of hyperbolic periodic orbits.

There are complexities in the small associated with the effect of the non-linear terms on the invariant curves of an elliptic fixed point. It turned out that some invariant curves survive and some do not and, in between the ones that do survive, there is an infinity of other points describing periodic orbits of arbitrarily long period. Poincaré did not obtain any result on the survival of the curves - that had to wait half a century for Kolmogorov - but, in the last months of his life (1912), he published a conjecture usually called Poincaré's Geometric Theorem which would have been a step toward the proof that there are an infinite number of periodic orbits in any neighborhood of an elliptic fixed point. Since in general some of these are again elliptic, we have a structure of worlds within worlds forever. This structure constitutes a third main theme of the modern theory of dynamical systems and will be treated in Gallavotti's lectures.

Finally, here are a few remarks about 4), bifurcation theory. This theory can be regarded, in the context of dynamical systems, as a response to the question: how does the phase portrait of a flow change when the vector field which defined its ODE varies smoothly? Poincaré considered this question in general, but also in the special context of a homogeneous rotating fluid moving under its own gravitational attraction. This special problem has a famous history beginning with Newton who predicted the flattening of the earth at the poles, over the protests of the experimentalists of his generation. (The matter was settled in Newton's favor by the polar expedition of Maupertuis who was congratulated by Voltaire for having "aplati les pôles et les Cassini".) One of the most remarkable discoveries in mathematical physics in the nineteenth century was Jacobi's of 1834: the existence of symmetry breaking solutions of this problem in which the spinning fluid mass is an ellipsoid with unequal principal axes. This family of solutions splits off (= bifurcates) from the axially symmetric family discovered by Newton and studied by McLaurin, when the ellipticity e is 0.81267. Poincaré found an infinite number of other bifurcations into pear-shaped objects beginning with the Jacobian family. Unfortunately, the physical significance of these solutions is not clear since in 1922 they were proved unstable by Henri Cartan. For modern accounts see the reviews of Chandrasekhar [10] and Tassoul [11]. Despite this brilliant physical discovery, it is Poincaré's geometric outlook on bifurcation theory which has had the most influence on twentieth century developments, as a perusal of Chapter VI Bifurcation Theory in V.I. Arnold's Chapitres Supplémentaires... should convince you.

Birkhoff's Contribution

Birkhoff made himself famous by proving Poincaré's Geometric Theorem, but he was very well prepared. He had studied Poincaré's writings on dynamical systems in great depth. (Marston Morse remarked that Poincaré was Birkhoff's true teacher.) It is not an exaggeration to say that in the two decades after Poincaré's death, Birkhoff made the deepest contribution to the program of study of dynamical systems which Poincaré had laid out. Here is a taste of what Birkhoff did.

First, there was the Geometric Theorem [12]:

Theorem

Let ϕ be an area preserving mapping of an annulus $0 < a \leq r \leq b$ in the $\{x,y\}$ plane with $r = \sqrt{x^2+y^2}$ and suppose that ϕ increases angles on the circle $r = b$ and decreases angles on $r = a$. Then ϕ has at least two fixed points in the annulus.

This is the grandfather of a whole tribe of twist theorems, some of them also proved by Birkhoff which assert the existence or non-existence of invariant curves in the annulus.[13]

A second contribution of Birkhoff was a solution of the normal form problem for Hamiltonian systems of differential equations. This is not a corollary of the solution of the normal form problem for general ODE, because by hypothesis the conjugacy is here restricted to be a canonical transformation given by a formal power series in the canonical variables. The result yields a normal form for the Hamiltonian which is a sum of independent one-dimensional harmonic oscillator Hamiltonians. If the formal series converges, the normal form implies that the solutions lie on invariant n-dimensional tori. However, it happens that the convergence or divergence can depend on the amplitude of the oscillations. This is an alternative way of looking at the question mentioned above in connection with elliptic fixed points of Poincaré mappings: some of the invariant circles, displayed in Figure 3, will be destroyed and some persist, deformed, when the effect of non-linear terms is taken into account.

Birkhoff devoted much effort to the question of survival of these invariant tori especially in the case of Hamiltonian systems of two degrees of freedom. (These were also a favorite of Poincaré and much of the work of his great treatise Les Méthodes Nouvelles de la Méchanique Celeste is carried out for this special case.) He was able to show the existence of zones of instability corresponding to the disappearance of layers of invariant tori.[14] Such a zone of instability is bounded by curves invariant under the twist mapping, but there is no invariant curve encircling the origin within the zone. In fact, Birkhoff showed that there are points arbitrarily close to the outer boundary of the zone, which are mapped under repeated twists arbitrarily close to the inner boundary, and similarly there are points arbitrarily close to the inner boundary mapped arbitrarily close to the outer boundary.

As a by-product of the construction of zones of instability, he found the first example of what would nowadays be called a strange attractor; he called it a "remarkable curve".[15] To this end, he took a twist mapping acting on a zone of instability between two invariant curves and modified it so that it squeezed the zone into a sub-region of smaller area. The iterates of the modified mapping then squeeze the zone to a limiting set of zero area and great complexity. Orbits of points lying in the zone converge to this set and as a consequence exhibit chaotic behavior. In recent years, computer studies have produced evidence that such chaotic behavior is a generic feature of dynamical systems and the occurrence of strange attractors has been proposed as an explanation of such important physical phenomena as turbulence.[16]

Birkhoff (and P.A. Smith) introduced the notion of metric transitivity of a flow with invariant measure to describe the situation in which there is no partition of a space $M = M_1 \cup M_2$ into sets M_1 and M_2 invariant under the flow and neither of measure zero.[17] This definition was historically important because it was later shown to be a necessary and sufficient condition for the equality time average = phase average in statistical mechanics. I will discuss this in more detail in the following section. (It is now customary to reverse the emphasis and say that the measure μ is ergodic with respect to a flow i.e. the flow is metrically transitive with measure μ, if μ is ergodic relative to the flow.)

Finally, it should be remarked that Birkhoff obtained a large number of results on the detailed properties of orbits in non-integrable flows whose ultimate significance remains to be determined. Suffice it to say that the conventional wisdom is that if one is interested in the really hard problems of dynamical systems it is worth the time to study Birkhoff's papers in detail. This point is especially worth making since, if Poincaré has been little appreciated among physicists, Birkhoff's work on dynamical systems has been more or less unknown.

Ergodic Theory 1931-1970

A new stage in ergodic theory was reached when Koopman (1931) introduced the idea of studying a measure-preserving flow by examining the induced flow on the Hilbert space of square integrable functions defined on the phase space. More explicitly, the definitions are as follows. Suppose the flow on the phase space M is given by $x \to T^t x = x(t)$ where

$$T^0 = 1 \qquad T^s T^t = T^{s+t} \qquad -\infty < s,t < \infty$$

Introduce a scalar product into the linear space of complex-valued functions on M

$$(f,g) = \int \overline{f(x)}\, g(x)\, d\mu(x)$$

and define the Hilbert space, $L^2(M,d\mu)$, of all square integrable functions f i.e. those whose norms are finite: $(f,f) < \infty$.

If μ is invariant under the flow $t \to T^t$ and one defines with Koopman

$$(U^t f)(x) = f(T^{-t}x)$$

for $f \in L^2(M,d\mu)$, then U^t is a unitary operator continuous in t and satisfies

$$U^0 = 1 \qquad U^s U^t = U^{s+t}$$

Thus, $t \to U^t$ is a continuous unitary one parameter group. Stone's theorem had recently been proven as well as the spectral theorem for self-adjoint operators, so Koopman knew that the U^t could always be written

$$U^t = \exp itL$$

where L is a self-adjoint operator in $L^2(M,d\mu)$, the so called Liouville operator in the present case. It was natural to attempt to express properties of the flow in terms of spectral properties of L. As an example of an elegant result of this character, I quote

Theorem

A flow in a phase space with finite invariant measure is ergodic if and only if 0 is a non-degenerate eigenvalue of L.

Shortly after Koopman's paper, von Neumann proved the Mean Ergodic theorem and Birkhoff the pointwise ergodic theorem.

Theorem (Mean Ergodic Theorem)

$$\lim_{\tau\to\infty} \frac{1}{\tau} \int_0^\tau (U^t f)(x)dt$$

exists in the sense of convergence in $L^2(M,d\mu)$ for every $f \in L^2(M,d\mu)$. If $\mu(M) < \infty$, the limit is $P_0 f$ where P_0 is the projection onto the subspace of functions invariant under the flow. If μ is ergodic, this subspace consists of constant functions and the limit is the phase average

$$[\int d\mu]^{-1} \int f(x)\ d\mu(x)$$

Theorem (Pointwise Ergodic Theorem)

If $f \in L^1(M,d\mu)$

$$\lim_{\tau\to\infty} \frac{1}{\tau} \int_0^\tau (U^t f)(x)dt = \bar{f}(x)$$

exists for almost all X, and is invariant under the flow.

These results roused much interest among mathematicians because for the first time the ideas of statistical mechanics were given a precise mathematical sense in the context of general functional analysis. The physists were in general less than enthusiastic. They were willing to take the existence of the limit for granted and were unable to see how the assumption of ergodicity (= metric transitivity) was any more transparent physically than what was already to be found in the Ehrenfests' review. There was some justice in this attitude in the then existing state of knowledge, but, regrettably, it led to a long-standing prejudice against ergodic theory as physically irrelevant. The efforts of mathematicians to understand what local behavior of a flow would give rise to ergodicity were ignored by most physicists working in statistical mechanics (with the outstanding exception of N.S. Krylov), although such investigations could be regarded as responding to basic physical questions.

How then does a flow manage to be ergodic? This question was answered by G. Hedlund and E. Hopf in 1934. They found simple systems for which ergodicity could be proved, and then analyzed what made the proof of ergodicity possible. The systems they studied were geodesic flows on compact surfaces of constant negative curvature. I will not give any indication of the simple and elegant formalism involved (Poincaré's upper half plane...) but only note that it permitted them to express in a very explicit way the sensitivity to initial conditions that had been known since the work of Hadamard at the turn of the century. Hopf then showed that as long as the curvature stayed negative the restriction to constant curvature was inessential. The essential mechanism can be expressed in terms of the action the flow induces on tangent vectors to the energy surface. The tangent space at each point splits into a sum of subspaces, a one-dimensional one along the flow and two others, one called the contracting or stable subspace and the other the expanding or unstable subspace. Tangent vectors in the stable subspace contract exponentially in length as $t \to +\infty$ while those in the unstable subspece expand exponentially. This is the basic mechanism used by Anosov in his definition of what he calls C-systems and what most everyone else calls Anosov systems. They form a very general class for which ergodicity holds. You will hear more about it and related matters from Oscar Lanford under the heading of hyperbolic sets.

In addition to determining a local behavior of a measure preserving flow which accounts for ergodicity, the mathematicians of the 1930's classified measure-preserving flows according to the thoroughness with which they scramble the points of phase space. In this way, there arose a hierarchy:

ERGODIC ← WEAKLY MIXING ← MIXING ← K-SYSTEM ← BERNOULLI SYSTEM

The last two, K-system and Bernoulli system, the most scrambling of all, have been added here, although they only were defined in the 1950's and 1960's.

A simple way of comparing the notions of ergodicity, weak mixing, and mixing can be based on the following theorem.

Theorem

A measure-preserving flow in a finite measure space

a) is mixing if and only if for all $f,g \in L^2(M,d\mu)$

$$\lim_{t\to\infty} (f,U^t g) = \frac{(f,1)(1,g)}{\int d\mu}$$

b) is weakly mixing if and only if for all $f,g \in L^2(M,d\mu)$

$$\lim_{\tau\to\infty} \frac{1}{\tau}\int_0^\tau \left|(f,U^t g) - \frac{(f,1)(1,g)}{\int d\mu}\right| dt = 0$$

c) is ergodic if and only if for all $f,g \in L^2(M,d\mu)$

$$\lim_{\tau\to\infty} \frac{1}{\tau}\int_0^\tau \left[(f,U^t g) - \frac{(f,1)(1,g)}{\int d\mu}\right] dt = 0$$

Evidently b) and c) require that the limit of a) be arrived at only in some average sense. Since a) implies b), which implies c) we have the first three levels in the hierarchy. There are elementary examples which show that no pair of these three classes coincide.

In order to understand the next level in the hierarchy it is useful to look at a general and important problem of ergodic theory: conjugacy for flows. (There is an analogous definition for cascades.) Given two flows $T_1{}^t$, $T_2{}^t$ in phase spaces M_1 and M_2 with invariant measures μ_1 and μ_2, respectively, when are they conjugate in the sense that there exists a one-to-one mapping of M_1 on to M_2 under which sets of μ_1 measure zero correspond to sets of μ_2 measure zero and

$$T_2{}^t = \phi\, T_1{}^t\, \phi^{-1} \ ?$$

Two flows which are conjugate in this sense are not essentially different as measure-preserving flows although they may be very different in other ways because ϕ may not respect other properties of the systems. For example, if the flows are continuously differentiable, one could insist that ϕ is continuously differentiable. That would give a much finer classification than the conjugacy just defined. It is easy to see that spectral properties of the Liouville operator are invariant under conjugation so that they provide one invariant suitable for labelling conjugacy classes. However, already in the 1930's there were constructed cascades, the Bernouilli shifts, which all have the same spectral properties, but no one succeeded in showing them conjugate. The problem remained unsolved for two decades until Kolmogorov (1958) introduced a notion of entropy now usually called K-S entropy and showed that it is a conjugacy invariant which can be used to distinguish the Bernouilli shifts from each other. (The S is for Sinai who made essential simplifications in the theory.) It can also be defined for flows. I will not give the definition of the K-S entropy but will only remark that it would be more natural to regard it as a rate of production of entropy, a measure of the rate of destruction of information as the flow scrambles the phase space. A K system can be defined as one whose flow scrambles every finite non-trivial partition of the phase space at a strictly positive rate as measured by the K-S entropy.

The Bernouilli systems are a sub-class of the K systems which are the ultimate in random behavior. They are defined as K systems which are conjugate to some Bernouilli shifts, if cascades (or to an analogous Bernouilli flow, if flows). The grand theorem of the ergodic theory of the 1960's was

Theorem (Ornstein)

All Bernouilli systems with the same K-S entropy are conjugate.

Once this extraordinary result was discovered it became natural to conjecture that several of the types of smooth flows known to be K systems are Bernouilli systems. That, in fact, has been established in some cases e.g.

Theorem

Every Anosov system is a Bernouilli system.

The point of this somewhat lengthy digression into general ergodic theory is to provide preparation for the following remarks.

1) If one takes the framework of the general theory of measure-preserving flows as providing a mathematical realization of Boltzmann's program and accepts the idea that the utter randomness of a Bernouilli system is what Boltzmann would have asked for, then there is essentially only a one-parameter family of such systems and any of them can be realized by Anosov systems. Here "essentially" means "up to conjugacy". From this point of view, there is only a one-parameter family of systems in classicaly statistical mechanics and the aspirations of the founders are perfectly realized in them. It may be that N hard spheres in a box provide one such although a complete proof for general N has not been published.

2) It is much more reasonable, from a physical point of view, to introduce more physical structure into the theory. Then there will be many conjugate flows which are not physically equivalent. This will reduce the embarrassment of having only a one-parameter family of systems - there will be many distinguishable.

3) The flows defined by most of the models of classical statistical mechanics, for example N particles in a box interacting via a Lenard-Jones potential, are almost certainly not ergodic, not to speak of Bernouilli. Almost certainly means that there is no published proof but most of the experts regard the statement as likely to be correct. Thus, if one is to understand the foundations of classical statistical mechanics, it will be necessary to go into the complications to which Poincaré and Birkhoff devoted so much attention.

4) If, as just indicated, the typical systems of classical statistical mechanics are not ergodic, one has to face the basic general question: to what extent is the traditional apparatus of equilibrium statistical mechanics (Gibbsian ensembles etc.) justified? There are three traditional reasons often given for the legitimacy of the traditional methods, despite the non-ergodicity of the flow.

A (Thermodynamic Limit) The results of classical statistical mechanics may be valid in the thermodynamic or bulk limit (number of degrees of freedom, $N \to \infty$; volume, $V \to \infty$; $N/V \to \rho < \infty$). This might happen because the relative phase volume of the non-ergodic portion of the flow approaches zero in the thermodynamic limit, so that observables become insensitive to the non-ergodic portion.

B (Macroscopic Observables) There might be a restricted class of observables, called macroscopic, to which classical statistical mechanics would apply in the thermodynamic limit. In that limit, the macroscopic observables might be insensitive to the non-ergodic portion of the flow even if its relative phase volume does not go to zero.

C (Grain of Dust or Heat Bath) The idealization that the system under study is isolated should be made more realistic by the inclusion of coupling to outside systems. Then the extreme sensitivity of the non-ergodicity to initial conditions might result in an average behavior consistent with classical statistical mechanics.

Of course, A, B, and C have no claim to novelty; their general significance has been discussed since the earliest days of statistical mechanics. One should also not overlook the possibility that one may need essential modifications of the traditional classical statistical mechanics in order to bring it into harmony with the facts of dynamics. There are plenty of interesting open problems here. The first exploratory work in this area seems to indicate that A is not correct.[18]

K(olmogorov) A(rnold) M(oser) Theory

Although, as we have seen, Birkhoff was able to establish many elements of the picture of the behavior of a flow in a neighborhood of an elliptic periodic orbit including the existence of an infinite number of periodic orbits in any neighborhood, he got nowhere with the survival of invariant tori. However, there was progress on a simpler problem by C.L. Siegel as contained in the following theorem.

Theorem (Siegel Circle Theorem)

Let f be a holomorphic function in some neighborhood of the origin and vanishing there. Let the first term of the Taylor series at the origin for f be az. Then a sufficient condition that there exists an invertible holomorphic function ϕ leaving the origin fixed such that

$$[\phi^{-1} \circ f \circ \phi](z) = az$$

is that there exist $A, B > 0$ such that for all positive integers n

$$[a^n - 1]^{-1} \leq An^B$$

This result broke new ground on the problem of Poincaré's thesis described above, and displayed the fact that the convergence of the formal power series for ϕ depends on the arithmetical properties of the number a. That is not so implausible since it is precisely those arithmetical properties which determine how small the "small divisors" get in the denominators of the perturbation series as a function of the order, but the problems had waited a long time for a solution.

The analogous result for elliptic periodic orbits of analytic Hamiltonian systems was announced by Kolmogorov in his address to the International Congress of Mathematicians in Amsterdam in 1954. (It is translated into English in an Appendix of Abraham and Marsden's book Foundations of Mechanics). A sketch of the proof was given in Reference 19 and a detailed treatment was provided by V. Arnold in Reference 20. As you will hear from Gallavotti, the result states that for analytic Hamiltonians sufficiently close to an integrable Hamiltonian under analogous arithmetic conditions the invariant tori of the unperturbed problem survive, somewhat deformed, and constitute a set of positive measure. Furthermore, the fraction of the total phase volume swept out by the invariant tori approaches one as one gets closer to the periodic orbit. Moser gave an improved version of the theorem in which the Hamiltonian was only assumed to have a finite number of derivatives.[21]

With the KAM theorem one has nearly arrived at where we are now. The main features of the extraordinary structure of the phase portrait or a Hamiltonian flow seem to have been established. It remains to try to understand how it all fits together. You will hear about that from Newhouse and Lanford.

Final Remarks

In closing, I would like to comment on a sociological question: How was it that after ignoring the development of analytical mechanics for half a century, physicists took it up again and cultivated it to its present state in which it is dreadfully fashionable? I will list three papers as influential.

R. May Simple Mathematical Models with Very Complicated Dynamics Nature 261 (1976) 459-467

J. Gollub and H. Swinney Onset of Turbulence in a Rotating Fluid Phys. Rev. Letts. 35 (1975) 927-930

M. Feigenbaum Quantitative Universality for a Class of Nonlinear Transformations Jour. Stat. Phys. 19 (1978) 25-52

May's paper showed that the transition to chaos is a feature of the solutions of one-dimensional difference equations used to model population dynamics. He also assembled a mass of convincing experimental evidence for the phenomenon. Gollub and Swinney showed that in Couette flow (fluid flow between two cylinders, one rotating) the transition from laminar to turbulent flow is characterized by a change from discrete to continuous spectrum in the power spectrum of the autocorrelation function. This was in disagreement with the conventional wisdom as expressed by Landau and Feynman. It said that turbulent flow should differ from laminar simply in having a large number of incommensurate discrete frequencies. Feigenbaum recognized that in the period doubling bifurcations of mappings of the unit interval into itself there are universal features treatable by the methods of the renormalization groups which have had such success in the theory of second order phase transitions.

Of course, these papers had influential antecedents of which I will mention three. The Lorenz model, a system of three ODE's with a right-hand side polynomial of second degree in the unknowns, was put forward to convince meteorologists that stochastic behavior could arise from simple deterministic models. It was abstracted from the theory of convection in a layer of fluid heated below by throwing away all but three Fourier coefficients of the unknown function (E.N. Lorenz Deterministic Nonperiodic Flow Jour. Atmos. Sci. 20 (1963) 130-141).

There were also the calculations of Hénon and Heiles who showed convincingly that in a problem of coupled oscillators one could get a transition from regular to apparently ergodic flow by varying the energy (M. Hénon and F. Heiles The Applicability of the Third Integral of Motion: Some Numerical Experiments Astron. Jour. 69 (1964) 73-79).

The paper, D. Ruelle and F. Takens On Turbulence Commun. Math. Phys. 20 (1971) 167-192; 23 (1971) 343-4, showed that strange attractors are a generic feature of the solutions of smooth ODE's in three or more variables. It pointed out that strange attractors because of their prickly shape offer an explanation of the appearance of chaotic motions; to approach such an irregular set without running into it one has to move in a highly irregular fashion.

Last, but not least, one should contemplate the influence of the computer which, used in an exploratory and illustrative manner, produced fascinating information on the phase portraits of concrete flows.

REFERENCES

1. H. Poincaré, La Théorie Cinétique des Gaz, Revue Générale des Sciences (1894) 516
2. A. Rosenthal, Beweis der Unmoglichkeit ergodischer Gas systeme, Ann. der Physik (4) 42 (1913) 796-806
3. P. and T. Ehrenfest, Begriffliche Grundlagen der statischen Auffassung der Mechanik, Encykl. d. math Wiss 4. 2 II, Heft 6; Nr. 10a
4. E. Fermi, Beweis, dass ein mechanisches normal system im allgemeinen quasi-ergodisch ist, Phys. Zeits. 24 (1923) 261-264
   I thank G. Benettin for explaining Fermi's work to me. Further details and generalizations of Fermi's result are contained in G. Benettin, G. Ferrari, L. Galgani, and A. Giorgilli, An Extension of the Poincaré-Fermi Theorem as the Nonexistence of Invariant Manifolds in Nearly Integrable Dynamical Systems, Nuovo Cim. 72B (1982) 137-148
5. Sur les Propriétés des Fonctions Définies par Les Equations aux Différences Partielles, Oeuvres de Henri Poincaré Tome I XLIX-CXXIX
6. Y. Choquet-Bruhat, C. DeWitt-Morette, M. Dillard-Bleick, Analysis-Manifolds and Physics, North Holldand, 1977
7. H. Dulac, Detérmination et Intégration d'une certaine Classe d'Equations Différentielles Ayant pour Point Singulier un Centre, Bull. Sci. Math. (2) 32 (1908) 230-252
8. V.I. Arnold, Chapitres Supplémentaires de la Théorie des Equations Différentielles Ordinaires
9. M. Hénon, A Two-Dimensional Mapping with a Strange Attractor, Commun. in Math. Phys. 50 (1976) 69-77
10. S. Chandrasekhar, Ellipsoidal Figures of Equilibrium, Yale University Press 1969
11. J.-L. Tassoul, Theory of Rotating Stars, Princeton University Press 1978

12. G.D. Birkhoff, Proof of Poincaré's Geometric Theorem, Trans. Amer. Math. Soc. 14 (1913) 14-22
13. G.D. Birkhoff, An Extension of Poincaré's Last Geometric Theorem, Acta Math. 47 (1926) 297-311
Surface Transformations and Their Dynamical Applications Trans. Amer. Math. Soc. 18 (1917) 1-119
14. G.D. Birkhoff, Sur l'existence de régions d'instabilité en Dynamique, Ann. de l'Inst. Henri Poincaré 2 (1932) 369-386
15. G.D. Birkhoff, Sur quelques courbes remarquables, Bull. Soc. Math. de France 60 (1932) 1-26
16. For reviews of this and related subjects see Nonlinear Dynamics and Turbulence, Ed. G. Barenblatt, G. Iooss, and D. Joseph, Pitman, 1983.
17. G.D. Birkhoff and P.A. Smith, Structure Analysis of Surface Transformations Jour. de Math Pures et Appl. 7 (1928 345-379
18. G. Benettin and L. Galgani, Transition to Stochasticity in a One-Dimensional Model of a Radiant Cavity, Jour. of Stat. Phys. 27 (1982)
C.E. Wayne, The KAM Theory of Systems with Short-Range Interactions, to appear
19. A.N. Kolmogorov, On the Conservation of Conditionally Periodic Motions under Small Perturbations of the Hamiltonian Dokl. Akad. Nauk USSR 98 (1954) 527-530
20. V. Arnold, Proof of A.N. Kolmogorov's Theorem on the Preservation of Quasi-Periodic Motions under Small Perturbations of the Hamiltonian, Russ. Math Surveys 18 (1963) 9-36
21. J. Moser, On Invariant Curves of Area-Preserving Mappings of an Annulus, Nach. Akad. Wiss. Göttingen Math. Phys. Klasse II (1962) 1-20

Stability and Nonlinear Character of Ordinary Differential Equations in Nonlinear Problems in Nonlinear Problems Ed. Lander Winn of Wisconsin Press, Madison 1963

SELECTED BIBLIOGRAPHY

General Texts on Mechanics

V.I. Arnold Mathematical Methods of Classical Mechanics Springer-Verlag

R. Abraham and J. Marsden Foundations of Mechanics W.A. Benjamin

W. Thirring A Course in Mathematical Physics I Classical Mechanics, Springer-Verlag

C.L. Siegel and J. Moser Lectures on Celestial Mechanics Springer-Verlag

G. Gallavotti Elements of Mechanics, Springer-Verlag

Mechanics (More Detailed but Generally Relevant)

V.I. Arnold and A. Avez Ergodic Problems of Classical Mechanics, W.A. Benjamin

J. Moser Stable and Random Motions in Dynamical Systems Ann of Math Studies #77

S. Newhouse Lectures on Dynamical Systems Progress in Math 8, Birkhauser

J. Guckenheimer and P. Holmes Nonlinear Oscillations, Dynamical Systems and Bifurcations of Vector Fields Springer-Verlag

A.J. Lichtenberg and M.A. Lieberman Regular and Stochastic Motion, Springer-Verlag

Ergodic Theory

V.I. Arnold and A. Avez Ergodic Problems of Classical Mechanics, W.A. Benjamin 1968

Ya. Sinai Introduction to Ergodic Theory, Princeton Math. Notes 1977

I.P. Kornfeld, S.V. Fo min, Ya. Sinai, Ergodic Theory, Springer-Verlag (1982)

R. Bowen Equilibrium States and the Ergodic Theory of Anosov Diffeomorphisms, Lecture Nores in Math 470, Springer-Verlag

D. Ruelle Ergodic Theory of Differentiable Dynamical Systems, Publ Math. IHES

Prerequisites for the Lectures of R. Helleman

Self-Generated Chaotic Behavior in Nonlinear Mechanics pp 165-233 in Fundamental Problems in Statistical Mechanics, Ed. E.G.D. Cohen, North Holland has a very complete bibliography.

One Mechanism for the Onsets of Large-Scale Chaos in Conservative and Dissipative Systems pp 95-126 in Long Time Prediction in Dynamics Ed. C.W. Horton et al., John Wiley

Differentiable Dynamical Systems, Smooth Mappings and Bifurcation Theory

P. Collet and J.-P. Eckmann, Iterated Maps on the Interval as Dynamical Systems, Birkhauser

V. Arnold Chapitres Supplémentaires de la théorie des équations différentiales ordinaires Mir (1980)

J. Palis, Jr. and W. de Melo, Geometric Theory of Dynamical Systems, Springer (1980)

M. Shub Stabilité globale des systèmes dynamiques, Asterisque 56 (1978)

A. Katok Dynamical Systems with Hyperbolic Structure Trans. Amer. Math. Soc. 2 (1980) 116

Z. Nitecki Differentiable Dynamics, M.I.T. Press (1971)

Conference Proceedings

Dynamical Systems CIME Lectures by Guckenheimer, Moser, Newhouse Birkhauser

Symposium in Pure Math. Vol 14 (Global Analysis) Amer Math. Soc.

Nonlinear Dynamics Annals of NY Academy of Sciences 359 (1980)

Topics in Nonlinear Dynamics, Amer. Inst. Physics Conf. Proceedings #46

APPLICATIONS OF SCALING IDEAS TO DYNAMICS

Leo P. Kadanoff
The James Franck and Enrico Fermi Institutes
The University of Chicago
Chicago, IL 60637 USA

Lecture I ROADS TO CHAOS: COMPLEX BEHAVIOR FROM SIMPLE SYSTEMS

Hydrodynamical systems often show an extremely complicated and apparently erratic flow pattern (Fig. 1). These turbulent flows are so highly time-dependent that local measurements, say of one component of the velocity, would show a very chaotic appearance like that in Fig. 2e. However, there is also an underlying regularity in which the notion can be analyzed (see Fig. 1 again) as a series of large swirls containing smaller swirls, and so forth.

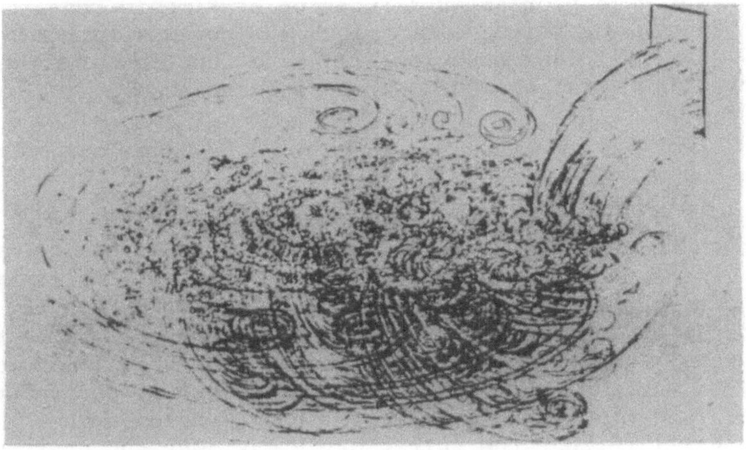

Fig. 1 A picture of a turbulent flow pattern by Leonardo da Vinci.
Note how the large swirls break up into smaller ones.

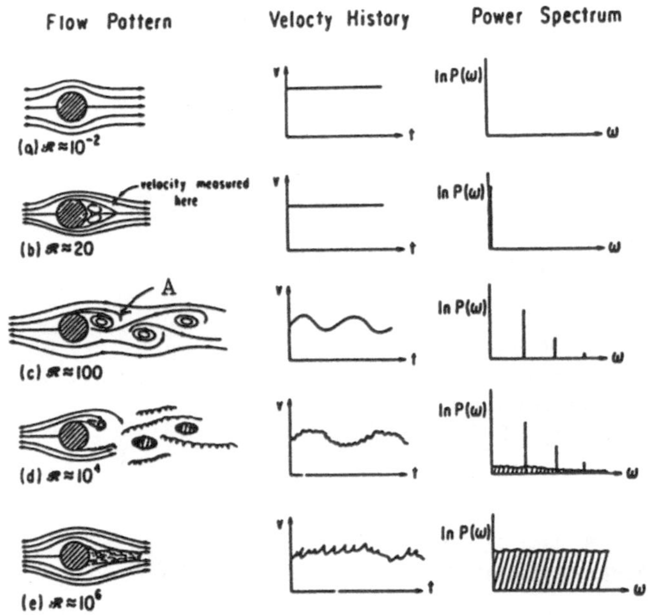

Fig. 2 Simplified view of patterns of hydrodynamic flow. The earlier rows show patterns of behavior which tend to be characteristic of smaller values of r; the lower rows show behavior which only appears for high values of r. The flow patterns of the first column are observed by measuring the velocity at a point, for example, the point marked with a cross in row b. The velocity is plotted as a function of time in the second column. The third column plots the power spectrum $P = |V(\omega)|^2$ (see equation (1)) as a function of frequency. The first column is drawn from reference 1.

One approach to understanding this turbulence is to ask how it arises. If one puts a body in a stream of a fluid--for example, a piece of a bridge sitting in the stream of a river--then for very low speeds (Fig. 2a) the fluid flows in a regular and time-independent fashion.[1] As the speed is increased (Fig. 2b), the motion gains swirls but remains time-independent. Then as the velocity, measured in a dimensionless fashion by the Reynolds number r, increases still further the swirls may break away and start moving downstream. This induces a time-dependent flow pattern--as viewed from the bridge. The velocity measured at a point--say point "A" on the drawing of Fig. 2c--gains a periodic time-dependence. As r is increased still further, the swirls begin to induce irregular internal swirls as in the flow pattern of Fig. 2d. In this case there is a partially periodic and partially irregular velocity history (see the second column of Fig. 2d). Raise r still further and a very complex velocity field is induced, and the v(t) looks completely chaotic as in Fig. 2e.

These different flow patterns can be characterized by looking at the Fourier-transforms of the velocity field:

$$V(\omega) = \frac{1}{\sqrt{T}} \int_0^T dt\, e^{2\pi i\omega t}\, v(t) \tag{1}$$

Then the power spectrum, $p(\omega) = |V(\omega)|^2$, shows a spike at zero frequency for the time-independent flows, Figs. 2a and 2b. In the periodic region (Fig. 2c), additional spikes appear at frequencies equal to one over the period, and the harmonics which are integer multiples of this frequency. As the motion becomes partially chaotic as in Fig. 2d, a broad slowly varying background appears behind the spectral lines. Finally, the fully chaotic flow has a power spectrum which is apparently continuous.

We would, of course, like to understand this transition to turbulence in hydrodynamical systems. Unfortunately, after many years of study, we still do not have a fully satisfactory approach to this problem. In this article, I would like to describe an extremely simplified model which shows a kind of transition to chaos and to discuss how the features of this model can be, and have been, observed in hydrodynamic systems. The spirit of this approach is similar to the one used in the theory of critical phenomena in condensed matter physics. To understand a complicated phase transition--i.e., a change in behavior of a many-particle system--choose a very simple system which shows a qualitatively similar change. Study this simple system in detail. Abstract the features of the behavior of the simple system which are "universal" --that is, appear to be independent of the details of the system's makeup. Apply these universal features to the more complex problem.

Our simple problem is so simple that one might, at first glance. imagine that it contains nothing of interest. But it has an amazingly intricate and regular structure. Consider a dynamical system characterized by one variable, x. At time 0, the value of this variable is x_0; at discrete later times $t = j\tau$ it has the value x_j. The major assumption is that the value of the variable at one step x_j determines the value at the next . Mathematically we write

$$x_{j+1} = R(x_j) \tag{2}$$

where $R(x)$ is a function which describes the dynamics. Our job is to find time histories of the system.[2] That is we start from x_0, find $x_1 = R(x_0)$, $x_2 = R(x_1)$,... and see the patterns in the sequence $x_0, x_1, x_2 \ldots$. One simple visualizable model for such a system is an island containing an insect population which breeds in the summer and leaves eggs which hatch next summer. Then x_j is the ratio of the actual population to some reference population in the summer of the jth year. Our model is the statement that the population next summer x_{j+1} is determined by the population this summer via the relation

$$x_{j+1} = rx_j - sx_j^2 \tag{3}$$

Here there are two terms. The first rx_j represents the natural growth rate of the population; the term sx_j^2 represents a reduction of this natural growth caused by overcrowding of the insects. When $r > 1$, the first term would simply express an increase in the population by a factor r in each year. The other term represents the reduction in the population growth caused by, for example, competition for resources (or perhaps shyness) of the insects when the population is large. By rescaling x_j, i.e., by letting x_j be replaced by $(r/s)x_j$, one can convert this equation into the standard form:

$$x_{j+1} = r\, x_j(1-x_j) \tag{4}$$

We wish to examine the long term behavior of the population variables x_j based upon equation (4). In particular, we are interested in how this behavior depends upon the growth rate r. We can think of r as being akin to the Reynolds number in the hydrodynamic example. We wish to keep the insect population ratio in the interval between 0 and 1 and to do this we limit our examination to the region $0 < r < 4$.

First study the small r behavior. If $r < 1$, the insects are living in such an inhospitable environment that their population will diminish each year. Their population pattern is shown in Figure 3a. If for example, $r = 1/2$ and one starts from $x_0 = 1/2$, then $x_1 = 1/8$ and each succeeding x_j is less than $2^{-(j+2)}$. The population simply dies away to zero, for all starting values. This result is summarized in Fig. 4 in which we plot eventual population values as a function of r. For $0 < r < 1$, the eventual population is zero. Roughly speaking, we might think of this behavior as akin to the laminar (smooth) flow in Fig. 2a.

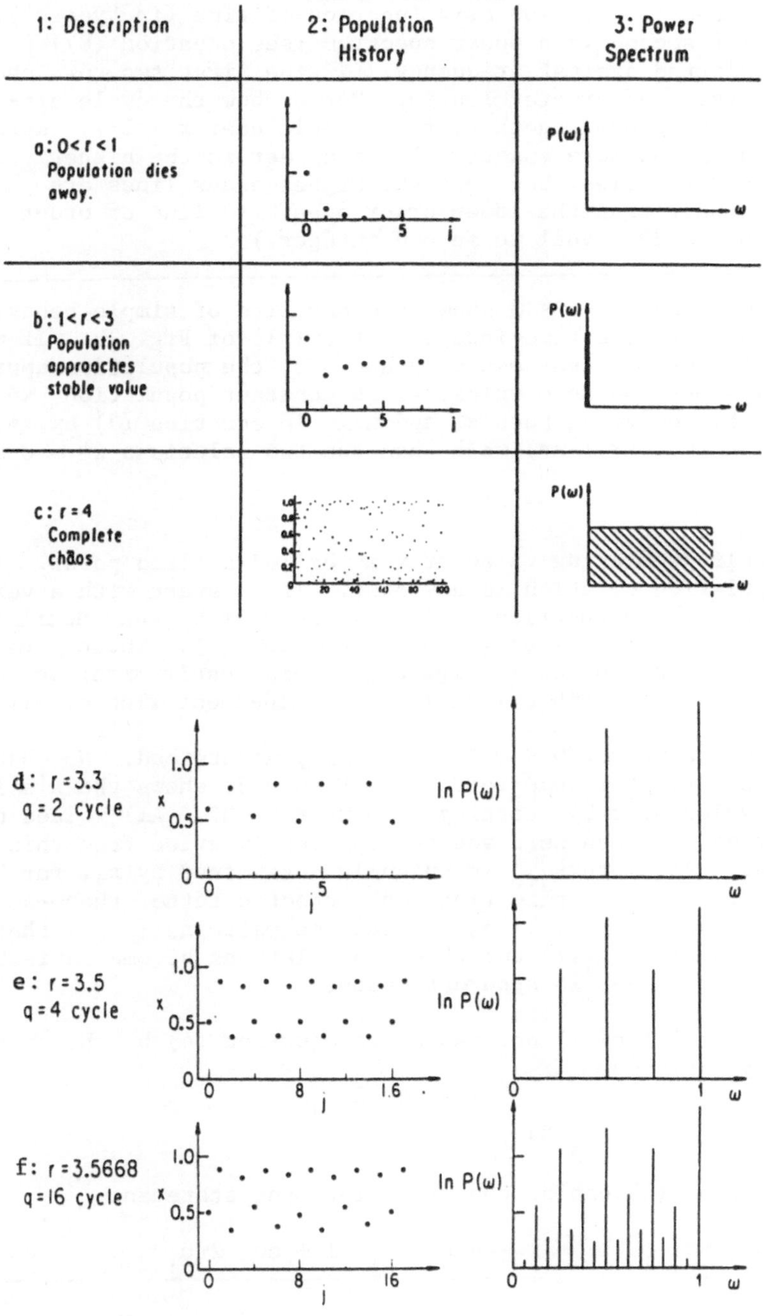

Fig. 3 (Legend on following page)

Fig. 3 Patterns of population behavior. The first column gives insect population as a function of time (in "years"). The second is a power spectrum (see equation (8)) plotted against frequency ω. The first two rows show cycles of greater length. Notice how the cycle elements group together, for example near $x = 1/2$. Note also how more spectral lines appear in the higher order cycles, but that the higher order lines are weaker than the lower order ones. (A line of order S has $2^S\omega$ equal to an odd integer.)

The region $1 < r < 3$ shows another kind of simple behavior, perhaps akin to the time-independent swirls of Fig. 2b. If we start with any x_0 between zero and one, the population approaches a constant but non-zero value. This constant population x^* can be found by replacing both x_j and x_{j+1} in equation (3) by x^*. Then x^* obeys $x^* = r\, x^*(1-x^*)$ which has the two solutions $x^* = 0$ and

$$x^* = 1 - r^{-1} \tag{5}$$

Such a self-generating value of x is called a fixed point. The zero population solution is unstable. If we start with a very low population, the population will increase year by year until it settles down to the value given in equation (5). These final populations are plotted in Figure 4. This result might be considered to be compatible with the time-dependent flow of Fig. 2b.

Thus the region $0 < r < 3$ is easily understood. No chaos has arisen so far. Now jump to $r = 4$. Figure 3c shows the x's induced at this value of r by starting from $x_0 = 0.707$. All values of x in the range between zero and one apparently arise from this starting point. Although x_{j+n} is uniquely determined by x_j, for large n the pattern of determination looks chaotic rather than--as it is--deterministic. For small n, one can see patterns (e.g., that small x_j produces small x_{j+1}) but these correlations become invisible as $n \to \infty$. What we see is apparent chaos.

For $r = 4$ (only!) one can solve equation (4) by the simple change of variables

$$x_j = \frac{1 - \cos 2\pi\theta_j}{2} \tag{6}$$

Then equation (4) can be converted into the statement

$$\frac{1 - \cos 2\pi\theta_{j+1}}{2} = 4\,\frac{1 - \cos 2\pi\theta_j}{2}\,\frac{1 + \cos 2\pi\theta_j}{2} = \frac{1 - \cos 4\pi\theta_j}{2}$$

which has as one solution $\theta_{j+1} = 2\theta_j$ or

$$\theta_j = 2^j \theta_0 \tag{7}$$

One can see the chaos in the solution quite directly. Since x_j is related to cos $2\pi\theta_j$, adding an integer to θ_j (or changing its sign) leads to the very same θ_j . Hence if one writes θ_j in ordinary base 10 notation as θ_j = 11.2693... one can simply throw away the 11. Better yet, if one writes θ_0 as a "decimal" base 2, as for example

$$\theta_0 = 1/2 + 1/8 + 1/16 + 1/64 + \ldots$$

$$= 0.101101 \ldots$$

then the multiplication by 2 is simply a shift in the "decimal" point, so that

$$\theta_1 = .01101 \ldots$$

$$\theta_2 = .\ 1101 \ldots$$

$$\theta_3 = .101 \ldots$$

$$\theta_4 = .01 \ldots$$

Thus, if we start out with any θ_0 the θ_j's produced will depend on the jth and higher digits in θ_0 . This gives us one possible definition of chaos: For large j the dynamical variable x_j has a value which is extremely sensitive to the exact value of x_0. In particular, in our case, if we have two different starting values x_0 and x_0' which differ by a small number ε then if we generate a sequence of resulting populations x_j and x_j' based respectively upon x_0 and x_0' then after j steps, the difference grows to the value $2^j\varepsilon$.

In fact, the calculation represented in the picture of Fig. 3c is, in some sense, incorrect. It was calculated upon a computer with 16 decimal digits. After about 50 steps an initial error of 10^{-16} grows to be an inaccuracy of order 1. Consequently, all data after step 50 is wrong.

This might be an appropriate point to mention that one of the major sources of modern stochastic theory is the work of the meteorologist Lorenz.[3] As an analog to weather-forecasting, he studied systems like this r = 4 system, in which the final state is an extremely sensitive function of the initial state. For this kind of system, as the prediction period grows longer both the initial data needs and the computational power required will grow exponentially. True, long-range, detailed weather prediction is in practical terms impossible.

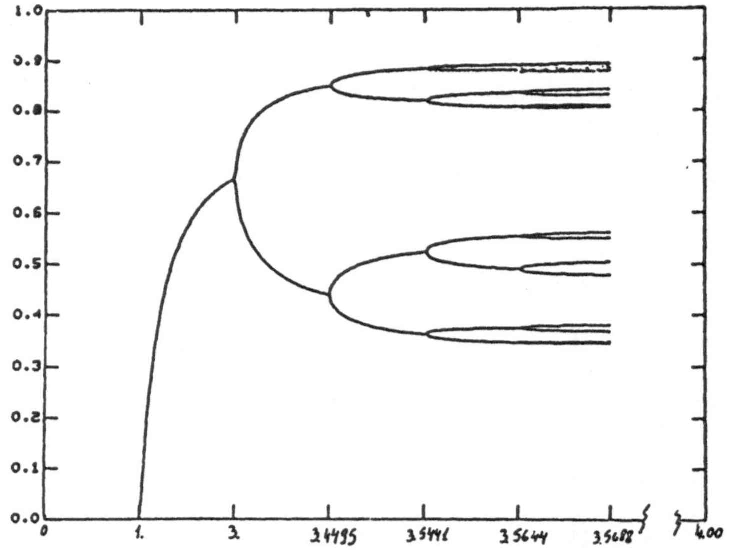

Fig. 4 Possible values of x_j as $j \to \infty$ as a function of r. The solid lines give the stable x-values as a function of r. Note the non-linear nature of the scale in r, which has been chosen to emphasize the successive period doublings as $r \to 3.5699$. There is a break in scale and then $r = 4$ behavior--in which all populations occur--is shown.

The system at $r = 4$ is chaotic in another sense. For almost any randomly chosen x_0 or θ_0 the set of resulting θ_j will be uniformly distributed between 0 and 1. Correspondingly, we will get a set of x_j's in which the probability that x_j will have the value between y and y + dy is given by $p(y) \sim (y(-y))^{-\frac{1}{2}}$ for almost any starting point. Thus the time average for this chaotic system is the same for almost every starting point.

We have inserted on the right-hand side of Fig. 4 this distribution of x's for this value of r by showing all x's between 0 and 1 as possible long-term values.

A final definition of chaos is that the power spectrum

$$p(\omega) = \frac{1}{N}\left| \sum_{j=1}^{N} x_j \, e^{2\pi i \omega j} \right|^2 \qquad (8)$$

has broad spectral features. For $r = 4$, and large N the power spectrum can be calculated exactly. For almost any x_0 it is perfectly flat.

There are some special x 's which generate exceptional patterns of x_j. Since θ_j can be flipped in sign and shifted by an integer without changing x_j one can always choose θ_j to be in the interval between 0 and 1/2. Thus, the recursion relation for θ_j can be written as

$$\theta_{j+1} = \begin{cases} 2\theta_j & \text{for } 0 \le \theta_j \le 1/4 \\ 1 - 2\theta_j & \text{for } 1/4 \le \theta_j \le 1/2 \end{cases} \tag{9}$$

Any choice of θ_0 which makes it a rational number will lead to a recurring pattern of θ_j and x_j. For example, if we take $\theta_0 = 1/3$ this is a fixed point in that all subsequent θ_j are also 1/3. If one starts with $\theta_0 = 1/5$ then the subsequent θ_j are $\theta_1, \theta_2, \ldots =$ 2/5, 1/5, 2/5, 1/5,... . Thus we have a cyclic behavior with a length of period $q = 2$. One period three solution is 1/2, 2/9, 4/9. Equation (9) has periodic solutions with all possible periods.[4]

Now we have a long-term solution which is time-independent for $0 < r < 3$ and which is quite chaotic for $r = 4$. Next increase r from 3 and observe the first hints of chaos which arise. As r increases just above 3 the fixed point at $x^* \approx 2/3$ becomes unstable. What happens, see Fig. 3d is that a cycle of length 2 becomes the stable behavior. The insects start from a low population. They reproduce avidly. Hence the population next year will be low. Odd years will have high x's; even years low x's. In the Bible is is recorded that Joseph[5] predicted such a period-behavior with a basic time step of seven years. In our model the exact values of x for the two cycles are

$$x = \frac{1}{2}(1 + r^{-1}) \pm \frac{1}{2}\sqrt{(1 + r^{-1})(1 - 34^{-1})}$$

These $q = 2$ cycles remain stable over the range between $r = 3$ and $r = 1 + \sqrt{6} = 3.4495$. Call the value of r at which the $q = 2$ cycle becomes unstable r . At larger values of r between r_2 and r_4 the stable behavior is a four cycle, as shown in Fig. 3e. The basic period of the behavior has doubled once more. Above r_4 a $q = 8$ cycle appears and remains stable between r_4 and r_8 whereupon a $q = 16$ cycle appears. These stable behaviors are shown on Fig. 4 as doublings in the number of x-values which remain present as $j \to \infty$. Successive doublings continue until at $r_c = 3.5699\ldots$ at which there apperas a cycle of infinite length.

To see the beginnings of the onset of chaos in this model, look at the power spectrum defined by equation (8). When the fixed point is the stable behavior, the power spectrum is a spike at zero frequency. When the period two cycle appears, another frequency, $\omega = 1/2$, appears in the spectrum. See Fig. 3d. This frequency is, of course, equal to the inverse period of the motion. At $q = 4$, $\omega = 1/4$ and $\omega = 3/4$ also enter, as in Fig. 3e. As the period increases, more and more spectral lines enter until at $q = 2$ there are an infinite number of lines.

Of course, a spectrum with an infinite number of lines is not the same as a broad band spectrum. Even at $r = r_\infty$ there is no fully developed chaos of the type which occurs at $r = 4$. In this model, the development of an infinite number of lines through successive period doublings is a major step toward the production of chaos.

To see the remaining steps turn to Fig. 5. This picture is, like Fig. 4 a depiction of the x_j's which arise after a huge number of iterations of an initial x_0. Our job is to understand how this picture changes between $r = r_c$ and $r = 4$. We shall discuss the predominat features of this development, leaving out small regions--like the one marked three cycle-- in which stable cycles once again dominate the picture.

Between $r = 4$ and $r = r_1'$ an initial x_0 generates points which move erratically through the entire band of permitted x-values. As r decreases, this band narrows slightly to between $r(1 - r/4)$ and $r/4$ but it is otherwise qualitatively unchanged. The spectrum contains no sharp peaks. However, at $r = r_1'$ the behavior changes. The band splits into two. Between r_1' and r_2', on even steps the point lies in the lower band; on odd steps, in the upper. There is then a broad spectrum, produced by the erratic motion within each band and a sharp peak at $\omega = 1/2$ produced by the regular motion from band to band. Then at r_2' the motion changes once more. There are four bands which we can number from bottom to top as 1, 2, 3, 4 and the motion goes from band 1 to 3 to 2 to 4, which is, not accidentally, exactly the same ordering as the motion in the $q = 4$ cycle. As n decreases beyond r_4' there are eight bands, then beyond r_8' sixteen, and so forth. When there are $q = 2^n$ bands, the motion returns to a given band after q steps but the exact point at which it returns is chaotic in exactly the same sense as there is chaos at $r = 4$. In this region of 2^n bands there are sharp spectral lines at $\omega = 2^{-n}$ times an integer together with a broad background produced by the erratic behavior in the band. To the naked eye this erratic behavior looks very much the same as the $r = 4$ chaotic motion. The only differences are ones of scale. In this 2^n-band case, the erratic motion is confined to a narrow region, inside a given band.

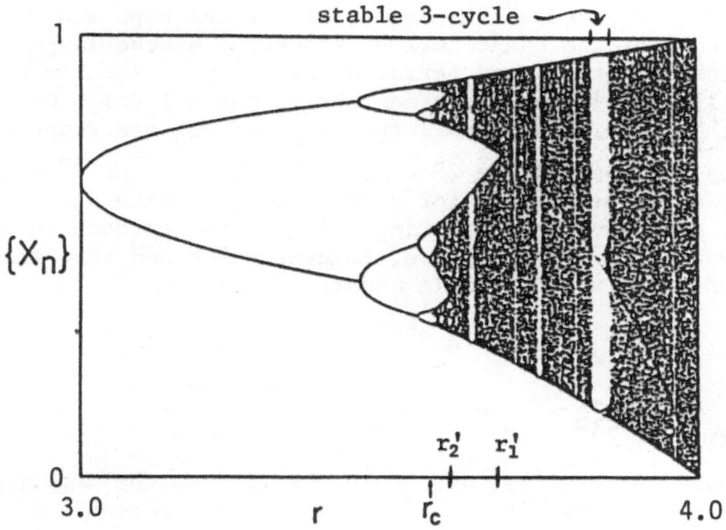

Fig. 5 A continuation of Fig. 4. Here are plotted on a linear scale between r = 3 and r = 4 points which arise during 20,000 iterations of a starting x_0. The period doubling occurs between r = 3 and $r = r_c = 3.5699$. After that a large number of bands appear which as r increases merge together to form $2^n \ldots 4,2$ and finally one band. (After Huberman, private communication.)

Furthermore, the motion only returns to the band every 2^n steps, so that there is a change in the time-scale as well. (A reader who is acquainted with renormalization and scaling, might wonder whether this scale invariance might be used to build a renormalization group analysis of the period doubling process. It can, and has.[6])

This process of successive band-splittings enables the system to interpolate smoothly between the full chaos at r = 4 and the 2^∞ cycle at $r = r_c$. As r approaches r_c from above, we get more and more bands until at $r = r_c$ there are 2^∞ bands.

All of this so far has applied to a single simple model given by equation (4). However, it is important to notice that the general nature of the processes of period doubling and band splitting are independent of the details of the model and will occur with any mapping of the form $x_{j+1} = rf(x_j)$, with f(0) = f(1) and f being a smooth function with a single maximum in the interval 0 to 1.

Furthermore, Feigenbaum[6] has looked at several maps and demonstrated that some of the quantitative properties of the large q behavior of band splitting and period doubling apply equally well to almost all maps with a smooth maximum in which $f''(x) < 0$ at the maximum. In particular recall that r_{2n} and r'_{2n} are respectively the values of r at which a 2^n cycle first appears or 2^n bands merge. (See Figs. 4 and 5 for a depiction of these bands and cycles.) As $n \to \infty$ these limiting values both approach the same limiting value r_c. For large n, Feigenbaum showed that they approached r_c in a very simple manner, namely

$$r_{2n} - r_c = A\,\delta^{-n}$$

$$r'_{2n} - r_c = A'\,\delta^{-n}$$

Here A and A', naturally, depend in detail upon the mapping function f(x). However, the exciting and surprising result of his work was that δ is universal: it does not change as we change the mapping function f. A similar result is obtained for the splitting of the x_j values. When one chooses in an appropriate fashion two neighboring x_j's which lie in a 2^n cycle, one finds that the separation between these values decreases with n as $B\,\alpha^{-n}$, where α is universally equal to 2.503... . Feigenbaum has also given a renormalization group treatment which verifies this universality and generates these numbers.

If these quantitative large n results apply to all maps, might they not apply to real dynamical systems? In particular might not x_t refer to some dynamical property at time t and $f(x_t)$ describe how the dynamical property at time t determined its value at time $t + \tau$?

Non-linear electrical circuits have been shown[6a] to give a very similar behavior to the one described above. As a control parameter, r, in the circuit is varied the circuit trace patterns in which successive period doublings occur. Moreover, the observed values of α and δ are the same as the values mentioned above.

It is not surprising that simple circuits, which can be described in terms of a few variables, show identical behavior to the mapping problems. But, what of real hydrodynamical systems. They are much more complex. Can they show this behavior also? One class of studies which could make contact with this theory of dynamical behavior is the experimental work on the Rayleigh-Bénard instability in fluid systems. When an enclosed fluid is heated from below, at low heating rates, no flow occurs. At highter rates time-independent convection is set up. At higher rates yet, a periodic time dependence appears. At still higher rates the time-dependence looks very chaotic--with a broad band spectrum. In a small system containing helium at low temperatures, Libchaber and Maurer[7] observed a series of successive period doublings. When they adjusted the heating rate very carefully, they could see the power spectrum shown in Fig. 6. Notice the qualitative similarity between this spectrum for a real hydrodynamic system and the spec-

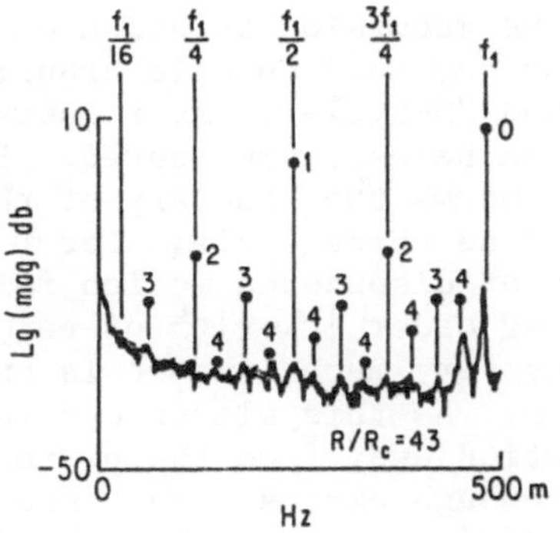

Fig. 6 A power spectrum observed by Libchaber and Maurer for a convective flow in a small cell contining helium. The labels show the order of the various lines.

trum shown in Fig. 3f. This connection, is however, more than qualitative. The relative heights of the weaker spectral lines are predicted to be universally determined by the properties of the high-n band splittings. In particular, one can determine the quantity α from these heights. There is a quite satisfactory agreement between this experimental value of α and the one calculated by Feigenbaum's renormalization group analysis. Hence, one route to chaos in one real system may be said to be largely understood.

However, this is only the beginning of the story--not the end. Libchaber and Maurer's cells are rectangular in cross section. Ahlers and Behringer[8] have done a series of parallel experiments on cylindrical cells. They observe a different route to chaos. In fact, the period doubling route seems to be rather rare. Are there other relatively universal routes to chaos observable in real systems? Can they also be analyzed in terms of very simple models? We do not know, but there are a large number of workers trying to find out.

Since the experiments show additional roads to chaos, it is sensible to look back at the mapping problems described by equation (2) and see whether they, too, have additional paths to interesting behavior. In fact, the recursion relation we have been studying, equation (4) does show one more chaotic transition. Notice the region of Fig. 5 marked "3-cycle". As r decreases and moves out of that region, the motion becomes disordered. For r just above the critical value, which marks the boundary of this region, the long-term stable motion is the three cycle. For r just below this value, r_c, there are periods of disordered motion followed by long periods (which have a length of order $|r_c - r|^{-\frac{1}{2}}$ of very orderly motion in which the system looks very much like it is undergoing cyclical motion of period three. As this almost cyclical motion progresses there is a gradual motion away from the period three cycle elements. Finally, you get far enough away so that once again you have a period of apparently random motion. This kind of motion: long orderly periods mixed with bursts of disorder (see Fig. 7) is called intermittency and has been studied in some detail.[9] It is also observed in experimental systems, but to date the detailed correspondence between the model systems and the real ones has not been fully worked out.

In the examples mentioned heretofore, the model problems all exhibit chaos because the mapping function f(x) has a maximum at some value of x. A map which does not have a maximum, for example,

$$x_{j+1} = x_j + \Omega - \frac{k}{2\pi} \sin 2\pi x_j \tag{10}$$

for $|k| < 1$ cannot show any chaotic structure. These are "no-passing" systems. Imagine that if you start with two points x_0 and x_0' and go through steps to construct x_j from x_0 and x_j' from x_0'. The no-passing property is the statement that if $x_0 < x_0'$ then it must be true that $x_j < x_j'$. For $|k| < 1$ there are two kinds of stable motions, both being smooth and unchaotic. For some values of Ω, the system will fall into a cycle of length q in which x advances by p units in the q steps. In this motion $x_{j+q} = x_j + p$ and the average rate of advance of x, $p/q \equiv w$ is a rational number. This motion may be described as commensurate in the sense that the

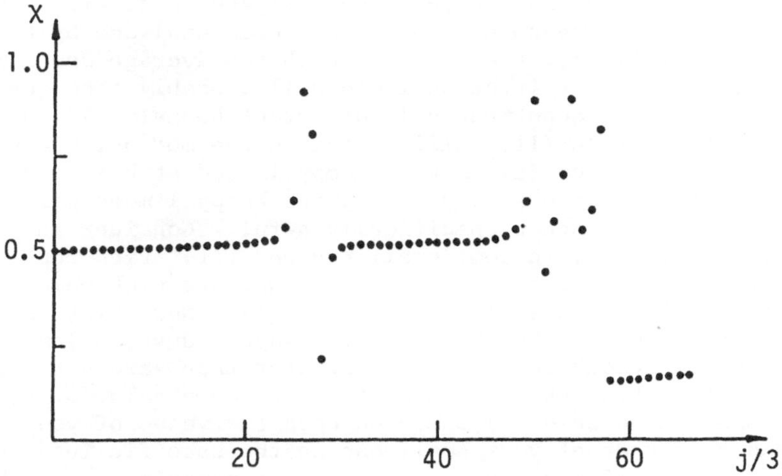

Fig. 7 Another road to chaos: Intermittency. At a closely neighboring value of r there is a three cycle with x = .5, .96, .16 ... . Here we see in a picture of every third step at r = 3.8282. In this case, the motion gets "stuck" for comparatively long periods near three values. The chaos is in the bursts of "unpredictable" behavior in between the ordered motion, in the range of variation of the trapping times, and in the stochastic motion among the three different trapping points.

cycles are commensurate with the period of sin $2\pi x$ in equation (10). On the other hand, Ω may also be chosen so that the average rate of advance per step, w, is irrational. In this incommensurate motion, if one starts from $x_0 = 0$ then the subsequent motion will look like

$$x_j = j\,w + \phi(jw) \tag{11}$$

where $\phi(t)$ is a periodic function of t with period 1.

As k passes through unity, the cyclical or commensurate motion persists. Near the Ω-values which produced commensurate motion for $k = 1 - \varepsilon$, for ε small but > 0, there is also orderly cyclical motion for $k = 1 + \varepsilon$. However, infinitesimally close to each Ω value which rpoduces an incommensurate motion at $k = 1 - \varepsilon$(as $\varepsilon \to 0$ from positive values) there is, for $k = 1 + \varepsilon$, a domain of Ω in which the motion is chaotic. The incommensurate motion becomes unstable to chaos at k = 1.

This instability has not yet been analyzed in detail. However, the $k = 1$ incommensurate motion has been analyzed by two groups,[10] especially for the case in which the average speed is $w = (\sqrt{5} - 1)/2$. (Other irrational w's will probably show qualitatively similar but quantitatively different behavior.) They conclude that equation (11) still describes the motion, but that the continuous function (t) is very bumpy indeed at $k = 1$, while for $|k| < 1$, it was quite smooth. By quite bumpy I mean something rather specific and rather specifically awful. Consider the derivative of ϕ, $\phi'(t)$ in some small region of t. Pick the interval to be as small as you like. Furthermore pick some big number (say 10^{50}) and a small one (say 10^{-50}). Now I let k approach closer to one, but keep $k < 1$ always. Just so long as $k < 1$, $\phi'(t)$ is smooth and is always greater than zero but not infinite. It is true that I can always find some value of k, close to one, in which $q'(t)$ takes on both the value of your big number and also that of your small one in the specific interval you have chosen. If you choose more extreme numbers, I just have to go closer to $k = 1$. Clearly, we have reached a situation in which $\phi(t)$ exists and describes a more or less physical problem, but the function in question is, at $k = 1$, not differentiable anywhere.

This strange mathematical behavior can be seen experimentally It results in a power spectrum which contains an infinite number of discrete lines which are bunched together and pile up toward $\omega = 0$.

Experimentalists will, no doubt, be looking for power spectra of this character structure to perhaps observe the onset of chaos in the theoretically predicted manners. Also, theorists will, of course, be looking in their models and at experimental data hoping to see new forms of the onset of chaos.

Acknowledgements

This paper was written while I was in Israel enjoying the hospitality and support of the Israel Academy of Sciences and Humanities of Tel-Aviv University, and of the Weizmann Institute. It has been my pleasure to learn about dynamical systems from M. Feigenbaum and my students David Bensimon, Shoudan Liang, Subir Sarkar, Scott J. Shenker, Chao Tang, Michael Widom, and Alber Zisook. Mr Bensimon has helped in the production of some of the figures shown here.

References

1. R.P. Feynman. R.B. Leighton and M. Sands The Feynman Lectures on Physics (Addison-Wesley, 1964) Vol. II, Chapter 41.

2. R.B. May, Nature 261, 459 (1976) is a review of this subject.

3. E.N. Lorenz, J. Atmos. Sci. 20, 130 (1963).

4. See, for example, P. Collet and J.-P. Eckmann, Iterated Maps on the Interval as Dynamical Systems (Birkhauser, Boston, 1980).

5. Genesis:41.

6. M.J. Feigenbaum, J. Stat. Phys. 19, 25 (1978); 21, 669 (1979).

6a. See, for example P.S. Lindsay, Phys. Rev. Letts. 47, 1349 (1981) and J. Testa, J. Perez and C. Jeffries, Phys. Rev. Letts. 48, 715 (1982).

7. A. Libchaber and J. Maurer, J. de Physique 41, C3, 51 (1980).

8. G. Ahlers and R.P. Behringer, Progr. Theor. Phys. Suppl. 64, 186 (1978).

9. P. Manneville and Y. Pomeau, Phys. Lett. 75A, 1 (1979).

10. M.J. Feigenbaum, L.P. Kadanoff, and S.J. Shenker, Physica 5D, 370 (1982); D. Rand, S. Ostlund, J. Sethna and E.D. Siggia, Phys. Rev. Letts. 49 132 (1982).

APPLICATIONS OF SCALING IDEAS TO DYNAMICS

(Notes by R. de la Llave and J. Sethna for Lectures II, III, and IV)

Leo P. Kadanoff

The James Franck and Enrico Fermi Institutes
The University of Chicago
Chicago, IL 60637

Lecture II FROM PERIODIC MOTION TO UNBOUNDED CHAOS: INVESTIGATIONS OF THE SIMPLE PENDULUM

The problem we will discuss in this lecture has a long history. The basic work in the problem was done by A. Kolmogorov, V. Arnold and J. Moser; more recent work has been done by J. Greene, B.V. Chirikov, D. Escande, F. Doveil, and R. MacKay. I have worked on this problem in collaboration with S.J. Shenker; parts of this and related work were done in collaboration with M. Widom, A. Zisook, M. Feigenbaum and D. Bensimon.

The phenomenology of Hamiltonian systems is quite different from that of dissipative systems (which we'll discuss in the next lecture). In this lecture we shall analyze in detail the breakdown of a KAM curve and the onset of unbounded chaotic motion in a particular map. First, let us give three physical systems to motivate the study of this map.

First consider a pendulum (Figure 1)

$$m\ell\ 2\pi\ddot{\theta} = -\ mg\ \sin\ (2\pi\theta) \tag{II.1}$$

in which we choose units of θ so that $\theta = 1$ corresponds to 360°. Let us act upon this system periodically by modulating g, the force due to gravity (say, by wiggling the support of the pendulum)

$$g = g_0 + g_1\ \sin\ (\omega t) \tag{II.2}$$

We get equations

$$\begin{aligned} \dot{r} &= -k(t)\ \sin(\theta/2\pi) \\ \dot{\theta} &= r \end{aligned} \tag{II.3}$$

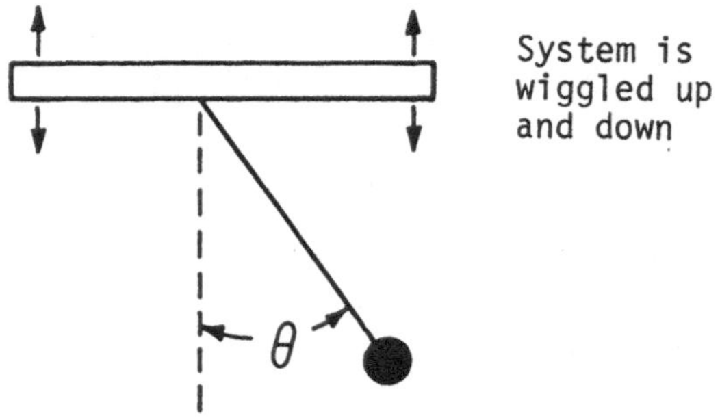

Fig. 1 Our first model.
A pendulum is accelerated up and down.

where r is the velocity of the pendulum. We can now use a trick due to Poincaré to transform this differential equation into a map. Observe the pendulum once each period of the force; let $r_j = r(t_j)$ and $\theta_j = \theta(t_j)$ where $t_j = (2\pi/\omega)_j$. Since the phase of the external force at time t_j is independent of j, one can integrate the equations of motion (II.3) over this period, expressing the new state of the pendulum in terms of its state one period earlier:

$$\begin{aligned} r_{j+1} &= F(r_j, \theta_j) \\ \theta_{j+1} &= G(r_j, \theta_j) \end{aligned} \tag{II.4}$$

In general, F and G will be some nonlinear functions periodic in θ. The simplest model which seems to capture the physics of this system is the standard map

$$\begin{aligned} r_{j+1} &= r_j - k/2\pi \sin(2\pi\theta_j) \\ \theta_{j+1} &= \theta_j + r_{j+1} \end{aligned} \tag{II.5}$$

also known as the Chirikov-Taylor model. (E. Fradkin and B. Huberman have studied this periodically modulated pendulum and have indicated how Eq. (II.4) can be converted to Eq. (II.5) in several limiting situations.)

The second system we'll use to motivate this map is an accelerator model (Figure 2). Envision a particle moving around a circular track, accelerated each time it enters a small box; the acceleration is provided by an AC field θ_j of the box at the time the particle enters; the equations (II.5) describe the state of the particle as it enters the box for the $j+1^{st}$ time in terms of its state as it entered the time before. One can reexpress Eq. (II.5) in the form

$$\theta_{j+1} - 2\theta_j + \theta_{j-1} = - k/2\pi \sin(2\pi\theta_j) \qquad \text{(II.6)}$$

which as a discrete version of Eq. (II.1) perhaps makes the connection to the pendulum problem more clear.

Finally, consider a solid state model of a one-dimensional array of atoms adsorbed on a periodic substrate (Figure 3). The jth atom feels a force from the springs connecting it to its two neighbors, and from the gradient of the potential energy at its position on the periodic substrate. If we choose $\theta_j = x_j/a$ to be the position of the jth atom x_j divided by the substrate lattice constant a, then

$$k_{spring} [(\theta_j-\theta_{j+1}) + (\theta_j-\theta_{j-1})] = k_{substrate} \sin(2\pi\theta_j) \qquad \text{(II.7)}$$

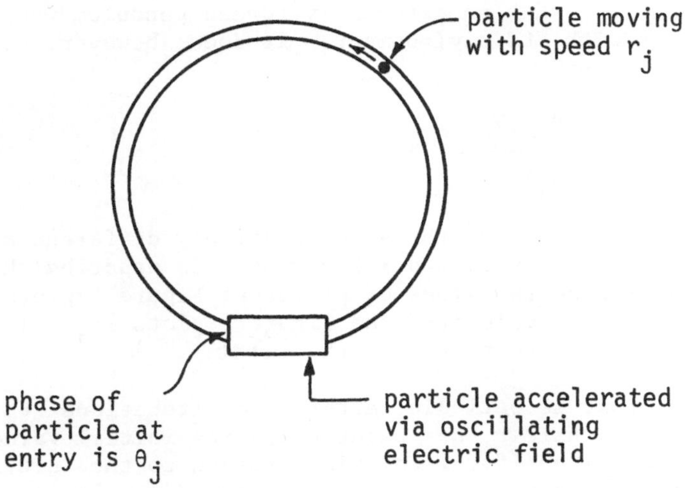

Fig. 2 The second model. A simplified accelerator.

describes a static configuration of atoms. This is of the form (II.6). The breakdown of the KAM surface has a physical meaning here as a pinning of the density wave of adsorbed atoms; this problem has been studied by S. Coppersmith, D. Fisher, S. Aubry, and others.

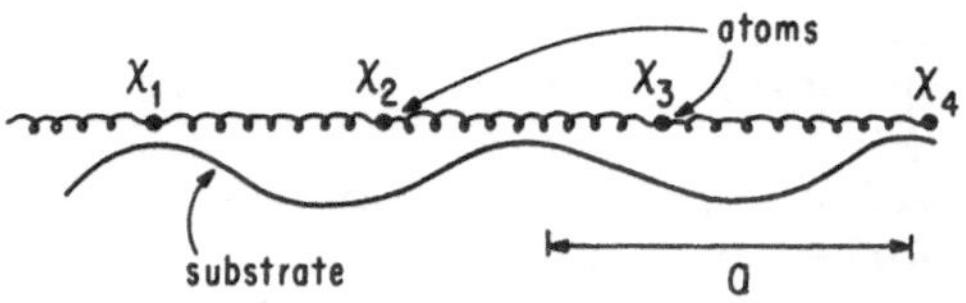

Fig. 3 A solid state example. Particles on a wavy surface.

The standard map (II.3) has no conserved energy; as noted above, it can represent an externally forced pendulum which exchanges energy with its environment. It does, however, obey Liouville's theorem

$$\frac{\partial(r_{j+1},\theta_{j+1})}{\partial(r_j,\theta_j)} = 1 . \tag{II.8}$$

The map possesses at least three qualitatively different kinds of orbits: periodic, chaotic, and KAM curves. To describe these orbits I shall draw two kinds of pictures: Figure 4 plots θ_j vs j to show the "average velocity" and Figure 5 plots (r_j,θ_j) in phase space to show the nature of the orbit.

The top orbit is periodic; after three iterations θ_j has increased by two while r_j has returned to its initial value; since θ and $\theta - 1$ are identified, this is a return to the initial state The orbit progresses q = 2 units in its period p = 3, so its average speed W is p/q = 2/3.

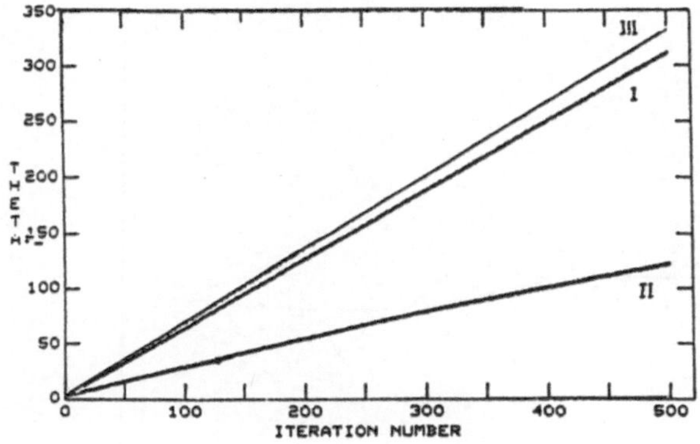

Fig. 4 Value of versus j for the three kinds of orbits. On this scale, seems to simply increase linearly.

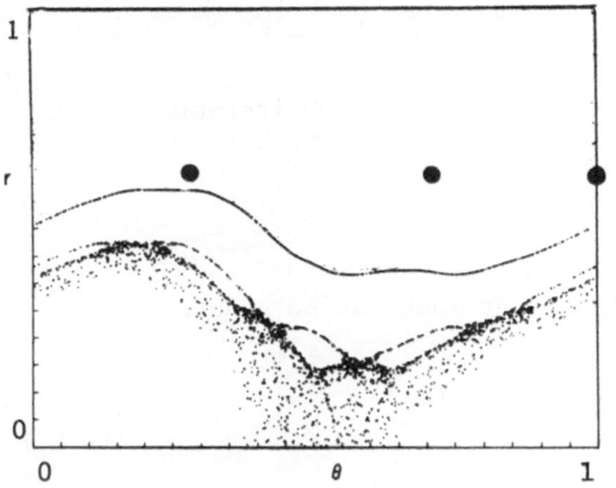

Fig. 5

Three kinds of orbits in the r - θ plane.

The bottom orbit is chaotic. Chaotic orbits appear to fill areas which we shall argue are bounded by the KAM curves. The middle curve is an example of such a KAM curve. All the points on the orbit fall on a smooth periodically climbing curve; it has a well-defined average speed of $W = (\sqrt{5}-1)/2$.

For large values of k, the chaotic region becomes unbounded in the vertical direction along r (Figure 6). For many (but not all) initial conditions the motion in this regime appears diffusive, with a diffusion constant

$$D = \lim_{n\to\infty} \frac{(r_{j+n}-r_j)^2}{n} \qquad \text{(II.9)}$$

which depends on K (Figure 6). Numerical studies due to Chirikov indicate that D grows from zero at a critical value (now known to be given by $k_c = 0.971\ldots$) like $(k-k_c)^{2.56}$. On the other hand, for

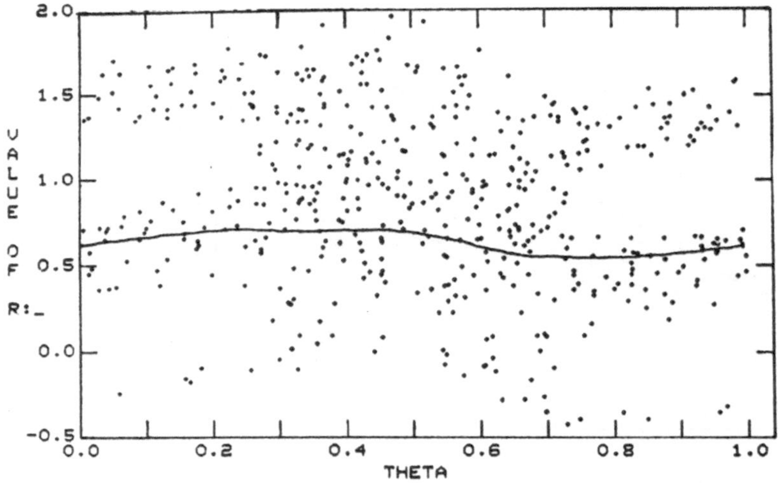

Fig. 6 Unbounded Chaos for $k > k_c$. Here 500 iterations of a single starting point are plotted when k = 1.8.

large values of k, $D \sim k^2$. The latter fact can be understood; write

$$r_{j+n} - r_j = \sum_{\ell=j}^{j+n} \left(\frac{k}{2\pi}\right) \sin(2\pi\theta_\ell)$$
$$\theta_{j+1} = \theta_j + r_{j+1} \tag{II.10}$$

For large k, the successive θ_j's can be considered virtually independent random numbers mod one, since the r_j change by a number of order k after each iteration. Thus

$$\begin{aligned}\overline{(r_{j+n}-r_j)^2} &= (k/2\pi)^2 \ \overline{(\Sigma \sin 2\pi\theta_\ell)^2} \\ &\approx (k/2\pi)^2 \ \Sigma \overline{(\sin 2\pi\theta_\ell)^2} \\ &\approx (k/2\pi)^2 \ n/2 \\ &= D/2\ n\end{aligned} \tag{II.11}$$

and $D = (k/2\pi)^2$.

Why is the chaos bounded for $k \leq k_c$? Let's start at k = 0, where the orbits of (II.3) lie on the straight horizontal lines r = W. For rational W, the orbits along r = W are periodic; for irrational W any orbit fills the entire line densely. Some of these latter orbits will form the KAM trajectories as we increase k; the rational orbits will destroy nearby irrational orbits to form chaotic regions.

For $k > 0$ but small, there are lots of KAM trajectories; in between any two, lies a chaotic region (in between any two irrationals lies a rational). We assert that the chaotic regions are confined by the persistence of horizontal KAM trajectories (as in Figure 5). Consider the chaotic region, for example, containing the fixed point $r = 0$, $\theta = 0$. The union of the images of a small neighborhood of the origin (Figure 7) under successive iterations of the map will in general form a very contorted open set. However, it must always include the origin, and cannot intersect a KAM surface (since points on a KAM surface are images only of other points on the surface). Thus no point in the neighborhood can ever leave the region bounded by horizontal KAM surfaces.

As we continue to increase k, more and more KAM curves break up, and the chaotic regions merge (Figure 8) until at $K = K_c$ the last remaining horizontal KAM curves disappear and unbounded vertical chaotic diffusion ensues. Empirically, the last surfaces to go have $W = (\sqrt{5}-1)/2$, the inverse of the Greek's golden mean

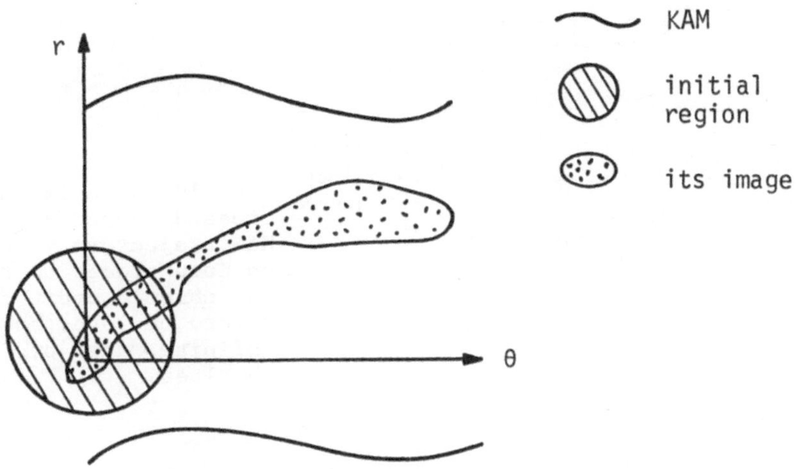

Fig. 7 Image of initial region including origin. Note that the image can never cross a KAM curve.

(up to an integer). That is, at k_c (Figure 9) only isolated horizontal curves are left amid the chaos, and these have become very crinkled. For $k > k_c$ these curves also break up; gaps form in them, changing the curve into a Cantor set (Figure 10).

The remainder of this lecture is devoted to describing the detailed mechanism for the break up according to the lines of the research carried on by Scott J. Shenker and myself.

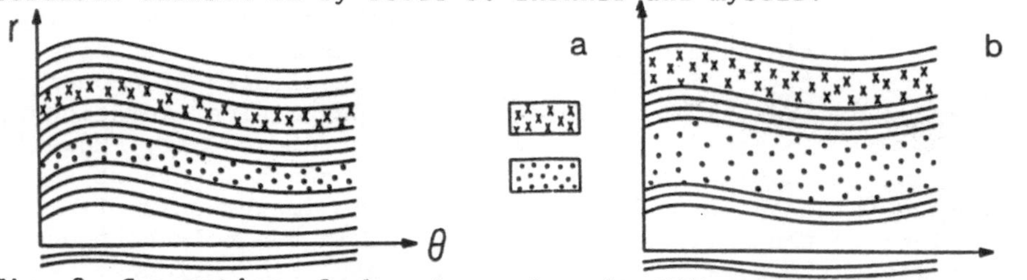

Fig. 8 Separation of chaotic regions by KAM trajectories. The x's and dots represent two different chaotic regions, separated by many KAM trajectories. In Fig. 8b $|k|$ is larger than in Fig. 8a. As $|k|$ increases, KAM trajectories disappear and chaotic regions merge.

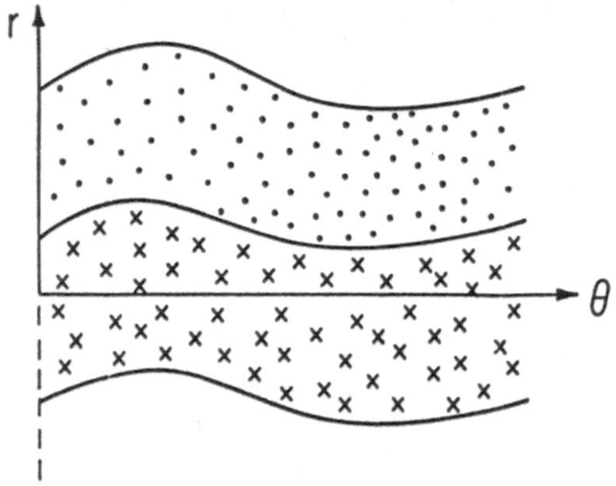

Fig. 9 At k_c, only a few well-separated KAM trajectories are left so that there are large chaotic regions.

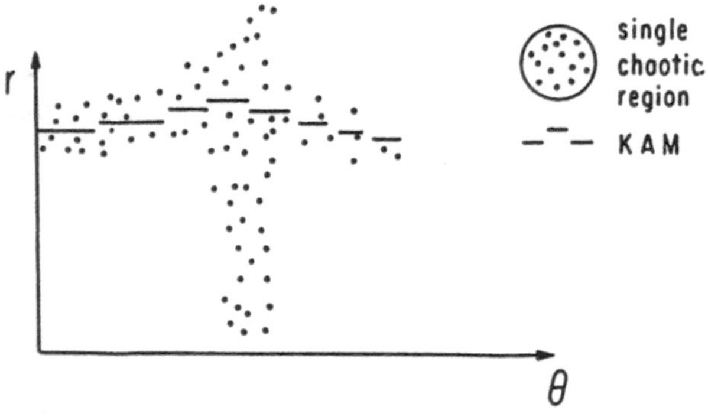

Fig. 10 As k is increased beyond k_c, the KAM trajectory breaks up into pieces so that the chaotic orbit can spread out over an infinite range of r-values.

How can we study this transition in detail? Greene has observed that these last KAM curves can be thought of as a limit of the particular periodic cycles with $W_n = p_n/q_n$ and $q_n = p_{n+1} = F_n$, the Fibonacci numbers. (F_n satisfies $F_0 = F_1 = 1$, $F_{n+1} = F_n + F_{n-1}$). In Figure 11, for example, we see two periodic cycles with W = 2/3 above the KAM surface, and two below with W = 3/5. These periodic cycles alternately converge to the KAM surface with $W = (\sqrt{5}-1)/2$ from above and below. Indeed, the solid line in Figure 11 actually is composed of two cycles with q = 4181 = F_{18}, as we can see in the expanded scale of Figure 12. Here note how magnificiently smooth the KAM surface appears on this length scale.

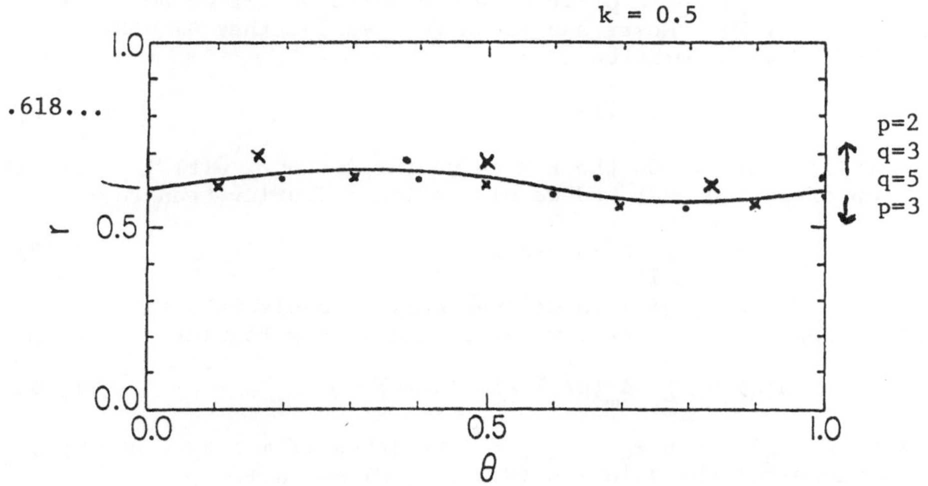

Fig. 11

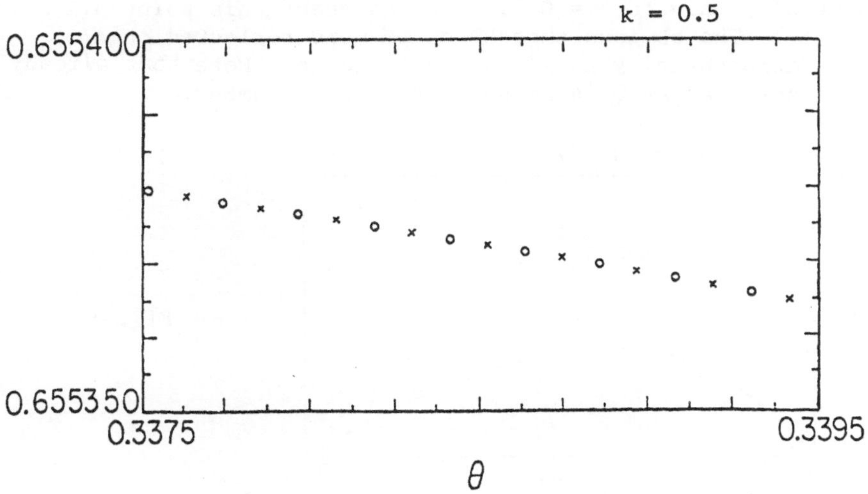

Fig. 12

Why do we choose precisely this sequence of periodic orbits to look at $W = (\sqrt{5}-1)/2$? Perhaps there are better answers, but roughly this sequence provides an orderly progression, converging rapidly and uniformly in its properties to those of the surface. One can think of the $n + 1^{st}$ cycle as "almost" being the nth cycle followed by the $n - 1^{st}$ cycle. Since the convergence is alternating a proper weighting of the previous two cycles is a natural choice for the next; for $W = (\sqrt{5}-1)/2$ the proper weighting is equal.

At this point it will be useful to introduce the Moser representation. Remember that W is an average speed for θ : $\theta_j = \theta_0 + Wj + O(1)$ on a periodic orbit or a KAM surface. If we define a time $t_j = t_0 + Wj$. Moser has shown for small k that $\theta_j = \theta(t_j)$ where $\theta(t)$ can be written

$$\theta(t) = t + u(t) \tag{II.12}$$

with $u(t+1) = u(t)$. On the KAM curve for $k < k_c$, $u(t)$ is a smooth function of period one, and we may define a Fourier transform

$$u(t) = \sum_{\omega=1}^{\infty} A_\omega \sin(2\pi\omega t). \tag{II.13}$$

For cycles $W = p/q$, $u(t)$ is defined only on a discrete set of points. Nonetheless, one may define a discrete Fourier transform

$$u(t) = \sum_{\omega=1}^{q} A_\omega(p/q) \sin(2\pi\omega t) \tag{II.14}$$

Clearly, $A_\omega(W_n)$ with $W_n = F_{n-1}/F_n$ vanishes if $\omega \geq F_n$. However, (Greene asserts) the values $A_\omega(W_n) \to A_\omega(W)$ as $n \to \infty$.

In Figure 13 we see $u(t)$ and $\omega A_\omega(W)$ for $W = \sqrt{5}-1/2$, $k = 0.5$. (We plot $u(t)$ only to $t = 0.5$; u is odd about this point $u(1/2 +t) = u(1/2 -t)$.) The obvious smoothness of u is reflected on $A(\omega)$, which is exponentially small as ω gets large. Note that already A_5 is larger than A_4 ; 4 is not a Fibonacci number.

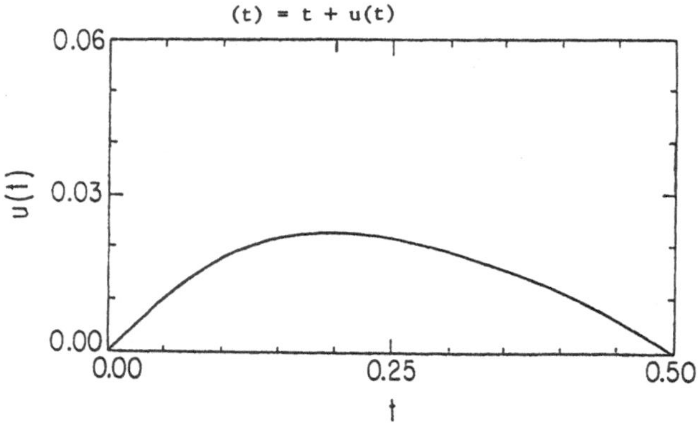

Fig. 13a

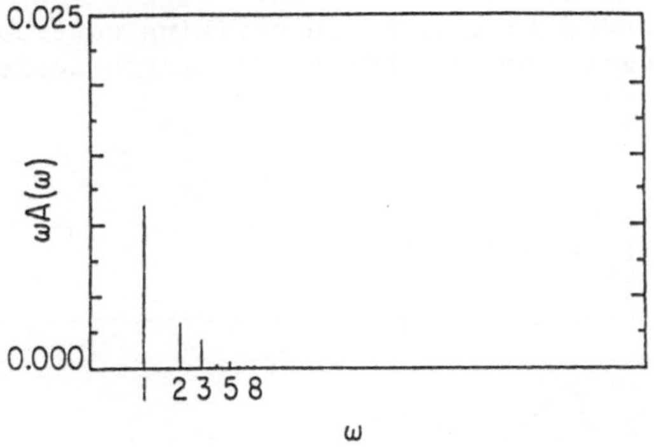

Fig. 13b

In Figure 14, we see u and ωA_ω for k = 0.9. New bumps have appeared in u, and the Fourier transform reflects this with prominent structure at the low Fibonacci numbers. (Notice that these peaks do not reflect numerical effects from our use of Fibonacci length orbits in approximating the KAM surface. These frequencies are naturally arising in the spectrum of the surface.)

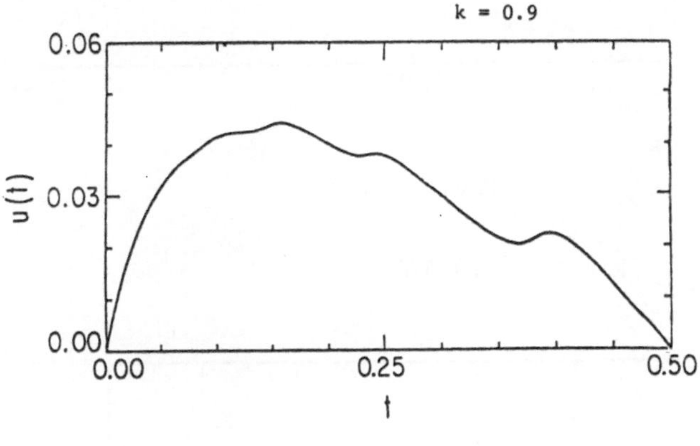

Fig. 14a

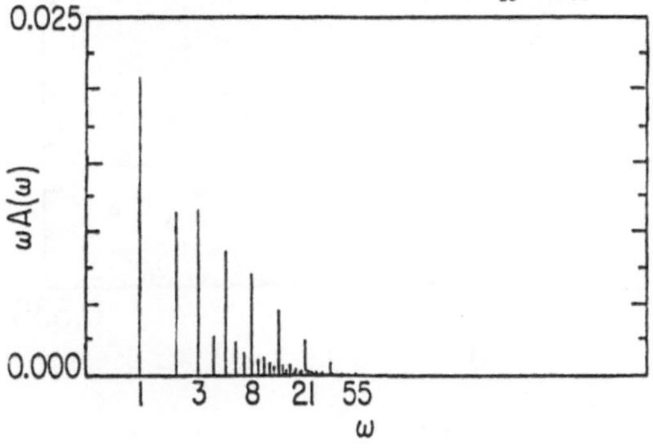

Fig. 14b

Finally, in Figure 15 we look at $k = k_c$. Figure 15a(1) is a graph of u(t) shown expanded in 15a(2). In striking contrast to Figure 12, it is very bumpy even on this small length scale.

$k = k_c = 0.9716354$

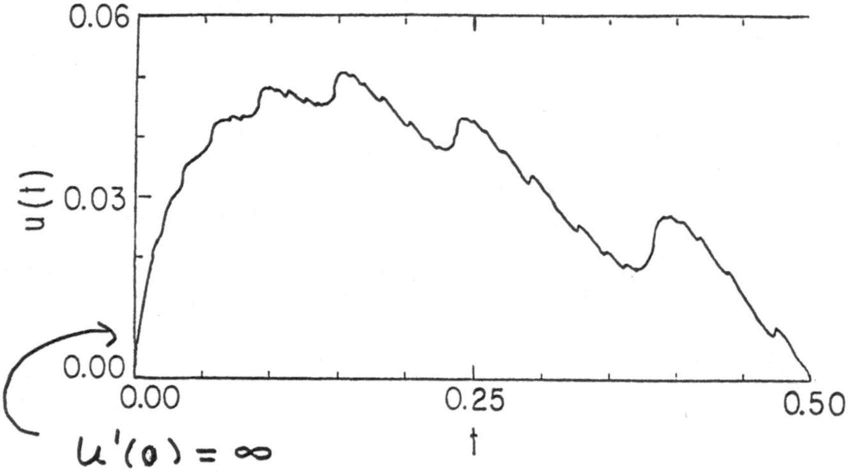

Fig. 15a(1)

$k = k_c = .97163$

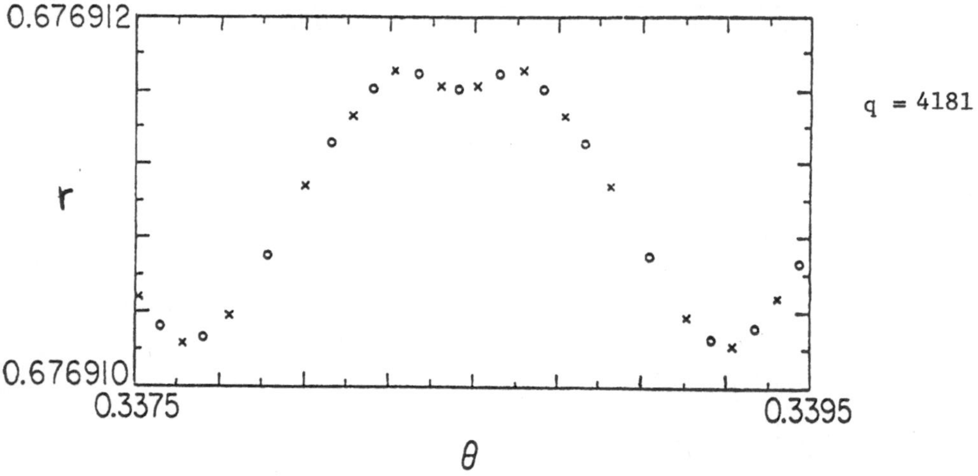

Fig. 15a(2)

A longer orbit than in 15a(2) would fill the gaps between these dots with more bumps like these. Figure 15b shows the spectrum $\omega \cdot A_\omega$ at k_c. One should ignore the low frequencies (long wavelengths) as they depend upon the details of the map and are not universal. One must also ignore the very high frequencies (short wavelengths) as the finite length of the orbit numerically introduces errors. The scaling region of the spectrum is expanded in Figure 15c.

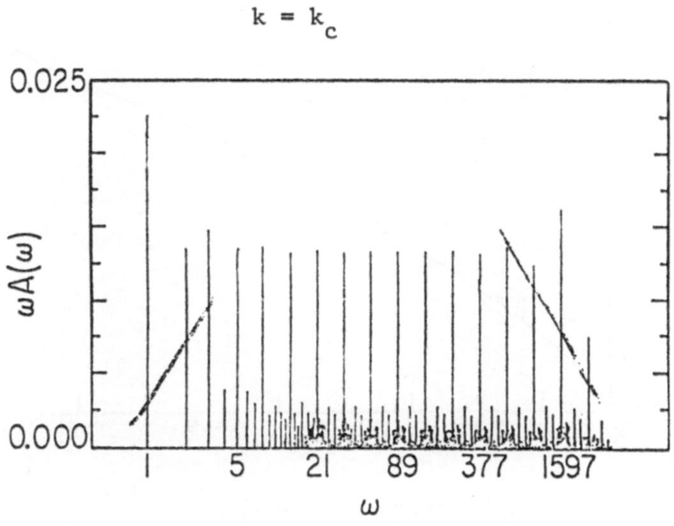

Fig. 15b

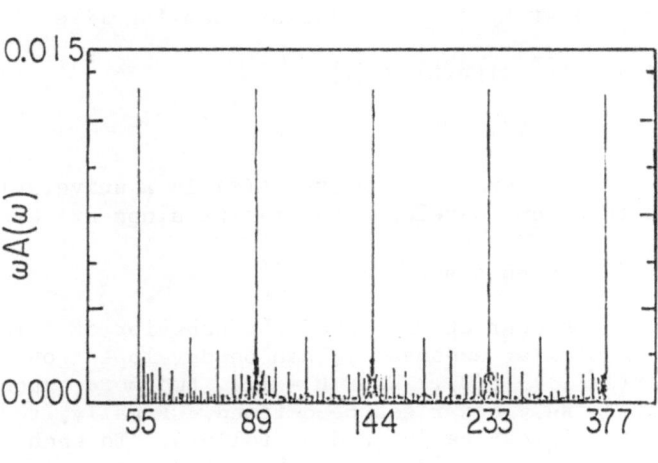

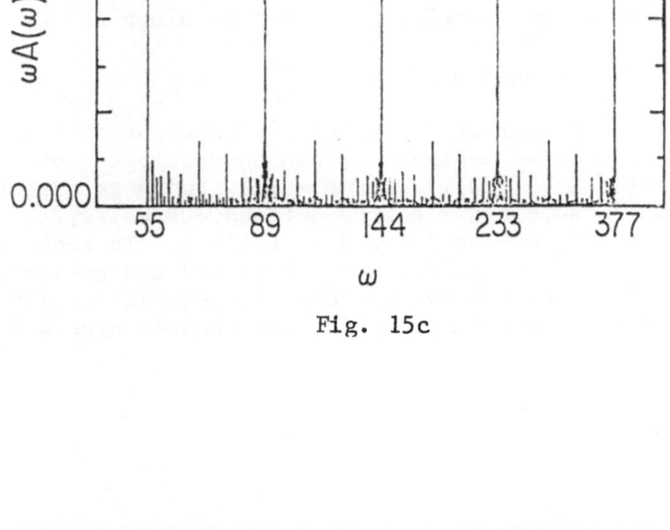

Fig. 15c

The big lines at the Fibonacci numbers have settled down . $|q_n A q_n|$ is clearly constant as $n \to \infty$. Also the smaller peaks are settling down, giving a beautiful self-similarity.

Let us look in more detail at $\Theta(t)$ near $t = 0$ (Figure 16). ($t = 0$ and $t = 1/2$ are special symmetry points of our map II.3.)

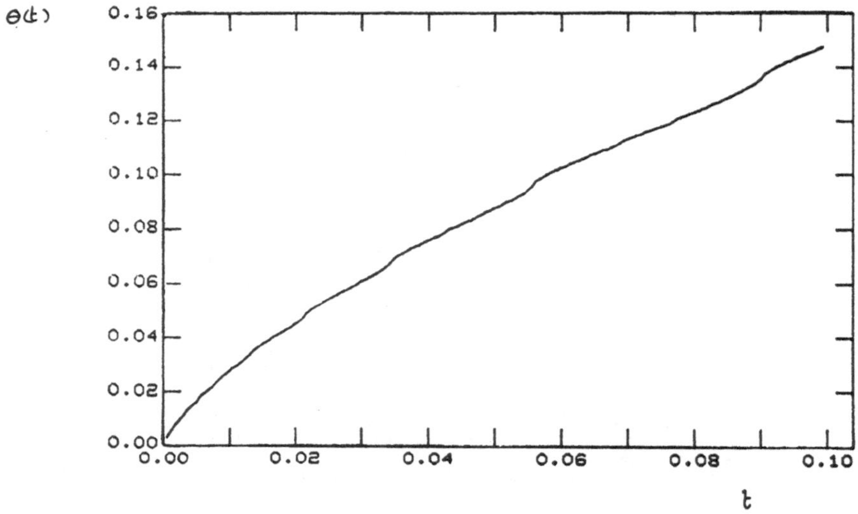

Fig. 16 Singular behavior at $k = k_c$.
Each bump in this plot of $\Theta(t)$ versus t marks an infinity in the slope.

The function $\Theta(t)$ at k_c is self-similar, scaling like

$$\Theta(t) = t^{x_0} \operatorname{sign}(t) \; \Theta(\tau)$$
$$\tau = \ln t / \ln W \qquad \text{(II.15)}$$

(LPK, Phys. Rev. Letters 1981) where $\Theta(\tau)$ is a universal function. Overall, the function develops an infinite slope at zero, with

$$\Theta(1/F_n) = \text{constant} \cdot (1/F_n)^{x_0} \qquad \text{(II.16)}$$

The exponent x_0 is characteristic of the behavior of Θ in this transition. A similar exponent y_0 can be developed governing the behavior of $r(1/F_n) - r(0)$. Near $t = 1/2$, two more exponents x_1 and y_1 governing Θ and r can be defined. Finally, two more exponents R_g and R_n can be defined as follows. To each Fibonacci ratio approximating the golden mean there are two periodic orbits - one stable (elliptic) and one unstable (hyperbolic). (In the figures the circles denote elliptic, the crosses hyperbolic orbits.)

The derivative of the q-times iterated map for a p/q cycle has two eigenvalues e^{λ} and $e^{-\lambda}$ (with λ pure imaginary for elliptic orbits). The residue is defined to be 1/4 $(2-e^{\lambda}-e^{-\lambda})$; its behavior for $W = F_{n-1}/F_n$ as $n \to \infty$ at $k - k_c$ is described by the exponents R_g and R_u.

These exponents are tabulated in Table I.

TABLE I

Potentially Universal Observables

W =	[1,1,1...]	[3,1,4,1...]	[2,2,2...]	[1,1,1...]
x_0	.7211	.72	.72	.72
x_1	1.093	1.09	1.10	1.09
y_0	2.329	2.33	2.34	2.33
y_1	1.957	1.96	1.96	1.95
R_s	.250	.25	.23	.25
R_u	-.255	-.25	-.23	-.26

(These are old numbers - now each is known to at least one more decimal point.) The first column shows the numerical values of these exponents for the breakdown of the last KAM curve in the standard map II.5. The last column shows the breakdown of the last curve in a different area preserving map of the plane; the agreement in the exponents shown has persisted using the better accuracy available today. The second column shows the breakdown of another KAM curve, whose winding number W has continued fraction which ends in [1,1,1...] [but begins with the digits of π]. Again, the exponents agree.

However, choosing $W = \sqrt{2}-1 = [2,2,2...]$ and varying k until this KAM curve breaks down, gives distinctly different values for R_s and R_u. The greater precision available today indicates that this difference is real, and extends to the other exponents as well. It is thought that the curve with winding number ending in ones separates chaotic regions at its breakdown, while (e,e,...) is surrounded by surviving curves (e.g. [2,2,2...,2,1,1,1...]) at its breakdown - its environment is quite different as it goes.

Thus we see universal behavior; the singularities at k_c are independent of the exact map, or (within limits) the exact W. We see scale invariant behavior (II.15) with structures which recur on all t-scales as $t \to 0$, 1/2, and which occurs again and again as the frequency W goes through Fibonacci numbers. We see behavior which is connected to Fibonacci numbers. We need an explanation for this behavior. In the next lecture I will describe the theory (the renormalization group) which has been used successfully to study this problem.

LECTURE III. THE MECHANICS OF THE RENORMALIZATION GROUP

The renormalization group was introduced into this kind of dynamics by M. Feigenbaum, J. Stat. Phys. 19 25 (1978); 21 669 (1979). This lecture is, of course, based upon the methodology he introduced.

Renormalization group methods have recently been found useful in several different contexts (Table II) in mechanics and related fields. In this lecture I hope to give you a feeling for the mechanics involved in these renormalization-group calculations. I'd like to explain the general backbone of practical steps involved in doing these calculations, using the Feigenbaum analysis of period doubling to illustrate features found common to all these problems. Overall, the steps fall into two stages. The first stage is a qualitative analysis of the behavior; the second is a quantitative implementation of a renormalization group.

TABLE II

Problems Attacked via R.G.

A. Period Doubling
   1. Dissipative Feigenbaum
   2. Conservative

B. Intermittency
   1. d = 1
   2. higher d

C. Quasi-Periodic to Chaotic Transition
   1. Hamiltonian   d = 2
   2. Circle map
   3. Complex plane   Siegel domain
   4. Schrödinger Equation

The first stage uses mainly computer exploration to gain a description of the behavior.

The first step is to qualitatively describe the transition: the phenomena and objects involved. In period doubling, one notices the period doubling cascade in the logistic map, the bands in the chaotic region, and the structure on all time and spatial scales at the transition.

A second step going past this intuitive understanding of the problem is obtaining scaling relations. These scaling relations are usually the most fruitful result of the analysis, they will largely determine the renormalization group equations as well as describe most of the interesting behavior. In the period doubling case, they correspond to noticing that, at the transition, $f^{2^n}(0) \approx \alpha^{-n}$ and, more generally, that defining g_n by

$$g_n(x) = \alpha^n f^{2^n}(\alpha^{-n}x) \qquad \text{(III.1)}$$

then the g_n approach a smooth function g as $n \to \infty$.

The third step in the exploratory analysis is to identify all the relevant symmetries of the problem. The reason why it is important to identify all the symmetries is because the renormalization group equation may act in different spaces depending on the symmetries of the problem and this can lead to different results. In the period doubling case, we are dealing with maps of the interval which are analytic and symmetric about a point, say zero, and have a quadratic maximum there.

In the fourth step, we test the universality of the observed scaling laws. In period doubling, different maps with the same symmetries have the same exponent (α is universal) but, maps with a non-quadratic maxima have different exponents.

Once this qualitative exploration is completed, publish. The second stage - implementing a renormalization group - will establish universality, compute exponents and, occasionally, lead to new insights about the problem. However, the return on effort invested is not high.

The first step in the second stage is to propose a renormalization group equation consistent with the scaling relations. In the period doubling case, Equation (III.1) immediately suggests the fixed point equation

$$f^* = \alpha f^*\left[f^*(\alpha^{-1}x)\right] \qquad \text{(III.2)}$$

The operation of composing a map with itself and scaling by α will be the renormalization group transformation for which we search for fixed points. Notice that the renormalization group equations usually follow immediately from the scaling laws - without finding the fixed point.

The second step is to describe the function space. In period doubling, the renormalization group transformation maps the space of analytic functions with a quadratic maximum at zero into itself. If you choose the wrong space, you may find spurious solutions (e.g. continuous but without derivatives).

The third step is to try to solve the fixed point equation. Unfortunately, this step is quite painful. The equations to be solved are usually functional equations involving compositions and scaling. There seem to be no systematic theories - or numerical methods - to deal with these problems. Indeed, there is no guarantee that experience gained in solving one functional equation will be useful for the next one. This step can easily take months. It is discouraging that the amount of information given by this step is not commensurately large. It may be that we can compute the scaling parameters much more accurately; however, we can find then to great accuracy (six or more figures) from more direct computations.

The last step is a linear perturbation analysis near the fixed point. This is the purpose of implementing the renormalization group - here one establishes the universality and finds the universal exponents. For the period doubling case, we iterate the renormalization group transformation

$$g_{n+1} = \alpha\, g_n \circ g_n(\alpha^{-1} x) \tag{III.3}$$

starting from a small perturbation of the fixed point

$$g_n = g + \psi_n \tag{III.4}$$

and we consistently keep only terms of first order in ψ .

Then we are led to

$$\psi_{n+1} = \alpha\, \psi_n(g(x/\alpha)) + \alpha\, g'(g(x/\alpha))\, \psi_n(x/\alpha)$$

This is a linear (functional) equation which is very accurately diagonalized once one knows the fixed point. When we do that, we find that there are eigenvalues

$$\delta = 4.669,\ \alpha,\ \ 1,\ \ldots.\text{(all simple)} \tag{III5}$$

and others of modulus less than one.

The eigenvalues with modulus less than one are technically called irrelevant; they describe perturbations which get damped asymptotically and do not change the critical behavior. The eigenvalues of modulus one (marginal eigenvalues) are dangerous because the linear theory does not suffice to describe their effects. The eigenvalues with modulus greater than one (relevant) describe perturbations which grow exponentially, and determine the scaling behavior near the transition.

The eigenvalue one, as noted above, needs special treatment.

A closer look at the problem reveals that, indeed, there is a one parameter family of solutions of the renormalization group equation. If $g(x)$ is a solution, so is $1/A\ g(Ax)$. So the eigenvalue one corresponds to these changes of coordinates and describes no new physics.

The eigenvalue α corresponds to changes of the origin (equivalently, to an eigenvector with a maximum not at zero). Thus, if we expand in the proper space (even functions) α will not show up as an eigenvalue.

The eigenvalue δ has no such simple description. We see that to produce a change of order one after 2^n iterations, it suffices to make a modification of the order δ^{-n} in perturbations:e.g. those which represent the deviation of r from r_c or of g_0 from the Feigenbaum point of infinite period doubling. Thus, if $f(x)$ is close in function space to the fixed point $g = f^*$, then except for disturbances with components in the relevant direction (eigenvalue δ) one retains exactly the same scaling behavior at long times.

This renormalization group analysis, when possible and successful, links closely several features expected in a theory of a transition to chaotic behavior. It describes the long time asymptotics, the oscillations on small spatial scales, with quantitative predictions independent of the concrete details of the model.

<u>Lecture IV</u> ESCAPE RATES AND STRANGE REPELLORS

Our group at Chicago has recently been shifting its attention away from these renormalization group treatments of the onset of chaos, towards the study of the chaotic region. Strange attractors seem very hard; I will describe instead our work on escape times for strange repellors.

This last lecture will be different in character from the previous ones. We will not present a finished theory. Rather we report on a numerical work (still in progress) with a heuristic explanation of the results. Our aim in presenting these results to a mathematical audience is to stimulate someone to sharpen our conjecture and to prove it. In the systems we have studied, the numerical evidence for our conjecture is compelling. However, we do not have a precise feeling about what conditions a system must satisfy for our result to hold, though we conjecture some sort of hyperbolicity will be an ingredient of them.

The quantity we study is an escape rate. Suppose f is a map of a manifold into itself, with an (unstable) invariant set contained in an open region R. Assume that when a point leaves R it never returns. We can imagine throwing a number (very big) N of points distributed according to Lebesgue measure and then following their trajectories under iteration of f. After n iterations only N_n remain in R (we call $\Gamma_n = N_n/N$ the "staying ratio") and we are interested in the large n behavior of this number.

One could imagine several types of long-time (large n) behavior for Γ_n. For example, some fraction of the points may never leave R (e.g. R contains an attractor), so $\Gamma_n \to \Gamma > 0$. Or, Γ_n could go to zero faster than exponentially (as it often will if R does not contain an invariant set). The only cases of interest to us are those in which Γ_n decreases to zero exponentially

$$\lim_{n\to\infty} \frac{\Gamma_{n+1}}{\Gamma_n} = e^{-\alpha} \tag{IV.1}$$

We call α the escape rate.

One way in which such systems may appear is when R is a neighborhood of a hyperbolic repellor. The following rather trivial example may clarify what we mean.

Suppose that we have a hyperbolic fixed point and we draw a square box very close to it (Fig.17). After a sufficiently large number of iterations

$$\Gamma_{n+k} \approx \prod_{|\Lambda|>1} \frac{1}{|\Lambda_j|^k} \Gamma_n \tag{IV.2}$$

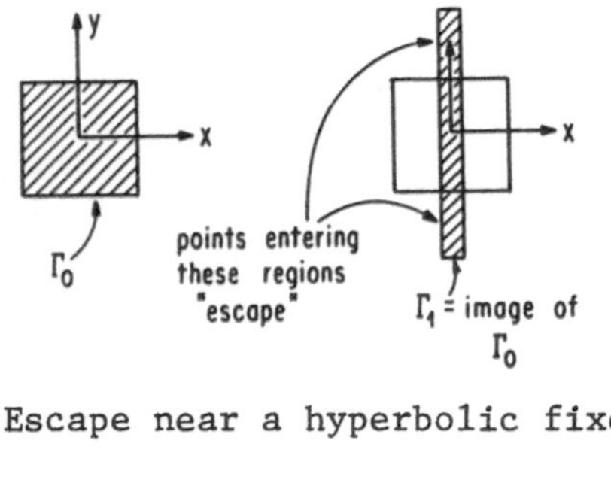

Fig. 17 Escape near a hyperbolic fixed point.

where Λ_j are the eigenvalues of the linearization at the fixed point. So that, in this case

$$e^{-\alpha} = \prod_{|\Lambda_j|>1} \frac{1}{|\Lambda_j|} \tag{IV.3}$$

Computing the eigenvalues and deciding which ones are bigger than others is a painful task. In the limit of large k we have

$$\prod_{|\Lambda_j|>1} \frac{1}{|\Lambda_j|^k} \simeq \frac{1}{|\det(1-\Lambda^k)|} \tag{IV.4}$$

so that we can compute the escape rate by

$$e^{-\alpha} = \lim_{k\to\infty} \frac{|\det(1-\Lambda^k)|}{|\det(1-\Lambda^{k+1})|} \tag{IV.5}$$

The fact that we have found numerically is a generalization of this, namely that the escape rate α can also be computed by

$$e^{-\alpha} = \lim_{n\to\infty} \frac{A_{n+1}}{A_n} \tag{IV.6}$$

where

$$A_n = \sum_{z\,\varepsilon\,\mathrm{FIX}\, f^n} \frac{1}{|\det(1-Df^n(z))|} \tag{IV.7}$$

where the sum is over all periodic points with period n which lie in the region R . This set is abbreviated as FIX f^n .

We can also obtain similar results for one-dimensional maps. In that case we take as our phase space the real line, as the map acting on it

$$f(x) = rx(1-x) \quad \text{with } r > 4 \tag{IV.8}$$

This is the range of parameters where most of the orbits go to infinity.

We choose our domain R to be the region [0,r/4]. One can convince oneself easily that all the points outside R rapidly iterate to infinity and never come back. After one iteration of f, the region [1,r/4] is mapped outside R, so $\Gamma_1 = (r/4-1)/(r/4)$ (Figure 18.)

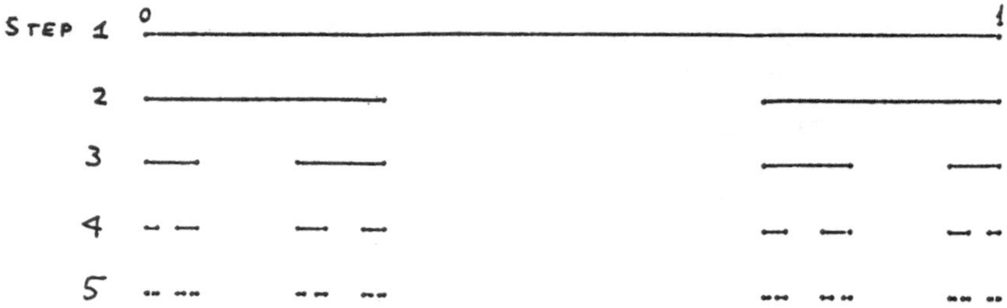

Fig. 18 All points with $x > 0$ rapidly map toward $-\infty$. The successive steps here show x-values which give positive images after 1,2,3,... iterations. These admissible x-values approach a Cantor set.

After a second iteration, $f^{-1}[1,r/4]$ leaves R, chopping a piece out of the center of the segment [0,1] remaining after the first iteration. In the nth iteration, the map chops an interval out of each remaining segment left after n-1 iterations, leaving a Cantor set of points that remain forever (our strange repellor).

The computation of the staying ratio can be done very efficiently in this example by keeping track of these intervals. The computation of the periodic orbits (and their derivations) on the Cantor set remaining can also be implemented very effectively. The numbers we obtain as limits of Γ_{n+1}/Γ_n and A_{n+1}/A_n agree to six decimal places for various values of r. The dependence of α on r seems to have a square root singularity for $r \gtrsim 4$.

$$\alpha(r) \simeq (r-4)^{\frac{1}{2}}$$

and for large r

$$\alpha(r) \sim \ln r$$

$\alpha(r)$ was calculated directly (via the intervals) and compared with $\ln A_n/A_{n-1}$ for $n = 2,3,\ldots,13$. Excellent agreement between the two calculations was obtained.

Before turning to our other examples, let us give a first attempt at a heuristic derivation of our results. Although our derivation is full of gaps, we hope it will be stimulating.

Consider the Frobenius-Perron operator P (discussed in Collet's lectures) which gives the time evolution of the probability density:

$$P[\rho](x) = \int d\bar{x}\; P(x,\bar{x})\; \rho(\bar{x}) \qquad (IV.9)$$

$$P(x',x) = \delta(x'-f(x)) \qquad (IV.10)$$

P maps a distribution of points to the distribution of the images of the points under f. The composition of two such operators is done by convolution

$$(PQ)(x',x) = \int d\bar{x}\; P(x',\bar{x})Q(\bar{x},x) \qquad (IV.11)$$

so that

$$P^n(x',x) = \delta(x_j'-f^n(x)) \qquad (IV.12)$$

Now, using the simple properties of the δ-function and the definition (IV.7) of the A_n, one can see

$$A_n = \int_R dx\; \delta(x-f^n(x)) = \int_R dx\; P^n(x,x) = \text{Trace } P^n(x) \qquad (IV.13)$$

On the other hand

$$\Gamma_n = \frac{\int_R dx' \int_R dx\; P^n(x',x)}{\int_R dy} = \langle|P^n|\rangle \qquad (IV.14)$$

If there were a spectral theorem for such operators - which probably there is not - then we could write

$$P^n(x',x) = \sum_\mu e^{-n\varepsilon_\mu}\, \psi_\mu(x')\phi_\mu(x) \qquad (IV.15)$$

(Notice that the operator is not symmetric so that the left eigenfunctions are not necessarily the same as the right eigenfunctions.)

so that

$$A_n = \sum_\mu e^{-n\varepsilon_\mu} I_\mu \qquad \text{(IV.16)}$$

$$\Gamma_n = \sum_\mu e^{-n\varepsilon_\mu} \gamma_\mu \qquad \text{(IV.17)}$$

for some weights I_μ and γ_μ clearly such a spectral result would imply our conjecture with α equalling the lowest eigenvalue ε_0 .

Now, while the spectral analysis of our operator P may not exist, it is reassuring to note that equation (IV.16) has been derived properly by R. Bowen and D. Ruelle (with the I_μ integers); one can hope that (IV.17) may also be proven.

Let us conclude with two more examples which we have investigated numerically - escape from a neighborhood of a Julia set, and escape in a three-dimensional area preserving map .

We will not try to describe Julia sets in detail here; they have been treated in detail by Professor Douady in this volume. Let us just remember that they form the boundary in the complex plane between the points whose orbits are bounded under iterations of a polynomial and those who escape to infinity.

As a quick first example of the Julia set repellor, consider the Julia set for the logistic map considered in the previous example. It can be shown that the Cantor set on the reals are the only points with bounded orbits. Since now each point on a periodic orbit has two unstable directions in the complex plane, the formula (IV.7) for A_n must be modified for use in C:

$$A_n = \sum_{z \in \mathrm{FIX}\, f^n} \frac{1}{\left|1 - \frac{df^n}{dz}\right|^2} \qquad \text{(IV.18)}$$

Also, the Γ_n are modified - points are scattered now in a neighborhood in the complex plane. Numerically, the escape rate agrees with the predicted value.

As a second example which we can analyze fully, consider the map $z' = z^2$, with R a thin annulus including $|z| = 1$. The Julia set is the circle $|z| = 1$, and half the points escape after each iteration, $\Gamma_n = 2^{-n}$.

The periodic points with period n in R (solutions of $z^{2^n} - z = 0$) are $z_k = e^{2\pi i k/(2^n-1)}$, $k = 0 \ldots 2^n-2$. The derivative

$$\frac{df^n}{dz} = \frac{d}{dz}\left(z_k^{2^n}\right) = 2^n z_k^{2^n-1} = 2^n ,$$

so

$$A_n = \sum_{k=0}^{2^n-2} \frac{1}{|1-2^n|^2} = \frac{2^n-1}{(1-2^n)^2} = \frac{1}{2^n-1} \qquad \text{(IV.19)}$$

$A_n \to \Gamma_n$ for large n.

Finally, we have numerically studied the map

$$f(z) = z^2 + C \qquad C \in \mathbb{R} \qquad \text{(IV.20)}$$

(see Figure 19 a,b,c).

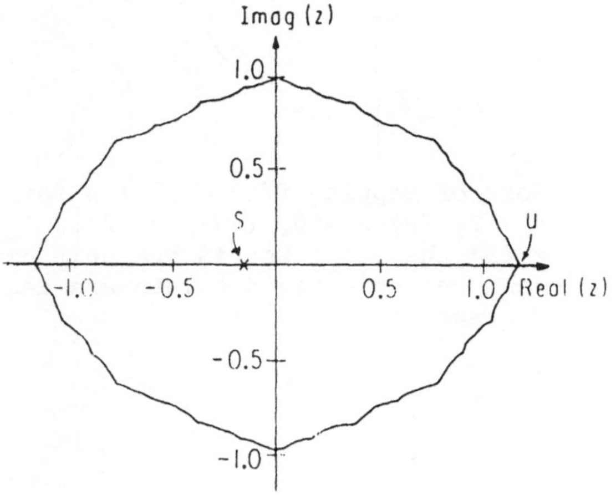

(a)

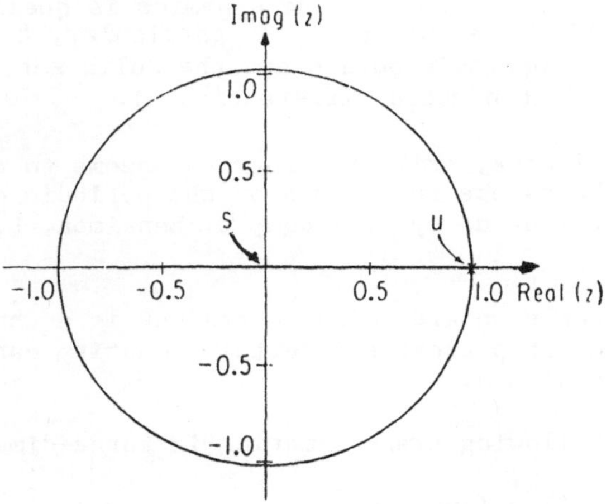

(b)

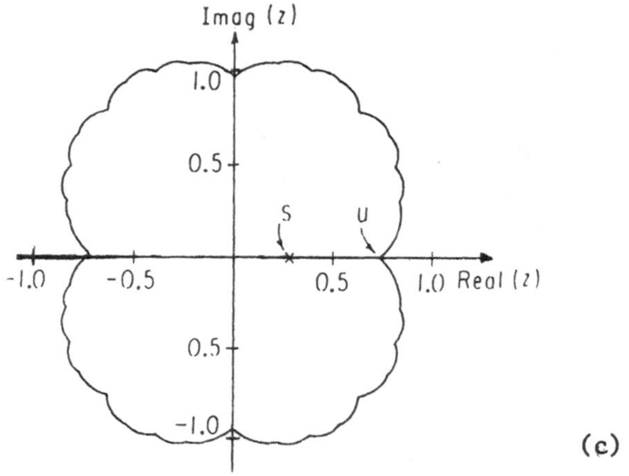

Fig. 19 Julia Sets of mapping $f(z) = z^2 + p$ for small p. (a) $p = -.2$, (b) $p = 0$, (c) $p = .2$. In case (b) the Julia Set is the unit circle. Stable and unstable fixed points are denoted s and u respectively.

We remark that, since the Julia set of $z' = z^2$ is hyperbolic, as proved in Lanford's lectures, the dynamics is qualitatively the same for all small values of p . In particular, there should be the same number of periodic points and the Julia set should be a continuous curve (but nowhere differentiable).

The numerical work, still in progress, seems to confirm the predictions of the escape rate based on the periodic orbits. This work is being carried out by M. Widom, D. Bensimon, L. Kadanoff and S.J. Shenker.

The last example we are going to present is a three-dimensional problem that, since it preserves a certain quantity, can be reduced to a two-dimensional one.

Define the following transformation in three-dimensional space.

$$(u',v',w') = (2uv - w, u, v) \qquad (IV.21)$$

Apart from the interest it has in this context we point out that this problem has been introduced for its relation with quasi-Schrodinger equations with quasi-periodic potentials (see M. Kohamoto, L. Kadanoff and C. Tang, Phys. Rev. Lett. 1983 for further details).

It is easy to check that this transformation preserves the usual euclidean volume and that, if we define the quantity

$$A(u,v,w) = u^2 + v^2 + w^2 - 1 + \\ - 2\,uvw \tag{IV.22}$$

then we have

$$A(u',v',w') = A(u,v,w)$$

(This function A should not be confused with the A_n introduced before.)

The most effective way of dealing with this three-dimensional first order recursion is to transform it to a third order one dimensional recursion putting

$$(u_j,v_j,w_j) = (y_j,\ y_{j-1},\ y_{j-2}) \tag{IV.23}$$

this recursion becomes

$$y_{j+1} = 2\,y_j\,y_{j-1} - y_{j-2} \tag{IV.24}$$

If we restrict ourselves to the surface A = 0, that is

$$0 = y_j^2 + y_{j-1}^2 + y_{j-2}^2 + \\ - 2\,y_j\,y_{j-1}\,y_{j-2} \tag{IV.25}$$

the problem becomes very easy to analyze in a certain region.

The surface A = 0 is qualitatively like a disc of radius one with four arms just touching but not really attached (see Figure 20).

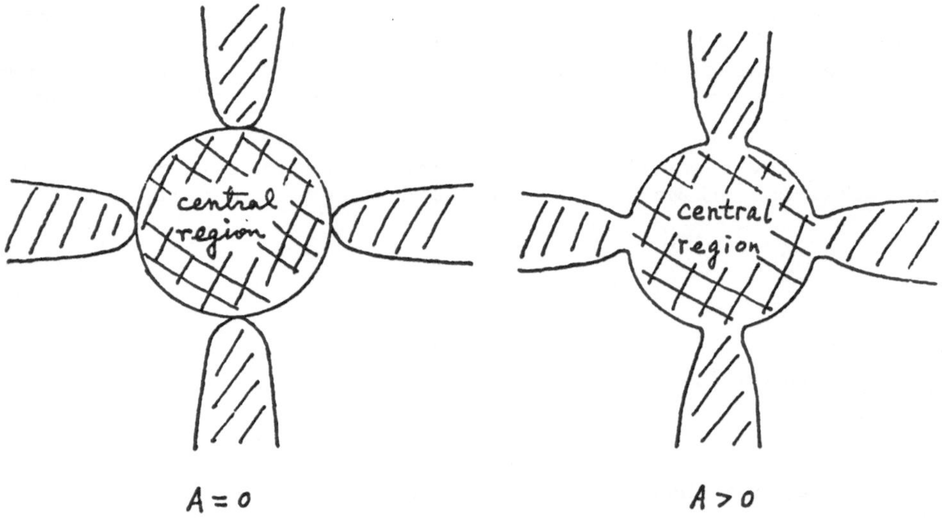

Fig. 20 Topology of the manifold (IV.22)

The dynamics of the central region can be described very conveniently in the θ_j variables defined by

$$y_j = \cos 2\pi\, \theta_j$$

Then, the recursion relation becomes

$$\theta_{j+1} = \theta_j + \theta_{j-1}$$

which is Anosov. One can easily find the periodic orbits which fill the surface densely. On the arms, the map (IV.21) generally moves the points off to infinity, even for $A > 0$, then the manifold becomes a sphere with four arms which go off to infinity.

Again, we just report that our numerical computations of the escape times seem to bear with the prediction from the periodic orbits. In this case, the analysis has been carried out by C. Tang and L. Kadanoff.

Acknowledgements

Sections II, III and IV come from a set of notes taken by R. de la Llave and J. Sethna. I would like to thank them for their very hard work; I suspect that the notes are an improvement upon the lectures. The figures were prepared by many people including: Scott Shenker, Subir Sarkar, David Bensimon and Chao Tang.

INTRODUCTION TO HYPERBOLIC SETS

*Oscar E. Lanford III**
IHES
91440 Bures-sur-Yvette
France

1. Preliminaries

One of the most illuminating general observations about dynamical systems is that it often happens that orbits starting very close together diverge exponentially. Exponential separation of orbits gives rise, notably, to the sensitive dependence on initial condition which accounts for the apparently stochastic behavior of deterministic dynamical systems. In these lectures, I will discuss systems which have a technically very strong version of the property of exponential separation of orbits, ones in which there are no neutral separations between nearby orbits so that each pair separates exponentially either forward or backward in time (and most separate both forward and backward) with strong uniformity assumptions on the rate of separation. Separation of orbits will not, however, be required everywhere in the state space of the system, but only in the neighborhood of some compact invariant set. Thus, the objects we will analyze are, roughly, compact invariant sets with the property that pairs of orbits starting out very near to each other and remaining near the set in question diverge exponentially in a uniform way. Such sets are called *hyperbolic sets.* (The above is intended only as a very general indication of what a hyperbolic set is and will be misleading if taken too literally; the formal definition is given in Section 3.)

The study of hyperbolic sets has led to a rich and deep mathematical theory which grew out of work of Anosov and Smale in the 1960's, with important contributions from many others. From the point of view of applications this investigation has not been as successful as might have been hoped; non-trivial attracting sets arising in examples studied so far generally do not satisfy the strong uniformity assumptions essential for this theory. This discouraging estimate of the practical usefulness of the theory of hyperbolic sets may change as more (and higher-dimensional) examples are studied. Even if it does not, the thoroughly non-trivial mathematical ideas developed in this theory are a promising starting point for the analysis of sensitive dependence on initial conditions in more general situations.

My objective in these notes is to present some of the basic elements of the theory of hyperbolic sets with a minimum of technical complexity and mathematical prerequisites. We will consider only mappings f and have nothing to say about flows despite the fact that the theory of hyperbolic flows is not an entirely routine extension of the theory for mappings. Our mappings will be assumed to be invertible and (at least)

*Preparation of these notes was supported in part by the National Science Foundation through Grant MCS81-07086 at the University of California, Berkeley.

continuously differentiable. We call the space on which they act M, and we will write all formulas for the case where M is an open set in m-dimensional Euclidean space, in order to avoid the distractions involved in the coordinate independent theory of more general manifolds. This convention will permit us to treat things like derivatives in a concrete and elementary way. In particular, $Df(x)$ will denote the $m \times m$ Jacobian matrix of first partial derivatives of the components of f at the point x.

I have not been able to include in these notes either historical remarks or an adequate bibliography. For these matters, as well as for a more systematic development of the theory sketched here, I refer the reader to Shub's excellent monograph [2].

2. Hyperbolic fixed points. Stable and unstable manifolds.

In this section we review some facts about hyperbolic fixed points of diffentiable mappings. In view of future applications, we call the reader's attention to the fact that the theory described in this section is valid for mappings on infinite dimensional spaces as well as finite dimensional ones. Indeed, beyond some care in the use of spectral theory, there is very little need to modify the treatment to gain for this useful extra generality.

Let L be a (bounded) linear operator on a Banach space E. We say that L is *hyperbolic* if the spectrum of L is disjoint from the unit circle, or, equivalently, if there is a direct-sum decomposition

$$E = E^s \oplus E^u$$

where E^s and E^u are closed linear subspaces, invariant under L, such that the spectrum of the restriction of L to $E^s (E^u)$ is strictly *inside (outside)* the unit circle. Intuitively, E^s is the space spanned by the eigenvectors (and generalized eigenvectors, if L is not diagonalizable) whose corresponding eigenvalues have modulus smaller than one. Note, however, that in order to get these eigenvectors it is generally necessary to extend the space E to allow for multiplication by complex numbers, whereas the above direct-sum decomposition requires no such extension.

We will refer to the vectors in E^s as *stable* or *contracting* vectors for L. E^s can be characterized as the set of vectors ξ such that

$$L^n \xi \to 0 \text{ as } n \to \infty,$$

or as the set of vectors ξ such that

$$\|L^n \xi\| \text{ remains bounded as } n \to \infty.$$

Furthermore, the convergence to zero of the $L^n \xi, \xi \in E^s$ is *exponential* and *uniform:* There are constants c, λ, with $\lambda < 1$, such that

$$\|L^n \xi\| \leqslant c \lambda^n \|\xi\| \text{ for all } \xi \in E^s.$$

It need not be true in general that

$$\|L \xi\| < \|\xi\| \text{ for all non-zero } \xi \in E^s,$$

and in this sense it is not quite precise to refer to E^s as the space of contracting vectors. It is however possible to replace the original norm with an equivalent one in such a way that, for some $\lambda < 1$ (not necessarily the same as the one above)

$$\|L \xi\| \leqslant \lambda \|\xi\| \text{ for all } \xi \in E^s. \tag{2.1S}$$

Similarly, assuming that L is invertible, E^u can be characterized as the set of vectors ξ

for which $\|L^{-n}\xi\|$ converges to zero, or, alternatively, remains bounded, and we can renorm so that

$$\|L^{-1}\xi\| \leqslant \lambda \|\xi\| \text{ for all } \xi \in E^u \tag{2.1U}$$

holds simultaneously with (2.1S). Note, however, that the set of vectors ξ such that $\|L^n\xi\| \to \infty$ as $n \to \infty$ is *not* E^u; it is rather the set of all vectors with non-zero E^u component.

A fixed point x_0 for the continuously differentiable mapping f is said to be a *hyperbolic* fixed point if $Df(x_0)$ is a hyperbolic linear operator. For the remainder of this section x_0 will denote a hyperbolic fixed point for f, and $E^s(E^u)$ the contracting (expanding) subspace for $Df(x_0)$. We will assume that the norm has been arranged so that there is a constant $\lambda < 1$ such that (2.1S),(2.1U) hold (with L replaced by $Df(x_0)$).

We now formulate a theorem asserting the existence of an invariant manifold for f, passing through x_0, which can be thought of as a non-linear generalization of the linear subspace

$$\{x_0 + \xi^s : \xi^s \in E^s\}$$

invariant under the linearization $Df(x_0)$ of f at x_0, and on which this linearization is contractive. We will use systematically, as coordinates for a neighborhood of x_0, pairs (ξ^s, ξ^u) with $\xi^s \in E^s$, $\xi^u \in E^u$; the corresponding point x is $x_0 + \xi^s + \xi^u$.

Theorem 2.1. Stable Manifold Theorem for Hyperbolic Fixed Points. *For $\epsilon > 0$ sufficiently small:*

1. To each $\xi^s \in E^s$ with $\|\xi^s\| < \epsilon$, there corresponds exactly one $\xi^u \equiv w_s(\xi^s) \in E^u$ such that, writing x for $x_0 + \xi^s + \xi^u$,

$$d(f^n(x), x_0) < \epsilon \text{ for all } n > 0.$$

2. For x as in 1., the sequence $(f^n(x))_{n \geqslant 0}$ not only remains near x_0 but converges exponentially to it. There exists $\lambda_1 < 1$ such that

$$d(f^n(x), x_0) \leqslant \lambda_1{}^n d(x, x_0)$$

3. The mapping w_s (from a neighborhood of 0 in E^s into E^u) is continuously differentiable; both w_s and Dw_s vanish at 0. If f is r times continuously differentiable, so is w_s.

We will not prove the Stable Manifold Theorem here. A proof can be found in Chapter 5 of Shub [2], and a sketch in the spirit of these notes in Lanford [1].

Some of the content of the Stable Manifold Theorem can be put into more geometrical language as follows: For $\epsilon > 0$, define

$$W^s_\epsilon = \{x : d(f^n(x), x_0) < \epsilon \text{ for } n \geqslant 0\}.$$

It is immediate from the definition that f maps W^s_ϵ into itself. The theorem then says that, for sufficiently small ϵ, W^s_ϵ is a manifold, as smooth as f, passing through x_0, whose tangent space at x_0 is exactly E^s.

The sets W^s_ϵ (ϵ small) are called *local stable manifolds* for f at x_0. We can also define the *(global) stable manifold* as:

$$W^s = \{x \in M : f^n(x) \to x_0 \text{ as } n \to \infty\}.$$

If $x \in W^s$ and $\epsilon > 0$ then $f^n(x) \in W^s_\epsilon$ for some n and conversely, i.e.,

$$W^s = \bigcup_{n=0}^{\infty} f^{-n} W^s_\epsilon.$$

Thus, W^s is made up of countably many pieces each of which is an imbedded disk with the same dimension as E^s. Nevertheless, taken as a whole, W^s can (and often does) accumulate on itself in a complicated way. The following example gives some idea of what the possibilities are:

Example: The quotient space $\mathbf{R}^2/\mathbf{Z}^2$ is one of the standard ways of representing a two-dimensional torus. We define a (particular) *linear automorphism* of the two dimensional torus by passing to the quotient modulo $\mathbf{Z}^2$ of the mapping

$$(x_1,x_2)\rightarrow(2x_1+x_2,x_1+x_2)$$

on $\mathbf{R}^2$. The image of the origin under passage to quotients is a fixed point; the derivative of the mapping at this fixed point (and everywhere else) has matrix

$$\begin{pmatrix} 2 & 1 \\ 1 & 1 \end{pmatrix}$$

Since neither of the eigenvalues $(3\pm\sqrt{5})/2$ of this matrix is on the unit circle, the fixed point is hyperbolic. The contracting space is one dimensional. In view of the simple linear form of the mapping, it is easy to see that the global stable manifold is the image under passage to quotients modulo $\mathbf{Z}^2$ of the linear subspace of $\mathbf{R}^2$ generated by the contracting eigenvector of this matrix, i.e., the line in the plane passing through the origin with slope $-(1+\sqrt{5})/2$. Since the slope is irrational, this image wraps densely around the torus (but without ever crossing itself.)

Local and global *unstable* manifolds are defined respectively as:

$$W_\epsilon^u=\{x:d(f^{-n}(x),x_0)<\epsilon \text{ for all } n\geqslant 0\}$$

$$W^u=\{x:f^{-n}(x)\rightarrow x_0 \text{ as } n\rightarrow\infty\}.$$

i.e., as the local and global stable manifolds for f^{-1}. By this last observation, the properties of unstable manifolds can easily be read off from the properties already formulated for stable manifolds.

3. Hyperbolic sets.

We consider a differentiable mapping f with differentiable inverse, acting on a state space M which we take to be an open subset of $\mathbf{R}^m$. For any n and any $x\in M$, $Df^n(x)$ is an $m\times m$ matrix. We will say that a vector $\xi\in\mathbf{R}^m$ is a *contracting* or *stable* vector for f at x if $\|Df^n(x)\xi\|$ goes to zero exponentially as n goes to infinity, i.e., if there are constants c,λ with $\lambda<1$ such that

$$\|Df^n(x)\xi\|\leqslant c\lambda^n\|\xi\| \text{ for all } n\geqslant 0.$$

(At this point, c and λ can depend in an arbitrary way on x and ξ.) For any fixed x, the set of contracting vectors forms a linear subspace of $\mathbf{R}^m$, which we will denote by $E^s(x)$. We now claim that c and λ may be taken to depend only on x, not on $\xi\in E^s(x)$. To see this: Let $\xi_1,\ldots,\xi_j$ be a basis for $E^s(x)$ with $\|\xi_i\|=1$ for each i from 1 to j. For each i, let c_i and $\lambda_i<1$ be such that

$$\|Df^n(x)\xi_i\|\leqslant c_i\lambda_i{}^n \text{ for all } n\geqslant 0.$$

Also, let D be a constant such that

$$\sum_{i=1}^{j}|a_i|\leqslant D\|\sum_{i=1}^{j}a_i\xi_i\|$$

for all $a_1, \ldots, a_j$. It is then easy to see that if we put

$$\lambda = \max\{\lambda_1, \ldots, \lambda_j\}$$
$$c = D\max\{c_1, \ldots, c_j\},$$

we have

$$\|Df^n(x)\xi\| \leqslant c\lambda^n\|\xi\| \text{ for all } n \geqslant 0 \text{ and all } \xi \in E^s(x)$$

Similarly, we define the space $E^u(x)$ of *expanding* or *unstable* vectors at x to be the set of vectors ξ such that there exist constants c,λ with $\lambda<1$ such that

$$\|Df^{-n}(x)\xi\| \leqslant c\lambda^n\|\xi\| \text{ for all } n.$$

Again, $E^u(x)$ is a linear subspace of $\mathbf{R}^m$ and c,λ can be taken to depend only on x, not on $\xi \in E^u(x)$.

It follows almost immediately from the definition of $E^s(x)$ and $E^u(x)$ (and the chain rule) that

$$Df(x)E^s(x) = E^s(f(x))$$
$$Df(x)E^u(x) = E^u(f(x)).$$

Furthermore, a λ which works for x will also work for $f(x)$ (but it may be necessary to take c larger at $f(x)$ than at x.)

We say that a point x is *hyperbolic* if

$$\mathbf{R}^m = E^s(x) \oplus E^u(x),$$

i.e., if every vector can be decomposed uniquely as the sum of a vector expanding at x and one contracting at x. From the preceding remark, if x is hyperbolic, so are $f(x)$ and $f^{-1}(x)$.

The definitions given up to now in this section are not standard and should not be taken completely seriously. The following definition, on the other hand, is standard (although expressed in a slightly unconventional way): A set $\Lambda \subset M$ is said to be a *hyperbolic set* for f if it is compact, invariant for f (i.e., $f\Lambda = \Lambda$), if each point $x \in \Lambda$ is hyperbolic, and if the constants c,λ can be taken to be independent of x, i.e., if there exist c,λ, with $\lambda<1$, such that

$$\|Df^n(x)\xi\| \leqslant c\lambda^n\|\xi\| \text{ for all } x \in \Lambda,\ \xi \in E^s(x),\ n \geqslant 0 \tag{3.1S}$$
$$\|Df^{-n}(x)\xi\| \leqslant c\lambda^n\|\xi\| \text{ for all } x \in \Lambda,\ \xi \in E^u(x),\ n \geqslant 0 \tag{3.1U}$$

The essential thing in this definition is the uniformity of the rate of decay of the contracting (respectively expanding) vectors under application of $Df^n(x)$ (respectively $Df^{-n}(x)$). It would therefore be more accurate to call such sets Λ *uniformly hyperbolic* rather than simply hyperbolic. In this connection, it may be illuminating to contrast uniform hyperbolicity with a weaker probabilistic kind of hyperbolicity. One natural generalization of the notion of hyperbolic set is that of a finite f-invariant measure μ such that μ-almost all points x are hyperbolic. This condition can be reformulated in the language of *Lyapunov characteristic exponents* as saying that all Lyapunov exponents are non-zero μ-almost everywhere. In contrast to the theory of uniform hyperbolicity, which is highly developed and rich and seems to be fairly complete, the investigation of probabilistic notions of hyperbolicity is just beginning and shows great promise for further development.

Historically, the investigation of uniformly hyperbolic sets began with the case in which the whole space M is hyperbolic (if we leave aside the relatively trivial case of hyperbolic fixed and periodic points.) If M is a hyperbolic set for f, we say that f is an *Anosov diffeomorphism.*

As we have already noted, the fields of subspaces $E^s(x)$ and $E^u(x)$ have a natural covariance property:

$$Df^{\pm 1}(x)E^s(x)=E^s(f^{\pm 1}x) \tag{3.2}$$

and similarly for $E^u(x)$. Because of this covariance, the uniform expansivity condition (3.1U) is equivalent to

$$\|Df^n(x)\xi\| \geqslant c^{-1}\lambda^{-n}\|\xi\| \text{ for all } x\in\Lambda,\ \xi\in E^u(x),\ n\geqslant 0 \tag{3.3}$$

which is a more direct-looking formulation of expansivity.

It turns out to be an automatic consequence of the definition of hyperbolic set that the subspaces $E^s(x)$ and $E^u(x)$ vary in a continuous way with x on any hyperbolic set. The notion of continuity of a field of subspaces needs to be defined, but this poses no problems; any reasonable definition will work. We adopt the following one: If x is a hyperbolic point for f, we let $P^s(x)$ denote the projection onto $E^s(x)$ along $E^u(x)$, i.e., the linear operator on $\mathbf{R}^m$ which is zero on $E^u(x)$ and the identity on $E^s(x)$, and we write $P^u(x)$ for $\mathbf{1}-P^s(x)$. We say that the splitting

$$\mathbf{R}^m=E^s(x)\oplus E^u(x)$$

varies continuously with x if $P^s(x)$ is a continuous matrix-valued function of x.

Proposition 3.1. *Let Λ be a hyperbolic set for f. Then $P^s(x)$ is a continuous function of x on Λ.*

Proof. We first argue that the condition

$$\|Df^n(x)\xi\| \leqslant c\lambda^n\|\xi\| \text{ for all } n \tag{3.4S}$$

characterizes vectors ξ in $E^s(x)$. If ξ is not in $E^s(x)$, then it is the sum of a component in $E^s(x)$ and a *non-zero* component in $E^u(x)$. By (3.3), applying $Df^n(x)$ to the $E^u(x)$ component gives an exponentially growing sequence of vectors; the $E^s(x)$ component gives an exponentially decreasing sequence; so $\|Df^n(x)\xi\|$ is (eventually) exponentially growing and so does not satisfy (3.4).

Thus

$$\{(x,\xi)\colon x\in\Lambda,\ \xi\in E^s(x)\}=\bigcap_{n\geqslant 0}\{(x,\xi)\colon x\in\Lambda,\ \|Df^n(x)\xi\|\leqslant c\lambda^n\|\xi\|\}.$$

The right-hand side is an intersection of closed sets in $\Lambda\times\mathbf{R}^m$, so the left-hand side is closed. Hence:

$$\text{If } x_n\to x_0 \text{ and } \xi_n\in E^s(x_n)\to\xi_0 \text{ then } \xi_0\in E^s(x_0). \tag{3.5}$$

A similar statement holds for the $E^u(x)$'s.

The idea is now as follows: Intuitively, what we have just shown says that the fields of subspaces $E^s(x)$ and $E^u(x)$ are upper semi-continuous; their only possible discontinuities are jumps where they become discontinuously bigger. But since their direct sum is always (on Λ) equal to $\mathbf{R}^m$, if one of them were to become bigger discontinuously, the other would have to become smaller discontinuously. As this is impossible, both must be continuous.

To make this argument precise, it suffices to show
1. If (x_n) is a sequence of points of Λ converging to a limit x_0, and if the $P^s(x_n)$ converge to a limit P_0, then $P_0 = P^s(x_0)$.
2. $P^s(x)$ is bounded on Λ.
Continuity then follows from a standard compactness argument.

To prove 1., we first note that P_0 is a projection since

$$(P_0)^2 = \lim_{n\to\infty} (P^s(x_n))^2 = \lim_{n\to\infty} P^s(x_n) = P_0,$$

From (3.5),

$$P_0\xi = \lim_{n\to\infty} P^s(x_n)\xi \in E^s(x_0) \text{ for all } \xi,$$

so the space onto which P_0 projects is contained in $E^s(x_0)$. Similarly,

$$(1 - P_0)\xi \in E^u(x_0) \text{ for all } \xi,$$

so the null space of P_0 is contained in $E^u(x_0)$. Since

$$\dim(E^s(x_0)) + \dim(E^u(x_0)) = m,$$

both inclusions must be equalities, i.e. $P_0 = P^s(x_0)$, as desired.

To prove 2: If $P^s(x)$ is not bounded, there must exist a sequence of points (x_n) in Λ and a sequence ξ_n of non-zero vectors such that

$$\frac{\|P^s(x_n)\xi_n\|}{\|\xi_n\|} \to \infty.$$

We normalize the ξ_n's by requiring that

$$\|P^s(x_n)\xi_n\| = 1;$$

then

$$\xi_n \to 0.$$

By passing to a subsequence, we can assume that the x_n converge to a limit x_0 and that the $P^s(x_n)\xi_n$ converge to a limit ξ_0. By the normalization condition, $\|\xi_0\| = 1$. Since

$$P^s(x_n)\xi_n \in E^s(x_n) \to \xi_0,$$

(3.5) implies that $\xi_0 \in E^s(x_0)$. On the other hand,

$$P^u(x_n)\xi_n = \xi_n - P^s(x_n)\xi_n \to -\xi_0$$

(since $\xi_n \to 0$), so $\xi_0 \in E^u(x_0)$. Thus, ξ_0 is a non-zero vector belonging to both $E^s(x_0)$ and $E^u(x_0)$; this contradicts

$$\mathbf{R}^m = E^s(x_0) \oplus E^u(x_0),$$

and hence proves that $P^s(x)$ must be bounded.

In its dependence on x, $P^s(x)$ is slightly more regular than just continuous; it can be shown to be Hölder continuous (provided that $Df(x)$ is). No matter how smooth f is, however, $P^s(x)$ need not be more than Hölder continuous. There exist for example analytic Anosov diffeomorphisms for which $P^s(x)$ is nowhere differentiable in x.

The presence of the constant c in the condition

$$\|Df^n(x)\xi\| \leqslant c\lambda^n \|\xi\| \text{ for all } x \in \Lambda,\ \xi \in E^s(x),\ n \geqslant 0$$

is a technical nuisance; it is easier to work with the stronger condition

$$\|Df(x)\xi\| \leqslant \lambda \|\xi\| \text{ for all } x \in \Lambda,\ \xi \in E^s(x)$$

(which, by the chain rule, is equivalent to the preceding one with $c=1$). It is necessary to use the former condition to make the definition of hyperbolic set independent of coordinate system. It is possible, however, to make the latter condition hold by adjusting the norm used, at the expense, in general, of making the norm x-dependent.

Proposition 3.2. *Let Λ be a hyperbolic set for f. Then there exist a constant $\lambda<1$ and a Riemannian metric for M (i.e., an x-dependent positive-definite scalar product $(\ ,\ )_x$ with associated norm $\|\xi\|_x=[(\xi,\xi)_x]^{1/2}$) such that, for all $x \in \Lambda$*

$$\|Df(x)\xi\|_{f(x)} \leqslant \lambda \|\xi\|_x \text{ for } \xi \in E^s(x) \tag{3.6S}$$

$$\|Df^{-1}(x)\xi\|_{f^{-1}(x)} \leqslant \lambda \|\xi\|_x \text{ for } \xi \in E^u(x). \tag{3.6U}$$

Such a metric is called an *adapted* or *Lyapunov* metric. We omit the proof of the above proposition; it is not difficult.

To keep things elementary and to avoid complicating the notation unnecessarily, we will consider for the remainder of these notes only the case in which M is an open subset of $\mathbf{R}^m$ and in which (3.6) holds for the standard (x-independent) scalar product.

4. Examples.

A set consisting of a single hyperbolic fixed point is a simple example of a hyperbolic set. In this section we describe briefly some less trivial examples.

1. Linear automorphism of the two-dimensional torus. (This example was already introduced in Section 2.) Represent the two-dimensional torus as $\mathbf{R}^2/\mathbf{Z}^2$, and let f denote the mapping on the quotient space induced by

$$(x_1,x_2) \rightarrow (2x_1+x_2, x_1+x_2)$$

on $\mathbf{R}^2$. $Df(x)$ is everywhere equal to the 2×2 matrix

$$\begin{pmatrix} 2 & 1 \\ 1 & 1 \end{pmatrix}$$

which has one eigenvalue greater than 1 and one between 0 and 1. The contracting and expanding subspaces $E^s(x)$, $E^u(x)$ do not vary with x; they are respectively the one-dimensional spaces spanned by the contracting and expanding eigenvectors of

$$\begin{pmatrix} 2 & 1 \\ 1 & 1 \end{pmatrix}.$$

The whole torus is a hyperbolic set, i.e., f is an Anosov diffeomorphism.

2. Smale horseshoe. The idea here is to look, not at the full structure of a mapping but at its action on a square Δ somewhere in the space (taken to be two-dimensional) on which the mapping acts. The mapping does not send the square into itself, but rather compresses it horizontally, stretches it vertically, bends it into a horseshoe shape, and lays it back over itself (extending out of top and bottom) as indicated in Figure 4.1. The set of points in Δ mapped back into Δ by one application of f consists of two roughly horizontal strips which we call Δ_0 and Δ_1.

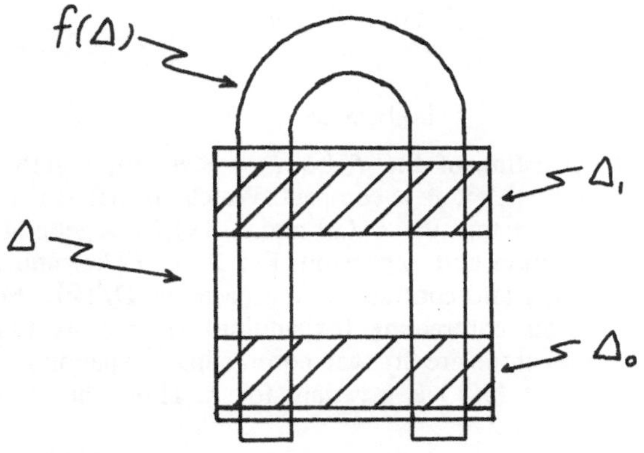

Figure 4.1. The Smale horseshoe

We now ask: What does the set Λ of points x such that $f^n(x) \in \Delta$ for all n (negative as well as positive) look like? It is not possible to answer this question on the basis of the qualitative description of f we have given, but it does become possible to answer it if we make the special assumption that f acts in a strictly affine way, with vertical expansion and horizontal contraction, on each of Δ_0 and Δ_1 (with, in addition, rotation through 180° on Δ_1). Under this assumption, there is a one-one correspondence between points of Λ and sequences

$$\mathbf{i} = (\cdots, i_{-1}, i_0, i_1, \cdots)$$

of 0's and 1's; the point x corresponding to the sequence $\mathbf{i}$ is determined by the condition

$$f^n(x) \in \Delta_{i_n} \text{ for all } n.$$

The set $\Lambda \subset \Delta$ is nondenumerable, closed, and totally disconnected. The orbit corresponding to a sequence $\mathbf{i}$ is periodic if and only if $\mathbf{i}$ is; since there are infinitely many periodic sequences, f has infinitely many periodic orbits in Λ. For each $x \in \Lambda$, $E^s(x)$ is the space of horizontal vectors and $E^u(x)$ the space of vertical vectors; it is not difficult to see that Λ is a hyperbolic set for f.

All these statements are proved in a straightforward and elementary way from the assumption that f is exactly affine on each of Δ_0 and Δ_1; the details can be found, for example, in Section 2 of Lanford [1]. It follows, however, from general theory we are going to develop that the qualitative properties of Λ (hyperbolicity, correspondence with sequences $\mathbf{i}$, existence of infinitely many periodic orbits) persist under small perturbations of f and hence are not specific to the assumed piecewise affine form.

3. Transverse homoclinic orbit. Let $\bar{x}$ be a hyperbolic fixed point for a differentiable mapping f on a two-dimensional space. Assume that $Df(\bar{x})$ has one-dimensional expanding and contracting subspaces, so W^s and W^u are both one-dimensional, i.e., are curves. Assume further that these curves intersect transversally (i.e., at non-zero angle) at some point $x_0 \neq \bar{x}$. (In general, a point x_0 other than $\bar{x}$ belonging to both W^s and W^u is called a *homoclinic point.* If, as we are assuming here, W^s and W^u intersect transversally at x_0, we speak of a *transverse* homoclinic point.) Since $x_0 \in W^s$,

$$f^n(x_0) \to \bar{x} \text{ as } n \to \infty,$$

and since $x_0 \in W^u$,

$$f^n(x_0) \to \bar{x} \text{ as } n \to -\infty.$$

Let Λ denote the set consisting of the $f^n(x_0)$, $-\infty < n < \infty$, together with $\bar{x}$. Because $f^n(x_0)$ converges to $\bar{x}$ as $n \to \pm\infty$, Λ is compact. We claim that Λ is a hyperbolic set. To check this, it is necessary to determine $E^s(x)$ and $E^u(x)$ for a general point $x \in \Lambda$ and to prove uniformity of contraction and expansion. For $x = \bar{x}$, $E^s(x)$ and $E^u(x)$ are just the (one-dimensional) expanding and contracting subspaces of $Df(\bar{x})$. For $x = x_0$, it is easy to see that the only possible contracting (expanding) vectors are those tangent to W^s (W^u) at x_0. That these vectors are in fact contracting (expanding) is not difficult to show, using the fact that W^s (W^u) is invariant for f. Thus, the point x_0 is hyperbolic. From the general relations:

$$Df(x)E^s(x) = E^s(f(x))$$

$$Df(x)E^u(x) = E^u(f(x))$$

it follows that each point $x_n = f^n(x_0)$ is hyperbolic. To complete the proof that Λ is a hyperbolic set, it is only necessary to show that the contraction and expansion are sufficiently uniform. This follows from a local analysis of the action of f in a neighborhood of $\bar{x}$ which we will leave as an exercise.

This example has a quite different flavor from the Smale horseshoe example discussed earlier. In this case, Λ contains no periodic orbit except for the fixed point $\bar{x}$, while the hyperbolic set obtained in the Smale horseshoe construction contains infinitely many periodic orbits. On the other hand, the set obtained from the Smale horseshoe construction is isolated in the sense that any orbit not in it must at some time leave Δ while, as we will see in the next section, there are periodic orbits which stay arbitrarily near to any transverse homoclinic orbit. Intuitively, the Smale horseshoe example seems to be "complete", and the homoclinic orbit example seems to be only a small piece of a larger hyperbolic set. To some extent, this intuitive distinction is made precise in the notion of locally maximal hyperbolic set to be developed in Section 12.

5. Shadowing

Let $\delta > 0$. A δ *pseudo-orbit* for f means a sequence of points (x_i) such that

$$d(x_{i+1}, f(x_i)) < \delta$$

for all (relevant) i. In other words, x_{i+1} is obtained from x_i by applying f, then making a jump which has length less than δ but which is otherwise arbitrary. (In the absence of a explicit statement to the contrary, our pseudo-orbits will always to be assumed to be defined for all integers i, negative as well as positive. The notion of pseudo-orbit does however make sense if the index set is any subinterval of $\mathbf{Z}$.)

One of the keys to the analysis of hyperbolic sets is the fact that, if δ is small enough, every δ pseudo-orbit near a hyperbolic set is followed for all i by a true orbit. The precise statement is as follows:

Theorem 5.1. Shadowing Theorem. *Let Λ be a hyperbolic set for f, and let $\epsilon > 0$. Then there exists $\delta > 0$ such that, for every δ pseudo-orbit (x_i) with $d(x_i, \Lambda) < \delta$ for all i, there is a y_0 such that*

$$d(x_i, f^i(y_0)) < \epsilon \text{ for all } i \in \mathbf{Z}. \tag{5.1}$$

Furthermore, if ϵ is small enough, y_0 is uniquely determined.

If (5.1) holds, one says that the orbit $f^i(y_0)$ ϵ-*shadows* the pseudo-orbit (x_i). The theorem as formulated can easily be deduced from the apparently weaker result in which the pseudo-orbit is required to be contained in Λ (rather than simply near it). On the other hand, it is a delicate matter whether the shadowing orbit $(f^i(y_0))$ is necessarily contained in Λ or not; this turns out to be the case if and only if Λ has a property called *local product structure* to be discussed in Section 12.

Before giving the proof of the Shadowing Theorem, we will indicate some of its uses. One immediate consequence of the uniqueness assertion of the theorem is

Corollary 5.2 *Let Λ be a hyperbolic set for f. There exists $\epsilon>0$ such that, if*

$$d(f^i(x_0),\Lambda)<\epsilon \text{ for all } i$$

and if $\bar{x}_0\neq x_0$, then

$$d(f^i(x_0),f^i(\bar{x}^0))>\epsilon \text{ for some } i.$$

In other words, two orbits near Λ cannot stay arbitrarily close together for all time. This property is called *expansivity*; a number ϵ with the indicated property is called an *expansivity constant.*

A typical use of the Shadowing Theorem is to prove the existence of periodic orbits. The idea is simply that it is often easy to construct directly periodic pseudo-orbits with arbitrarily small δ and that, if (x_i) is a periodic pseudo-orbit with a unique shadowing orbit, then the shadowing orbit must also be periodic. For a concrete example, consider the situation described in Example 3 of the preceding section. That is: Λ consists of a hyperbolic fixed point $\bar{x}$ of saddle type (in two dimensions) together with a single orbit $(f^i(x_0))$ where x_0 is a transverse homoclinic point for $\bar{x}$:

$$f^n(x_0)\to\bar{x} \text{ as } n\to\infty, \text{ so } x_0\in W^s(\bar{x})$$

$$f^n(x_0)\to\bar{x} \text{ as } n\to-\infty, \text{ so } x_0\in W^u(\bar{x})$$

$$W^s(\bar{x}) \text{ intersects } W^u(\bar{x}) \text{ transversally at } x_0.$$

We will use the Shadowing Theorem to show that there are periodic orbits arbitrarily near to x_0. For any large N we construct a pseudo-orbit $(z_n),0\leqslant n\leqslant N$ as follows: Put $z_0=x_0$, and make $z_n=f(z_{n-1})$ until $n=n_0\approx N/2$. At this point, z_n is very near to $\bar{x}$. Thus, by making only a small jump, we can take $z_{n_0+1}=f^{-(N-n_0-1)}(x_0)$, which is also very near to $\bar{x}$. Now go back to making $z_n=f(z_{n-1})$; this leads us back to $z_N=x_0$. We thus get a finite pseudo-orbit starting and ending at x_0; we repeat it periodically to get a periodic pseudo-orbit defined for all n. Given any $\epsilon>0$, we can by taking N big enough arrange that the pseudo-orbit constructed in this way is a δ pseudo-orbit for the δ related to ϵ as in the Shadowing Theorem. Let y_0 be a point whose orbit ϵ-shadows (z_n). Since (z_n) is periodic (with period N), the orbit of $f^N(y_0)$ shadows it exactly as well as does that of y_0, so, if ϵ is small enough to guarantee uniqueness of the shadowing orbit, it must be the case that

$$f^N(y_0)=y_0,$$

i.e., the shadowing orbit is also periodic. Since $z_0=x_0$, ϵ-shadowing implies that

$$d(x_0,y_0)<\epsilon.$$

Thus, as asserted, there are periodic points arbitrarily close to x_0

The above argument can easily be extended to prove the following: *For any* $\epsilon>0$, *there exists* N *such that, for any sequence* $\mathbf{i}=(\cdots,i_{-1},i_0,i_1,\cdots)$ *of 0's and 1's, there is a* y_0 *such that*

$$d(f^{nN}(y_0),x_0)<\epsilon \text{ if } i_n=1$$

and

$$d(f^{nN}(y_0),\bar{x})<\epsilon \text{ otherwise.}$$

In other words: There is an initial point y_0 whose orbit under f^N jumps back and forth from the vicinity of x_0 to the vicinity of $\bar{x}$ in any specified way. The preceding condition need not determine y_0 uniquely, but it is not difficult to specify a z_0 corresponding to each sequence $\mathbf{i}$ in such a way that $f^N(z_0)$ corresponds to the sequence obtained by shifting $\mathbf{i}$ one place to the left (and also in such a way that z_0 varies continuously with $\mathbf{i}$.) Thus we get

Proposition 5.3. *Let* $\epsilon>0$. *Then for any sufficiently large integer* N *there is a compact* $\Xi\subset\{x:d(x,\bar{x})<\epsilon \text{ or } d(x,x_0)<\epsilon\}$, *invariant for* f^N, *such that* f^N *on* Ξ *is conjugate to the left shift on the space* $\{0,1\}^{\mathbf{Z}}$ *of sequences of 0's and 1's.*

Such a Ξ is often called an *imbedded 2-shift.*

We are going to prove the Shadowing Theorem by reducing it to a fixed point problem in a Banach space. The idea is roughly as follows: Define an operator A on the space of sequences $\mathbf{x}$ in M by

$$A(\mathbf{x})_n=f(x_{n-1}).$$

A sequence $\mathbf{x}$ is an orbit if and only if

$$A(\mathbf{x})=\mathbf{x},$$

i.e., if and only if $\mathbf{x}$ is a fixed point for A. Similarly, if $\mathbf{x}$ is a δ pseudo-orbit,

$$\|A(\mathbf{x})-\mathbf{x}\|\equiv\sup_n\|A(\mathbf{x})_n-x_n\|\leqslant\delta.$$

What we need in order to prove the Shadowing Theorem is a result saying that near every point left approximately fixed by A there is a point left *exactly* fixed. We formulate in the next section a simple but powerful theorem along these lines, and in the following section we show how it can be applied to prove the Shadowing Theorem.

6. A fixed point theorem

Proposition 6.1. *Let* A *be a differentiable mapping defined on a neighborhood of* $\mathbf{0}$ *in a Banach space* X, *with values in* X, *and let* Γ *be a linear operator on* X *such that* $(\Gamma-\mathbf{1})^{-1}$ *is bounded. Assume that there are positive constants* ρ, κ *with* $\kappa<1$ *such that*

1. For $\|\xi\|\leqslant\rho$, $\|(\Gamma-\mathbf{1})^{-1}\|\cdot\|DA(\xi)-\Gamma\|\leqslant\kappa$.

2. $\|(\Gamma-\mathbf{1})^{-1}\|\cdot\|A(\mathbf{0})\|\leqslant(1-\kappa)\rho$.

Then A *has exactly one fixed point in* $\{\|\xi\|\leqslant\rho\}$.

Proof. We will apply the contraction mapping principle to

$$\Phi(\xi)\equiv\xi-(\Gamma-1)^{-1}[A(\xi)-\xi]$$

which has the same fixed points as A does. Condition 1. will imply that Φ is contractive on $\{\|\xi\|\leqslant\rho\}$; condition 2. that Φ maps this ball into itself. The verifications are almost immediate:

$$D\Phi(\xi)=1-(\Gamma-1)^{-1}(DA(\xi)-1)$$
$$=1-(\Gamma-1)^{-1}[(\Gamma-1)+(DA(\xi)-\Gamma)]=-(\Gamma-1)^{-1}(DA(\xi)-\Gamma)$$

From this and 1.,

$$\|D\Phi(\xi)\|\leqslant\kappa \text{ for } \|\xi\|\leqslant\rho,$$

which proves contractivity.

Now let $\|\xi\|\leqslant\rho$. Then

$$\|\Phi(\xi)\|\leqslant\|\Phi(\xi)-\Phi(0)\|+\|\Phi(0)\|$$
$$\leqslant\kappa\|\xi\|+\|(\Gamma-1)^{-1}\|\,\|A(0)\|\leqslant\kappa\rho+(1-\kappa)\rho=\rho.$$

7. Proof of the Shadowing Theorem

The proof of the Shadowing Theorem itself involves a certain amount of messy detail which distracts attention from the main ideas. We will therefore begin by proving a preliminary result which, although weaker than what we will eventually prove, displays the underlying argument in a clean way.

Proposition 7.1 *Let Λ be a hyperbolic set for f. Then there exists an $\epsilon>0$ such that, if $x_0\in\Lambda$ and if $x_0'\neq x_0$, then $d(f^n(x_0),f^n(x_0'))>\epsilon$ for some n.*

Note that this proposition proves less than expansivity since x_0 has to be in Λ (whereas, in Corollary 5.2, it is only required that $f^n(x_0)$ remain near Λ for all n).

Proof. Let $x_0\in\Lambda$; $x_n\equiv f^n(x_0)$, and let X denote the space of all bounded sequences $\xi=(\xi_n)_{n\in\mathbf{Z}}$ in $\mathbf{R}^m$, equipped with the norm

$$\|\xi\|=\sup_n\|\xi^n\|$$

Define an operator A on X by

$$A(\xi)_n=f(x_{n-1}+\xi_{n-1})-x_n.$$

We don't need to worry about the domain of definition of A beyond noting that it contains a neighborhood of $\mathbf{0}$. A sequence ξ is a fixed point for A if and only if $(x_n+\xi_n)$ is an orbit for f. In particular, $\mathbf{0}$ is a fixed point for A. We are going to use the uniqueness statement of the fixed point theorem of the preceding section to show that there is an $\epsilon>0$ such that A has no other fixed point of norm no larger than ϵ. *It is an essential part* *of the argument that, although A depends on the orbit (x_n), we show that there is an ϵ which works for all orbits.*

To apply the fixed point theorem, we have first to check that A is differentiable on a neighborhood of $\mathbf{0}$ in X. This is an elementary verification, which we omit. The derivative of A at $\boldsymbol{\xi}$ is the linear operator on X given by

$$(DA(\boldsymbol{\xi})\boldsymbol{\eta})_n = Df(x_{n-1}+\xi_{n-1})\eta_{n-1} \tag{7.1}$$

The next step is to choose a Γ, and in this case we can take it to be simply $DA(\mathbf{0})$. (In later applications of these ideas, finding the right Γ will be considerably more complicated.) The main thing we have to do is to establish a bound on $\|(\Gamma-\mathbf{1})^{-1}\|$ which is independent of (x_n); the rest of the argument will be relatively routine.

Let X^s be the subspace of X consisting of sequences $\boldsymbol{\eta}$ with $\eta_n \in E^s(x_n)$ for all n; similarly, let X^u be the space of sequences with $\eta_n \in E^u(x_n)$. We will refer to X^s (respectively X^u) as the space of *contracting* (respectively *expanding*) sequences. From (7.1) (with $\boldsymbol{\xi}=\mathbf{0}$),

$$(\Gamma\boldsymbol{\eta})_{n+1} = Df(x_n)\eta_n,$$

and, from the covariance of the splitting, Γ maps each of X^s and X^u to itself. Furthermore, if $\boldsymbol{\eta}\in X^s$,

$$\|\Gamma\boldsymbol{\eta}\| = \sup_n \|(\Gamma\boldsymbol{\eta})_{n+1}\| = \sup_n \|Df(x_n)\eta_n\|$$

$$\leqslant \lambda\cdot\sup_n \|\eta_n\| = \lambda\|\boldsymbol{\eta}\|,$$

i.e.,

$$\|\Gamma|_{X^s}\| \leqslant \lambda < 1.$$

Thus, on X^s, $(\Gamma-\mathbf{1})^{-1} = -\sum_{n=0}^{\infty}\Gamma^n$ and so $\|(\Gamma-\mathbf{1})^{-1}\| \leqslant (1-\lambda)^{-1}$. Similarly, for $\boldsymbol{\eta}\in X^u$,

$$(\Gamma^{-1}\boldsymbol{\eta})_n = Df^{-1}(x_{n+1})\eta_{n+1},$$

so

$$\|\Gamma^{-1}\boldsymbol{\eta}\| \leqslant \lambda\|\boldsymbol{\eta}\|,$$

so, on X^u,

$$(\Gamma-\mathbf{1})^{-1} = \Gamma^{-1}(\mathbf{1}-\Gamma^{-1})^{-1} = \sum_{n=1}^{\infty}\Gamma^{-n}$$

and

$$\|(\Gamma-\mathbf{1})^{-1}\| \leqslant \lambda/(1-\lambda) < (1-\lambda)^{-1}.$$

Since $\mathbf{R}^m = E^s(x_n)\oplus E^u(x_n)$ for all n, $X = X^s \oplus X^u$, and since $\Gamma-\mathbf{1}$ is invertible on each of X^s and X^u, it is invertible on X. Now comes an irritating complication. In general, $E^s(x_n)$ and $E^u(x_n)$ need not be orthogonal, so it need not be true that the norm of $(\Gamma-\mathbf{1})^{-1}$ on X is the larger of the norms of its restrictions to X^s and X^u. To cope with this, we let

$$D = \sup_{x\in\Lambda}\max\{\|P^s(x)\|, \|P^u(x)\|\}.$$

Then

$$\max\{\|\xi^s\|,\|\xi^u\|\}\leqslant D\|\xi^s+\xi^u\|$$

for all $\xi^s\in E^s(x)$, $\xi^u\in E^u(x)$, and for all $x\in\Lambda$. From this it follows at once that, if $\boldsymbol{\xi}^s\in X^s,\boldsymbol{\xi}^u\in X^u$,

$$D^{-1}\max\{\|\boldsymbol{\xi}^s\|,\|\boldsymbol{\xi}^u\|\}\leqslant\|\boldsymbol{\xi}^s+\boldsymbol{\xi}^u\|\leqslant 2\max\{\|\boldsymbol{\xi}^s\|,\|\boldsymbol{\xi}^u\|\}$$

and hence that

$$\|(\Gamma-\mathbf{1})^{-1}\|\leqslant 2D\max\{\|(\Gamma-\mathbf{1})^{-1}|_{X^s}\|,\|(\Gamma-\mathbf{1})^{-1}|_{X^u}\|\}\leqslant 2D/(1-\lambda)\equiv B. \tag{7.2}$$

This is what we needed: a bound on $\|(\Gamma-\mathbf{1})^{-1}\|$ independent of the orbit (x_n).

To complete the proof, we have to show how to choose ρ,κ with $\kappa<1$ so that

$$\|(\Gamma-\mathbf{1})^{-1}\|\cdot\|DA(\boldsymbol{\xi})-\Gamma\|\leqslant\kappa$$

if $\|\boldsymbol{\xi}\|\leqslant\rho$. (This is condition 1 of Proposition 6.1. In the case at hand, condition 2 is immediate since $A(\mathbf{0})=\mathbf{0}$.) Any $\kappa<1$ will do; suppose one has been chosen. Then take ρ small enough so that, if $d(x,x')\leqslant\rho$

$$\|Df(x)-Df(x')\|\leqslant\kappa/B$$

(using the uniform continuity of Df on a neighborhood of Λ; B is as defined in (7.2).) From

$$(DA(\boldsymbol{\xi})\boldsymbol{\eta})_{n+1}=Df(x_n+\xi_n)\eta_n$$

(and using $\Gamma=DA(\mathbf{0})$) it follows that

$$\|DA(\boldsymbol{\xi})-\Gamma\|\leqslant\sup_n\|DA(x_n+\xi_n)-DA(x_n)\|\leqslant\kappa/B\leqslant\kappa/\|(\Gamma-\mathbf{1})^{-1}\|$$

provided $\|\boldsymbol{\xi}\|\leqslant\rho$.

We have thus verified the hypotheses of the fixed point theorem of the preceding section; that theorem implies that A has no fixed point other that $\mathbf{0}$ in $\{\|\boldsymbol{\xi}\|\leqslant\rho\}$ so the proposition is proved with $\epsilon=\rho$.

The proof of the Shadowing Theorem begins in the same way as the above proof: Given a pseudo-orbit (x_n), we introduce the space X and the operator A exactly as before, i.e.,

$$A(\boldsymbol{\xi})_{n+1}=f(x_n+\xi_n)-x_{n+1}.$$

Again:

$$(DA(\boldsymbol{\xi})\boldsymbol{\eta})_{n+1}=Df(x_n+\xi_n)\eta_n. \tag{7.3}$$

The main problem is to construct an appropriate Γ, i.e., an operator approximating $DA(\mathbf{0})$ and for which we can estimate $\|(\Gamma-\mathbf{1})^{-1}\|$. Informally, the argument goes as follows: The first step is to construct a continuous extension of the splitting

$$\mathbf{R}^m=E^s(x)\oplus E^u(x)$$

to a neighborhood U of Λ. If, then, (x_n) is a pseudo-orbit with a very small δ staying very close the Λ, it will follow from continuity that, for each n, $Df(x_n)$ is contractive on $E^s(x_n)$ and expansive on $E^u(x_n)$. It need not, however, map $E^s(x_n)$ (respectively $E^u(x_n)$) to $E^s(x_{n+1})$ ($E^u(x_{n+1})$). We fix this by modifying $Df(x_n)$ slightly to get a new linear mapping $\Gamma^{x_n}_{x_{n+1}}$ which does preserve the splitting and show that the necessary modification can be made small enough so that $\Gamma^{x_n}_{x_{n+1}}$ is still contractive on $E^s(x_n)$ and expansive on $E^u(x_n)$. We take Γ to be defined by

$$(\Gamma\eta)_{n+1}=\Gamma_{x_{n+1}}^{x_n}\eta_n;$$

then Γ (respectively Γ^{-1}) maps X^s (respectively X^u) contractively to itself, and so $\|(\Gamma-\mathbf{1})^{-1}\|$ can be estimated as in the proof of Proposition 7.1. Since $\Gamma_{x_{n+1}}^{x_n}\approx Df(x_n)$, $\Gamma\approx DA(\mathbf{0})$. Further, if $\|\xi\|$ is small, $\|DA(\xi)-DA(\mathbf{0})\|$ will also be small, so we can satisfy condition 1 of the fixed point theorem on a small ball about $\mathbf{0}$. Condition 2 of that theorem will also be satisfied if δ is small enough, and the Shadowing Theorem will follow.

We turn now to the formal details.

1. *There is a neighborhood U of Λ and a bounded continuous projection-valued extension of $P^s(x)$ to U.* We will also denote the extension by $P^s(x)$.

Proof. $P^s(x)$ is a continuous matrix-valued function defined on Λ satisfying

$$(P^s(x))^2=P^s(x) \tag{7.4}$$

(equivalent to the fact that $P^s(x)$ is a projection.) First use the Tietze Extension Theorem to extend each of the matrix elements of $P^s(x)$ to a bounded continuous function defined on all of M. This gives a continuous matrix-valued extension $\tilde{P}^s(x)$ to all of M, but the projection condition (7.4) need not hold except on Λ. For $x\in\Lambda$, $\tilde{P}^s(x)$ *is* a projection, so its spectrum consists of the two points 0 and 1. By elementary perturbation theory, the spectrum of $\tilde{P}^x(x)$ varies continuously with x. Let U be a neighborhood of Λ on which the spectrum is contained in the union of two open disks, each of radius 1/3, centered respectively at 0 and 1. We can take, as our extension of $P^s(x)$ to U, the spectral projection for $\tilde{P}^s(x)$ corresponding to the part of the spectrum inside the disk of radius 1/3 and center 1.

We will also call the extension $P^s(x)$, and for general $x\in U$, we will write

$P^u(x)$ for $\mathbf{1}-P^s(x)$

$E^s(x)$ for the range of $P^s(x)$

$E^u(x)$ for the range of $P^u(x)$.

Warning: This notation is not consistent with the definition given in Section 3 of $E^s(x)$ and $E^u(x)$ for a general $x\in M$. The latter notation will not be used in this section.

2. For $x,x'\in U$, put

$$\Gamma_x^{x'}=P^s(x)Df(x')P^s(x')+P^u(x)Df(x')P^u(x'). \tag{7.5}$$

Note that

a. $\Gamma_x^{x'}$ *maps* $E^s(x')$ *to* $E^s(x)$ *and* $E^u(x')$ *to* $E^u(x)$.

b. $\Gamma_x^{x'}$ *is a continuous function of* x,x'.

c. If $x'\in\Lambda$ *and* $x=f(x')$ *then* $\Gamma_x^{x'}$ *reduces to* $Df(x')$.

3. Pick a λ_1 with $\lambda<\lambda_1<1$. *There exists* $\delta_1>0$ *such that, if* $d(x',\Lambda)\leqslant\delta_1$ *and* $d(f(x'),x)\leqslant\delta_1$ *then*

$$\|\Gamma_x^{x'}\xi\|\leqslant\lambda_1\|\xi\| \text{ for } \xi\in E^s(x')$$

$$\|(\Gamma_x^{x'})^{-1}\xi\|\leqslant\lambda_1\|\xi\| \text{ for } \xi\in E^u(x)$$

Proof. This follows easily from 2a.,2b., the continuity of $E^s(x')$ and $E^u(x)$, and

$$\|Df(z)\xi\|\leqslant\lambda\|\xi\| \text{ for } \xi\in E^s(z),\ z\in\Lambda$$

$$\|(Df(z))^{-1}\xi\| \leqslant \lambda \|\xi\| \text{ for } \xi \in E^u(f(z)),\ z \in \Lambda.$$

4. Let (x_n) be a pseudo-orbit in U. Define an operator Γ by

$$(\Gamma\boldsymbol{\eta})_{n+1} = \Gamma^{x_n}_{x_{n+1}} \eta_n, \tag{7.6}$$

and let

$$X^s = \{\boldsymbol{\eta} \in X : \eta_n \in E^s(x_n) \text{ for all } n\}$$

$$X^u = \{\boldsymbol{\eta} \in X : \eta_n \in E^u(x_n) \text{ for all } n\}$$

By 2a., Γ maps each of X^s and X^u to itself. Let λ_1, δ_1 be as in 3. If (x_n) is a δ_1 pseudo-orbit, and if $d(x_n,\Lambda) \leqslant \delta_1$ for all n, then 3. and the definition of Γ imply

$$\|\Gamma\| \leqslant \lambda_1 \text{ on } X^s$$

$$\|\Gamma^{-1}\| \leqslant \lambda_1 \text{ on } X^u.$$

By the argument used in the proof of Proposition 7.1, there is a constant D such that, for any pseudo-orbit as above,

$$\|(\Gamma - \mathbf{1})^{-1}\| \leqslant 2D/(1-\lambda_1) \equiv B \tag{7.7}$$

It is now easy to complete the proof of the Shadowing Theorem. We suppose, thus, that we are given $\epsilon > 0$; we want to find δ so that any δ pseudo-orbit staying within a distance δ of Λ is ϵ-shadowed by a true orbit. We will take δ smaller than the δ_1 of 3. and also small enough so that the pseudo-orbits we consider are necessarily in U; then (7.7) holds. Take any κ with $0<\kappa<1$. From the definition (7.6) of Γ, the formula (7.3) for $DA(\boldsymbol{\xi})$, and the continuity of $Df(x)$, it follows that if we take δ small enough we can guarantee that

$$\|\Gamma - DA(\mathbf{0})\| \leqslant \kappa/3B \tag{7.8}$$

Similarly, we can choose $\epsilon_1 < \epsilon$ so that $\|Df(z_1) - Df(z_2)\| \leqslant \kappa/3B$ whenever $d(z_1,z_2) \leqslant \epsilon_1$; then it follows from (7.3) that

$$\|DA(\boldsymbol{\xi}) - DA(\mathbf{0})\| \leqslant \kappa/3B \tag{7.9}$$

for $\|\boldsymbol{\xi}\| \leqslant \epsilon_1$.

Combining (7.7), (7.8), and (7.9), we get

$$\|(\Gamma - \mathbf{1})^{-1}\| \cdot \|\Gamma - DA(\boldsymbol{\xi})\| \leqslant B(\kappa/3B + \kappa/3B) < \kappa$$

for $\|\boldsymbol{\xi}\| \leqslant \epsilon_1$, so condition 1. of the fixed point theorem of Section 6 is verified. If, finally, we also take δ small enough so that

$$\delta < \epsilon_1(1-\kappa)/B \tag{7.10}$$

we get

$$\|(\Gamma - \mathbf{1})^{-1}\| \cdot \|A(\mathbf{0})\| \leqslant B\delta < \epsilon_1(1-\kappa),$$

which is condition 2 of the fixed point theorem (with $\rho = \epsilon_1$). Thus, A has exactly one fixed point with norm no larger than ϵ_1, i.e., there is one and only one y_0 such that

$$d(f^n(y_0), x_n) \leqslant \epsilon_1$$

for all n.

8. Extensions. Structural stability of hyperbolic sets.

There is much more to be extracted from the methods developed in the preceding section. One major extension starts from the observation that these methods imply the existence of shadowing orbits *not just for f itself but for any $\tilde{f}$ near enough to f.*

Theorem 8.1. *Let Λ be a hyperbolic set for f, and let $\epsilon>0$. Then there is a neighborhood W of f in the space of C^1 diffeomorphisms of M and a $\delta>0$ such that, if (x_n) is a δ pseudo-orbit for f with $d(x_n,\Lambda)\leqslant\delta$ for all n, and if $\tilde{f}\in W$, then there is a y_0 such that*

$$d(\tilde{f}^n(y_0),x_n)\leqslant\epsilon$$

for all $n\in\mathbf{Z}$. If ϵ is small enough, y_0 is uniquely determined.

Proof. The argument is a simple extension of the proof of the Shadowing Theorem; we indicate only the changes. We use the same Γ, and again take δ small enough so that (7.8) holds and ϵ_1 small enough so that (7.9) holds. We take W small enough so that

$$\|Df(x)-D\tilde{f}(x)\|\leqslant\kappa/3B$$

for all x if $\tilde{f}\in W$ Thus, if we define

$$\tilde{A}(\xi)_{n+1}=\tilde{f}(x_n+\xi_n)-x_{n+1}$$

we get

$$\|D\tilde{A}(\xi)-DA(\xi)\|\leqslant\kappa/3B$$

Combining estimates, we thus have

$$\|(\Gamma-\mathbf{1})^{-1}\|\cdot\|D\tilde{A}(\xi)-\Gamma\|\leqslant\kappa \text{ for } \|\xi\|\leqslant\epsilon_1.$$

Finally, instead of (7.10) we require

$$\delta\leqslant\epsilon_1(1-\kappa)/2B$$

and we require that W be small enough to guarantee that, if $\tilde{f}\in W$, then $d(\tilde{f}(x),f(x))\leqslant\delta$ for all x. From the latter condition,

$$\|\tilde{A}(0)-A(0)\|\leqslant\delta$$

and, as usual,

$$\|A(0)\|\leqslant\delta.$$

Hence

$$\|(\Gamma-\mathbf{1})^{-1}\|\cdot\|\tilde{A}(0)\|\leqslant B\cdot2\cdot\epsilon_1(1-\kappa)/2B=\epsilon_1(1-\kappa),$$

and Theorem 8.1 follows from the fixed point theorem of Section 6.

We get a striking result if we apply Theorem 8.1 to *exact* f-orbits on Λ: If ϵ is small enough, if W corresponds to ϵ as in the theorem, if $\tilde{f}\in W$, and if $x_0\in\Lambda$, then there is a unique y_0 such that

$$d(\tilde{f}^n(y_0),f^n(x_0))\leqslant\epsilon.$$

In other words, every f-orbit on Λ is ϵ-shadowed by an $\tilde{f}$-orbit. We write

$$y_0=\tilde{h}(x_0);$$

$\tilde{h}$ is a mapping of Λ into M satisfying $d(\tilde{h}(x),x) \leqslant \epsilon$ for all x. From the uniqueness of y_0, it follows immediately that

$$\tilde{h}(f(x)) = \tilde{f}(\tilde{h}(x)) \tag{8.1}$$

We will argue shortly that $\tilde{h}$ is one-one and continuous, and hence maps Λ homeomorphically onto a compact set $\tilde{\Lambda}$. The intertwining relation (8.1) implies that $\tilde{\Lambda}$ is invariant for $\tilde{f}$ and that $\tilde{f}$ on $\tilde{\Lambda}$ is topologically conjugate to f on Λ Thus:

Theorem 8.2. *Let Λ be a hyperbolic set for f. Then, for any $\tilde{f}$ sufficiently close to f in the C^1 topology, there is a homeomorphism $\tilde{h}$ of Λ onto a set $\tilde{\Lambda}$ invariant for $\tilde{f}$ such that*

$$\tilde{f} = \tilde{h} f \tilde{h}^{-1} \text{ on } \tilde{\Lambda}.$$

$\tilde{h}$ can be made as close as desired to the identity by making $\tilde{f}$ close to f.

Proof. It remains only to show that $\tilde{h}$, constructed above, is one-one and continuous. Take ϵ small enough so that any two distinct f orbits on Λ are somewhere separated by more than 2ϵ, and let $\tilde{f}$ belong to the corresponding W. Suppose

$$\tilde{h}(x_0) = \tilde{h}(x_0') \equiv y_0.$$

By the definition of $\tilde{h}$,

$$d(f^n(x_0), \tilde{f}^n(y_0)) \leqslant \epsilon$$
$$d(f^n(x_0'), \tilde{f}^n(y_0)) \leqslant \epsilon;$$

hence,

$$d(f^n(x_0), f^n(x_0')) \leqslant 2\epsilon$$

for all n. By the choice of ϵ, this implies $x_0 = x_0'$.

To prove that $\tilde{h}$ is continuous, it suffices, using compactness, to prove that if $x^{(j)} \to x$ and $y^{(j)} \equiv \tilde{h}(x^{(j)}) \to y$ then $\tilde{h}(x) = y$. For any j and any n,

$$d(f^n(x^{(j)}), \tilde{f}^n(y^{(j)})) \leqslant \epsilon.$$

Letting j go to infinity with n fixed, we get

$$d(f^n(x), \tilde{f}^n(y)) \leqslant \epsilon \text{ for all } n,$$

which means exactly that

$$y = \tilde{h}(x),$$

as desired.

Intuitively, this theorem expresses a *structural stability* property of hyperbolic sets. It says that, if Λ is a hyperbolic set for f and if $\tilde{f}$ is obtained by perturbing f slightly, then there is a set $\tilde{\Lambda}$ near Λ, invariant for $\tilde{f}$, such that the action of $\tilde{f}$ on $\tilde{\Lambda}$ is indistinguishable, from a topological point of view, from the action of f on Λ. Note that we have *not* proved, however, that $\tilde{\Lambda}$ is a hyperbolic set for $\tilde{f}$. This is in fact true, and will follow from a result to be proved in the next two sections.

If f is an Anosov diffeomorphism, i.e., if the whole state space M is a hyperbolic set for f, then the preceding result can easily be extended to show that f is structurally stable in the strict sense.

Corollary 8.3. Structural stability of Anosov diffeomorphisms. *Let f be an Anosov diffeomorphism. For any $\tilde{f}$ which is close enough to f in the C^1 topology, there is a* *homeomorphism* $\tilde{h}$ *of* M *onto itself such that*

$$\tilde{f}=\tilde{h}f\tilde{h}^{-1}.$$

Proof. All that needs to be shown is that the image of $\tilde{h}$ is all of M. This follows from general (if high-powered) topological considerations, but we can also prove it directly from shadowing. Given y_0, we have to show that there is an x_0 such that

$$d(f^n(x_0),\tilde{f}^n(y_0))\leqslant\epsilon \text{ for all } n,$$

i.e., such that the f-orbit of x_0 ϵ-shadows the $\tilde{f}$-orbit of y_0. But by making $\tilde{f}$ close enough to f in the C^0 topology, we can guarantee that the $\tilde{f}$ orbit of any y_0 is a δ pseudo-orbit (for f) with any pre-assigned δ, so the existence of x_0 follows directly from the Shadowing Theorem.

9. A Banach-space characterization of hyperbolicity.

One of the keys to the analysis of hyperbolic sets is the systematic exploitation of the notion of pseudo-orbit and shadowing. A second, which we have already seen in the proof of the Shadowing Theorem, is the translation of geometrical (or dynamical) questions in the finite-dimensional space M into questions about differentiable operators on Banach spaces. In this section, we develop this line of attack further by showing how to recognize a hyperbolic set in terms of properties of a related operator on a Banach space.

Let Λ be a compact invariant set for the C^1 diffeomorphism f. We let $X(\Lambda)$ denote the space of bounded but otherwise arbitrary mappings ξ from Λ to $\mathbf{R}^m$, equipped with the supremum norm, and we define an operator $f_*(\Lambda)$ on $X(\Lambda)$ by

$$(f_*(\Lambda)\xi)(x)=Df(f^{-1}(x))\xi(f^{-1}(x)).$$

We will refer to elements of $X(\Lambda)$ as *(bounded) vector fields* on Λ. It will usually be clear what set Λ we have in mind, and in this case we will frequently write X and f_* for $X(\Lambda)$ and $f_*(\Lambda)$ respectively. Recall that we defined in Section 2 a linear operator on a Banach space to be hyperbolic if its spectrum does not intersect the unit circle.

Theorem 9.1. Λ *is a hyperbolic set for* f *if and only if* $f_*(\Lambda)$ *is a hyperbolic linear operator on* $X(\Lambda)$.

Remark. The theorem remains true if $X(\Lambda)$ is replaced by the space of continuous vector fields of Λ. We will not give the proof of this stronger result. The main ideas involved are the same as in the proof of Theorem 9.1, but a bit more technical work is needed.

Proof. For the proof of this theorem, we need to work with the general, co-ordinate independent, condition

$$\|Df^n(x)\xi\|\leqslant c\lambda^n\|\xi\| \text{ for } \xi\in E^s(x)$$

and similarly for $E^u(x)$, rather than with the more specialized condition with $c=1$.

The proof that f_* is hyperbolic if Λ is a hyperbolic set is almost immediate. Assume Λ is hyperbolic, and write

$$X^s(\Lambda)=\{\xi\in X(\Lambda):\xi(x)\in E^s(x) \text{ for all } x\} \tag{9.1}$$

$$X^u(\Lambda)=\{\xi\in X(\Lambda):\xi(x)\in E^u(x) \text{ for all } x\}.$$

Then

$$X=X^s\oplus X^u,$$

and X^s and X^u are invariant subspaces for f_* (by the covariance of the splitting $\mathbf{R}^m = E^s(x) \oplus E^u(x)$.) By the chain rule,

$$(f_*{}^n \xi)(f^n(x)) = Df^n(x)\xi(x)$$

(where n may be negative as well as positive.) Hence, for $\xi \in X^s$,

$$\|f_*{}^n \xi\| = \sup_x \|f_*{}^n \xi(f^n(x))\| = \sup_x \|Df^n(x)\xi(x)\|$$

$$\leqslant c\lambda^n \sup_x \|\xi(x)\| = c\lambda^n \|\xi\|.$$

Thus,

$$\|(f_*|_{X^s})^n\| \leqslant c\lambda^n,$$

so the spectrum of $f_*|_{X^s}$ is contained in $\{|z| \leqslant \lambda\}$. Similarly, the spectrum of $f_*|_{X^u}$ is contained in $\{|z| \geqslant \lambda^{-1}\}$, so f_* is a hyperbolic operator on X.

Conversely, suppose f_* is a hyperbolic operator on X. Then there is a splitting

$$X = X^s \oplus X^u$$

into f_*-invariant subspaces and constants c, λ, with $0 < \lambda < 1$, such that

$$\|(f_*|_{X^s})^n\| \leqslant c\lambda^n$$

$$\|(f_*|_{X^u})^{-n}\| \leqslant c\lambda^n.$$

What we want to show, in essence, is that this splitting has the "local form" of (9.1).

Let $\xi \in X^s$. Then

$$c\lambda^n \|\xi\| \geqslant \|(f_*)^n \xi\| = \sup_x \|Df^n(x)\xi(x)\|.$$

In particular: For any fixed x, $\|Df^n(x)\xi(x)\|$ goes to zero exponentially with n, i.e., $\xi(x) \in E^s(x)$. Similarly, if $\xi \in X^u$, $\xi(x) \in E^u(x)$ for all x. Since X^s and X^u span X, it follows that $E^s(x)$ and $E^u(x)$ span $\mathbf{R}^m$ for every x.

Now let $x_0 \in \Lambda$ and let $\xi_0 \in E^s(x_0)$. Define a vector field ξ by

$$\xi(x) = 0, \ x \neq x_0$$

$$= \xi_0, \ x = x_0.$$

Then

$$\|f_*{}^n \xi\| = \|Df^n(x_0)\xi_0\|$$

(where the norm on the left is the norm on X while the norm on the right is on $\mathbf{R}^m$.) Since the right-hand side goes to zero as n goes to infinity, $\xi \in X^s$, so

$$\|Df^n(x_0)\xi_0\| = \|f_*{}^n \xi\| \leqslant c\lambda^n \|\xi\| = c\lambda^n \|\xi_0\|,$$

which establishes the uniformity in x of the rate of decay of vectors in $E^s(x)$ under application of $Df^n(x)$. A similar argument establishes the uniformity in the decay of vectors in $E^u(x)$ under the application of $Df^{-n}(x)$.

It remains to show that, for all x,

$$E^s(x) \cap E^u(x) = \{0\}.$$

This is proved by the same sort of argument: Let $\xi_0 \in E^s(x_0) \cap E^u(x_0)$; let $\xi(x)$ be ξ_0 for $x = x_0$ and 0 otherwise. Then $\|f_*{}^n \xi\|$ goes to zero as n goes to ∞, so $\xi \in X^s$, but also goes to zero as n goes to $-\infty$ so $\xi \in X^u$. Hence

$$\xi \in X^s \cap X^u = \{0\},$$

so $\xi_0 = 0$, as desired.

10. Stability of hyperbolicity

In Section 8, we saw that, if Λ is a hyperbolic set for f and if $\tilde{f}$ is near enough to f in the C^1 topology, then there is a set $\tilde{\Lambda}$ near Λ, invariant for $\tilde{f}$, such that $\tilde{f}$ on $\tilde{\Lambda}$ is topologically conjugate to f on Λ. We will show here, using the characterization of hyperbolic set established in the preceding section, that $\tilde{\Lambda}$ is also a hyperbolic set for $\tilde{f}$. In fact, we will prove something considerably more comprehensive: If $\tilde{f}$ is near enough to f, then *any* invariant set for $\tilde{f}$ which is near enough to Λ is hyperbolic. Besides the application mentioned above, this theorem can be used, for example, to prove hyperbolicity of the imbedded 2-shifts which, as we saw in Section 5, exist arbitrarily near to a transverse homoclinic orbit.

Theorem 10.1. *Let Λ be a hyperbolic set for f. Then there exists a C^1 neighborhood W of f and a neighborhood U of Λ such that, if $\tilde{f} \in W$ and if Ξ is a compact subset of U invariant for $\tilde{f}$, then Ξ is a hyperbolic set for $\tilde{f}$.*

Proof. We will combine the characterization of hyperbolic sets given in the preceding section, the constructions used in the proof of the Shadowing Theorem, and the following simple result from operator theory.

Proposition 10.2. *Let Γ be a hyperbolic linear operator and let*

$$B \equiv \sup_{\theta \text{real}} \| (\Gamma - e^{i\theta}\mathbf{1})^{-1} \|$$

Then any linear operator Γ_1 with

$$\|\Gamma_1 - \Gamma\| < B^{-1}$$

is also hyperbolic.

Proof of Proposition 10.2. Write

$$\Gamma_1 - e^{i\theta}\mathbf{1} = [\Gamma - e^{i\theta}\mathbf{1}] \cdot [\mathbf{1} + (\Gamma - e^{i\theta}\mathbf{1})^{-1}(\Gamma_1 - \Gamma)].$$

For any real θ, the first factor on the right is invertible by assumption and the second because

$$\| (\Gamma - e^{i\theta}\mathbf{1})^{-1}(\Gamma_1 - \Gamma) \| < B \cdot B^{-1} = 1$$

Returning to the proof of the theorem: As in the proof of the Shadowing Theorem, we find a continuous bounded projection-valued extension of $P^s(x)$ to a neighborhood U_1 of Λ. We define

$$P^u(x) = \mathbf{1} - P^s(x)$$

$$\Gamma_{x_2}^{x_1} = P^s(x_2) Df(x_1) P^s(x_1) + P^u(x_2) Df(x_1) P^u(x_1).$$

Pick λ_1 between λ and 1, and recall from step 3 of the proof of the Shadowing Theorem that there is a neighborhood U_2 of Λ, contained in U_1, and a constant $\delta_1 > 0$ such that, if x_1, x_2 are in U_2 and if $d(x_2, f(x_1)) \leq \delta_1$ then

$$\|\Gamma_{x_2}^{x_1} \xi\| \leq \lambda_1 \|\xi\| \text{ for } \xi \text{ in the range of } E^s(x_1) \tag{10.1}$$

$$\|(\Gamma_{x_2}^{x_1})^{-1}\xi\| \leqslant \lambda_1 \|\xi\| \text{ for } \xi \text{ in the range of } E^u(x_1).$$

Now let $\tilde{f}$ be a diffeomorphism such that

$$d(\tilde{f}(x), f(x)) \leqslant \delta_1 \text{ for } x \in U_2.$$

For any compact $\tilde{f}$-invariant set $\Xi \subset U_2$, we put

$$X = X(\Xi)$$

$$X^s = \{\xi \in X : P^s(x)\xi(x) = \xi(x) \text{ for all } x \in \Xi\}$$

$$X^u = \{\xi \in X : P^u(x)\xi(x) = \xi(x) \text{ for all } x \in \Xi\}$$

and we define an operator Γ on X by

$$(\Gamma\xi)(\tilde{f}(x)) = \Gamma_{\tilde{f}(x)}^{x}\xi(x). \tag{10.2}$$

Then X^s and X^u are invariant for Γ and, from (10.1),

$$\|\Gamma\xi\| \leqslant \lambda_1 \|\xi\| \text{ for } \xi \in X^s$$

$$\|\Gamma^{-1}\xi\| \leqslant \lambda_1 \|\xi\| \text{ for } \xi \in X^u$$

An argument used in the proof of Proposition 7.1 shows, from these bounds, that $\Gamma - \mathbf{1}$ is invertible on each of X^s and X^u and that both inverses have norm no larger than $(1-\lambda_1)^{-1}$. Applying that argument with Γ replaced by $e^{-i\theta}\Gamma$ shows that the same is true for $\Gamma - e^{i\theta}\mathbf{1}$ for any real θ.

Let

$$D = \sup_{x \in U_2} \left(\max\{\|P^s(x)\|, \|P^u(x)\|\}\right).$$

As in the proof of Proposition 7.1,

$$\|(\Gamma - e^{i\theta}\mathbf{1})^{-1}\| \leqslant 2D/(1-\lambda_1) \equiv B.$$

Now choose $U \subset U_2$ and $\delta_2 \leqslant \delta_1$ so that if $x_1 \in U$ and $d(x_2, f(x_1)) < \delta_2$, then

$$\|\Gamma_{x_2}^{x_1} - Df(x_1)\| \leqslant 1/3B.$$

(This is possible by step 2 of the proof of the Shadowing Theorem.) Finally, let W be a C^1 neighborhood of f small enough to ensure that any $\tilde{f} \in W$ satisfies

$$d(f(x), \tilde{f}(x)) \leqslant \delta_2 \text{ and } \|Df(x) - D\tilde{f}(x)\| \leqslant 1/3B \text{ for all } x.$$

If $\tilde{f} \in W$ and if Ξ is a compact $\tilde{f}$-invariant subset of U, then, with Γ defined as in (10.2), we have

$$\|(\Gamma - e^{i\theta}\mathbf{1})^{-1}\| \leqslant B \text{ for all real } \theta.$$

$$\|\Gamma - \tilde{f}_*(\Xi)\| = \sup_{x \in \Xi} \|\Gamma_{\tilde{f}(x)}^{x} - D\tilde{f}(x)\|$$

$$\leqslant \sup_{x \in \Xi} \|\Gamma_{\tilde{f}(x)}^{x} - Df(x)\| + \sup_{x \in \Xi} \|Df(x) - D\tilde{f}(x)\|$$

$$\leqslant 1/3B + 1/3B < 1/B$$

Proposition 10.2 now applies and shows that $\tilde{f}_*(\Xi)$ is hyperbolic, as desired.

11. Stable and unstable manifolds for hyperbolic sets.

We recall the principal results of Section 2. If x_0 is a hyperbolic fixed point for f, the *stable manifold* W^s of x_0 means the set of points x such that $f^n(x)$ converges to x_0 as $n \to \infty$. This set is a submanifold of M with dimension equal to that of $E^s(x_0)$. A useful way of characterizing a local piece of W^s is as the set of points x such that $d(f^n(x), x_0) < \epsilon$ for all $n \geqslant 0$, with ϵ a sufficiently small positive number.

These considerations generalize with straightforward changes to points x_0 which are not fixed or periodic but which do lie in a hyperbolic set. The stable manifold for such a point will be the set of x's such that

$$d(f^n(x), f^n(x_0)) \to 0 \text{ as } n \to \infty,$$

i.e., points whose forward orbits are asymptotic to the not-necessarily-constant forward orbit of x_0. Note that this means that stable manifolds are not, in general, invariant, but rather *covariant:* f maps the stable manifold of x_0 to the stable manifold for $f(x_0)$. As with fixed points, the main task is to analyze *local* stable manifolds.

Theorem 11.1. Stable Manifold Theorem for Hyperbolic Sets. *Let Λ be a hyperbolic set for f. For sufficiently small ϵ:*

1. For any $x_0 \in \Lambda$ and any $\xi^s \in E^s(x_0)$ with $\|\xi^s\| < \epsilon$, there is a unique element $\xi^u \equiv w_s(x_0, \xi^s)$ of $E^u(x_0)$ such that, writing x for $x_0 + \xi^s + \xi^u$,

$$d(f^n(x), f^n(x_0)) < \epsilon \text{ for all } n > 0.$$

2.

$$d(f^n(x), f^n(x_0)) \to 0 \text{ exponentially as } n \to \infty.$$

3. For fixed x_0, $w_s(x_0, \xi^s)$ is a continuously differentiable function of ξ^s which vanishes, together with its first derivative, at 0. *If f is r times continuously differentiable, $w_s(x_0, \xi^s)$ is r times continuously differentiable in ξ^s and the partial derivatives with respect to ξ^s are jointly continuous in x_0, ξ^s.*

Proof. We will not give a complete proof but only sketch an argument showing how to reduce this theorem to the Stable Manifold Theorem for Hyperbolic Fixed Points in an infinite dimensional space. The construction is, by now, familiar: Let X denote the space of bounded vector fields $\boldsymbol{\xi}$ on Λ. Define an operator A on X by

$$A(\boldsymbol{\xi})(f(x)) = f(x + \boldsymbol{\xi}(x)) - f(x).$$

(Intuitively: Write $h(x) = x + \boldsymbol{\xi}(x)$; then the h corresponding to $A(\boldsymbol{\xi})$ is fhf^{-1}.) The zero vector field is evidently a fixed point for A, and it is easy to check that A is continuously differentiable on a neighborhood of $\mathbf{0}$ in X and that

$$DA(\mathbf{0}) = f_*(\Lambda).$$

Since Λ is a hyperbolic set, we know from Theorem 9.1 that $f_*(\Lambda)$ is a hyperbolic linear operator, i.e., that $\mathbf{0}$ is a *hyperbolic* fixed point for A. Furthermore, from the proof of Theorem 9.1 we know what the contracting and expanding subspaces of $DA(\mathbf{0})$ are; they are respectively the spaces X^s (X^u) of vector fields with $\boldsymbol{\xi}(x) \in E^s(x)$ ($E^u(x)$) for all x.

We apply the Stable Manifold Theorem for Hyperbolic Fixed Points to A. Thus, for ϵ sufficiently small, to every $\boldsymbol{\xi}^u \in X^s$ with $\|\boldsymbol{\xi}^s\| < \epsilon$, there corresponds a unique $\boldsymbol{\xi}^u \equiv \mathbf{w}_s(\boldsymbol{\xi}^s) \in X^u$ such that

$$\|A^n(\boldsymbol{\xi}^s + \boldsymbol{\xi}^u)\| < \epsilon \text{ for all } n > 0. \tag{11.1}$$

It is easy to check that

$$A^n(\boldsymbol{\xi})(f^n(x))=f^n(x+\boldsymbol{\xi}(x))-f^n(x),$$

so (11.1) can be rewritten as

$$d(f^n(x+\boldsymbol{\xi}^s(x)+\boldsymbol{\xi}^u(x)),f^n(x))<\epsilon \text{ for all } n>0 \text{ and all } x\in\Lambda. \qquad (11.2)$$

In principle, the value of $\boldsymbol{\xi}^u$ at a particular x depends on the values of $\boldsymbol{\xi}^s$ on all of Λ. It is clear from the form of the determining equation (11.2), however, that different x's are completely independent. It thus follows from the above that, for any $x_0\in\Lambda$ and any $\xi^s\in E^s(x_0)$ with $\|\xi^s\|<\epsilon$, there is a unique $\xi^u\equiv w_s(x_0,\xi^s)\in E^u(x_0)$ such that, writing

$$x=x_0+\xi^s+\xi^u,$$

we have

$$d(f^n(x),f^n(x_0))<\epsilon \text{ for all } n>0,$$

and

$$\mathbf{w}_s(\boldsymbol{\xi}^s)(x)=w_s(x,\boldsymbol{\xi}^s(x)). \qquad (11.3)$$

This proves the first statement of the theorem.

If f is r times continuously differentiable, then so is A and hence, by the Stable Manifold Theorem for Hyperbolic Fixed Points, so is $\mathbf{w}_s$. It is not hard to show, using this fact and the form (11.3) of $\mathbf{w}_s$ that, for x_0 fixed, $w_s(x_0,\xi^s)$ is r times continuously differentiable in ξ^s.

It remains only to prove the joint continuity of the partial derivatives of $w_s(x_0,\xi^s)$ with respect to ξ^s. This can be done by repeating the above analysis for the operator obtained by letting A act on the space of *continuous* vector fields (which also has $\mathbf{0}$ as a hyperbolic fixed point).

As for the fixed point case, we can reformulate part of the above theorem in terms of the sets

$$W^s_\epsilon(x_0)=\{x\in M:d(f^n(x),f^n(x_0))<\epsilon \text{ for all } n\geqslant 0\}.$$

The theorem tells us that, if ϵ is sufficiently small, then for any $x_0\in\Lambda$, $W^s_\epsilon(x_0)$ is a submanifold of M, as smooth as f is, with dimension equal to that of $E^s(x_0)$, passing through x_0 and tangent there to $E^s(x_0)$; furthermore, in a natural sense, $W^s_\epsilon(x_0)$ varies in a continuous way with x_0.

We also define *global* stable manifolds:

$$W^s(x_0)=\{x\in M;d(f^n(x),f^n(x_0))\to 0 \text{ as } n\to\infty\}.$$

It is easy to show, by the same argument as we used in the fixed-point case, that $W^s(x_0)$ is a countable union of pieces each of which is a well-behaved submanifold of M. Also as in the fixed-point case, the global manifold can accumulate on itself in a complicated way.

We define local and global *unstable* manifolds $W^u_\epsilon(x_0)$ and $W^u(x_0)$ by

$$W^u_\epsilon(x_0)=\{x\in M:d(f^{-n}(x),f^{-n}(x_0)<\epsilon \text{ for all } n\geqslant 0\}$$

$$W^u(x_0)=\{x\in M:d(f^{-n}(x),f^{-n}(x_0))\to 0 \text{ as } n\to\infty\}$$

The theory of these objects is easily deduced by applying the theory of stable manifolds to f^{-1}.

12. Local product structure and local maximality

Let Λ be a hyperbolic set, ϵ a very small positive number, and let x,y be two points of Λ whose separation is much smaller than ϵ. We want to argue that, in this situation, $W^s_\epsilon(x)$ and $W^u_\epsilon(y)$ intersect in a single point. This is easy to see if x and y are the same; $W^s_\epsilon(x)$ and $W^u_\epsilon(y)$ are submanifolds of complementary dimension (i.e., whose dimensions add up to the dimension of the ambient space M) which intersect "at nonzero angle" at x because their tangent spaces there ($E^s(x)$ and $E^u(x)$, respectively) have only the zero vector in common. Thus, the intersection at x is isolated, and, if ϵ is small enough, the manifolds cannot bend enough to cross again. To deal with the case of y near but not equal to x, one uses an argument similar to one used in the proof of the Inverse Function Theorem (together with the continuity of the dependence of $W^u_\epsilon(y)$ on y).

The above argument is a sketch of a proof of the following result:

Proposition 12.1.
1. There exists $\eta>0$ such that, for any pair of points $x,y \in \Lambda$ $W^s_\eta(x)$ and $W^u_\eta(y)$ have at most one point in common.

2. Given η as in 1., there exists $\epsilon>0$ such that, for any pair of points x,y with $d(x,y)<\epsilon$, $W^s_\eta(x)$ and $W^u_\eta(y)$ do intersect; the point of intersection varies continuously with x,y.

We will write $[x,y]$ for the point of intersection of $W^s_\eta(x)$ and $W^u_\eta(y)$, with the understanding that it is defined only for pairs x,y with $d(x,y)<\epsilon$, where ϵ,η are as in the proposition. For a general pair of nearby points x,y in a general hyperbolic set Λ, $[x,y]$ may or may not be in Λ. We say that Λ *has local product structure* if $[x,y]\in\Lambda$ for every pair of points x,y with $d(x,y)$ sufficiently small.

The reason for this terminology is as follows: Suppose Λ has local product structure; pick $x_0\in\Lambda$; let $\epsilon>0$ be sufficiently small; and put

$$X = W^u_\epsilon(x_0) \cap \Lambda,$$
$$Y = W^s_\epsilon(x_0) \cap \Lambda.$$

Then the mapping

$$(x,y) \to [x,y]$$

maps $X \times Y$ homeomorphically onto a neighborhood of x_0 in Λ. Thus, in a neighborhood of each of its points, Λ admits local coordinates which map it into the product of a piece of stable manifold with a piece of unstable manifold.

We now turn to some apparently unrelated concepts. If f is any homeomorphism of a topological space to itself, and if Λ is a set invariant for f, we will say that Λ is a *locally maximal* invariant set if there is an open set U containing Λ such that there is no invariant set properly containing Λ and contained in U. The relation between U and Λ in this case is easily seen to be equivalent to the statement that any f-orbit which remains in U for all time must be in Λ, or that

$$\Lambda = \bigcap_{n=-\infty}^{\infty} f^{-n} U$$

i.e., that U is an *isolating neighborhood* for Λ.

At this point, the notion of local maximality may look very restrictive. It is, however, not vacuous; for example, a set consisting of a single hyperbolic fixed point is locally maximal. We note one immediate consequence of local maximality: If Λ is a locally maximal hyperbolic set, then the Shadowing Theorem can be sharpened to say that the shadowing orbit $(f^n(y_0))$ is actually *in* Λ if ϵ is small enough. This follows because, by definition, the shadowing orbit stays near Λ for all time, and an orbit which stays near a locally maximal set must be in that set.

The notion of local maximality connects in a very neat way with local product structure:

Theorem 12.2. *A hyperbolic set is locally maximal if and only if it has local product structure.*

It is almost immediate that a locally maximal hyperbolic set has local product structure. We will not prove the (more difficult) converse statement; for a proof, see Proposition 8.20 of Shub [2]. The theorem is surprising in the following respect: Local product structure (for a hyperbolic set) is a purely intrinsic property of the set and the action of the mapping on the set. To make this more evident, note that local product structure for a hyperbolic set Λ is equivalent to the following:

For each $\eta>0$, there is an $\epsilon>0$ such that, for any two points $x,y \in \Lambda$ with $d(x,y)<\epsilon$, there is a point $z \in \Lambda$ with

$$d(f^n(z),f^n(x))<\eta \text{ for all } n \geqslant 0$$

$$d(f^{-n}(z),f^{-n}(y))<\eta \text{ for all } n \geqslant 0.)$$

(Here, d can be any metric, not necessarily smooth or even Riemannian, inducing the topology of Λ.) Local maximality, on the other hand, is not an intrinsic property; it depends on how the set is imbedded in the ambient space. There is no real paradox, however; the property of hyperbolicity, necessary to deduce local maximality from local product structure, is a strong restriction on how the mapping acts on the ambient space in the vicinity of Λ.

13. Recurrence

In analyzing dynamical systems, it is useful to distinguish between orbits which are *transient,* i.e., which go away and never come back, and those which are in some sense recurrent. Unfortunately, there are a number of inequivalent ways of making this distinction precise, with different advantages and disadvantages. The elementary notion of recurrence—i.e., a point x is recurrent if it is an accumulation point of its forward orbit—does not seems to be very useful. We will discuss here three less obvious ways of making the distinction between recurrence and transience. For the purposes of this section, we will assume that that the space M on which our mapping f acts is compact (although this is inconsistent with our standing convention that M is an open subset of Euclidean space.)

1. Let x be a point of M. We define the ω *limit set* of x (written $\omega(x;f)$) to be the set of all accumulation points of the forward orbit of x, and we denote by $\tilde{\Omega}(f)$ the closure of the union over $x \in M$ of $\omega(x;f)$. It is easy to see that all periodic points of f are in $\tilde{\Omega}$ and that every forward orbit converges to $\tilde{\Omega}$. To say that x is in $\tilde{\Omega}$ is to say that the orbit of x is recurrent in a relatively strong sense, and hence to say that x is not in $\tilde{\Omega}$ is to say that the orbit of x is transient in a relatively weak sense.

2. A point $x \in M$ is *wandering* if there is an open set U containing x such that

$$f^{-n}U \cap U=\varnothing \text{ for all } n>0,$$

and a *non-wandering* point is one which is not wandering. We will denote by $\Omega(f)$ the set of all non-wandering points for f. By definition, the set of wandering points is open, so $\Omega(f)$ is closed. It is easy to see that

$$\tilde{\Omega}(f) \subset \Omega(f)$$

3. We say that a point x is *chain recurrent* if for every $\delta>0$ there is a (finite) δ pseudo-orbit starting and ending at x, or, equivalently, if x lies on a periodic δ pseudo-orbit. We let $R(f)$ denote the set of all chain-recurrent points for f. It is straightforward to prove that $R(f)$ is closed and contains $\Omega(f)$. To say that a point x is *not* chain recurrent is to say that x is transient in the strong sense that the orbit of x and the orbits of all points sufficiently near to it go away and never return, not only for the motion induced by the mapping f but even for motions obtained by small (possibly non-deterministic) perturbations on f.

The notion of chain recurrence is more subtle than might at first appear. Notably, we have

Proposition 13.1 *If $x \in R(f)$, then for each $\delta>0$ there is a δ pseudo-orbit in $R(f)$ (not just in the ambient space M) starting and ending at x.*

This proposition can be reformulated as the equality

$$R(f|_{R(f)})=R(f).$$

The corresponding equality for the non-wandering set does not hold in general.

Proof. We need some notation. For $\delta>0$ and $x \in M$, we write $R_\delta(x)$ for the set of points y such that there exist δ pseudo-orbits from x to y and from y to x. We also write $R(x;f)$ for the set of points y such that, for each $\delta>0$, there exist δ pseudo-orbits from x to y and from y to x. It is immediate that x is chain recurrent if and only if $x \in R(x;f)$ and that every point of $R(x;f)$ is chain recurrent. It is not difficult to see that, for any $\delta>0$ and any $x \in M$, $R_\delta(x)$ is open (Recall that a δ pseudo-orbit is defined by the condition that each $d(x_{n+1}, f(x_n))$ is *strictly* less than δ.) and that each $R(x;f)$ is closed.

Lemma 13.2. *As δ decreases to 0, $R_\delta(x)$ decreases to $R(x;f)$, i.e., if U is any open set containing $R(x;f)$, then $R_\delta(x) \subset U$ for all sufficiently small δ.*

Proof. Suppose not. Then there is an open set $U \supset R(x;f)$, a sequence (δ_n) decreasing to zero, and a sequence

$$y_n \in R_{\delta_n}(x) \setminus U.$$

By passing to a subsequence if necessary we can assume that the y_n converge to a limit y, and, since U is assumed to be open, y is not in U. We are going to argue that $y \in R(x;f)$; this will contradict $U \supset R(x;f)$ and thus establish the lemma.

What we have to do is to show that, for any $\delta>0$, there are δ pseudo-orbits from x to y and from y to x. We will show how to construct the second of these; the construction of the first is similar but slightly easier. Take $\epsilon>0$ small enough so that $d(f(z_1), f(z_2))<\delta/2$ whenever $d(z_1, z_2)<\epsilon$; then take n large enough so that $d(y_n, y)<\epsilon$ and also so that $\delta_n<\delta/2$. Then if

$$z_0=y_n, z_1, \ldots, z_j=x$$

is a δ_n pseudo-orbit from y_n to x,

$$y, z_1, \ldots, z_j$$

is a $\delta_n+\delta/2<\delta$ pseudo-orbit from y to x.

We can now complete the proof of Proposition 13.1. The idea is as follows: Given $x \in R(f)$ and $\delta>0$, we want to show that there is a δ pseudo-orbit in $R(f)$ starting and ending at x. Since $x \in R(f)$, there is such a pseudo-orbit in M, and, by the definition of $R_\delta(x)$, this pseudo-orbit lies in fact in $R_\delta(x)$. But Lemma 13.2 shows that, for δ small, $R_\delta(x)$ is not much larger than $R(x;f) \subset R(x)$ so the pseudo-orbit can be moved into $R(x)$ by a small perturbation.

To fill in the details, we first choose $\delta_1<\delta/3$ so that

$$d(z_1,z_2)<\delta_1 \text{ implies } d(f(z_1),f(z_2))<\delta/3.$$

Then, using Lemma 13.2, choose $\delta_2<\delta/3$ so that

$$y \in R_{\delta_2}(x) \text{ implies } d(y,R(x;f))<\delta_1.$$

Let

$$y'_0=x, y'_1, \ldots, y'_n=x$$

be a δ_2 pseudo-orbit starting and ending at x. For each $j=1,\ldots,n-1$, choose $y_j \in R(x)$ such that $d(y_j,y'_j)<\delta_1$, and put $y_0=y_n=x$. With these choices,

$$d(y_{j+1},f(y_j)) \leqslant d(y_{j+1},y'_{j+1})+d(y'_{j+1},f(y'_j))+d(f(y'_j),f(y_j))$$
$$<\delta_1+\delta_2+\delta/3<\delta/3+\delta/3+\delta/3=\delta,$$

so $y_0,\ldots,y_n$ is the desired δ pseudo-orbit in $R(x)$ starting and ending at x.

14. Global stability.

Up to now, we have studied hyperbolic sets semi-locally, i.e., without worrying about what is happening elsewhere in the state space M. We now take up the more global question: What can we say about a mapping f if we know that the set of all its recurrent points is hyperbolic? As we indicated in the last section, there are several alternative notions of recurrence and hence several variants of this question. In 1967 Smale formulated a condition on a diffeomorphism f, which he called *Axiom A,* the main content of which was the requirement that the non-wandering set of f is hyperbolic. He also found it necessary to require separately that the periodic points for f are dense in the non-wandering set. Even with this condition added, Axiom A was not strong enough to imply any reasonable kind of global stability for f; something further, called the *no-cycles condition,* was needed. It was observed after the fact that the two conditions in Axiom A and the no-cycles condition were, all taken together, equivalent to the single condition that the chain recurrent set $R(f)$ is hyperbolic. We will sketch here some of the consequences of this condition.

Proposition 14.1. *If $R(f)$ is hyperbolic, then periodic points are dense in $R(f)$*

Proof. By Proposition 13.1, each point $x \in R(f)$ lies on a periodic δ pseudo-orbit in $R(f)$ with δ as small as we like. Hence, by the Shadowing Theorem, there is a periodic orbit as close as we like to x. This periodic orbit is in $R(f)$ because *all* periodic points are chain recurrent.

Proposition 14.2. *If the closure of the set of periodic points of f is a hyperbolic set, it has local product structure.*

We make only a few remarks about the proof. The first idea is that, to prove local product structure, it is only necessary (by continuity) to show that $[x,y]$ is a limit of periodic points *for x,y themselves periodic points with d(x,y) small.* This latter statement is proved by a variant of the argument given in Section 5 to show that a transverse homoclinic point is a limit of periodic points. The construction of periodic pseudo-orbits passing through $[x,y]$ is done using that fact that, not only do $W^s_\eta(x)$ with $W^u_\eta(y)$ intersect (at $[x,y]$), but $W^u_\eta(x)$ with $W^s_\eta(y)$ also intersect (at $[y,x]$). The pseudo-orbits go from $[x,y]$ to x along $W^s_\eta(x)$, then from x to $[y,x]$ along $W^u_\eta(x)$, then from $[y,x]$ to y along $W^s_\eta(y)$, then back to $[x,y]$ along $W^u_\eta(y)$.

Putting together the above propositions, we see that, if $R(f)$ is hyperbolic, it has local product structure.

To close these notes we cite one of the high points of this circle of ideas:

Theorem 14.3. Ω-Stability Theorem. *Let f be a diffeomorphism of the compact manifold M such that $R(f)$ is hyperbolic. Then for any $\tilde{f}$ sufficiently close to f in the C^1 topology:*
1. $R(\tilde{f})$ is a hyperbolic set for $\tilde{f}$.
2. $R(\tilde{f})$ is close to $R(f)$.
3. The restriction of $\tilde{f}$ to $R(\tilde{f})$ is topologically conjugate to the restriction of f to $R(f)$.

Most of the elements of the proof are contained in the results we established in Sections 8 and 10 about structural stabiliy of hyperbolic sets and stability of hyperbolicity. To complete the argument, it is necessary to show that $R(\tilde{f})$ cannot be much larger than $R(f)$. This requires methods which we have not developed and for which we refer again to Shub's monograph.

References.

1. O.E. Lanford. Introduction to the mathematical theory of dynamical systems. To appear in the proceedings of the 1981 Les Houches school on Chaotic Behavior of Dynamical Systems, edited by R. Helleman and G. Iooss, to be published by North Holland.
2. M. Shub. Stabilité globale des systèmes dynamiques. Astérisque **56** (1978).

TOPICS IN CONSERVATIVE DYNAMICS

Sheldon Newhouse

Department of Mathematics
University of North Carolina
Chapel Hill, N.C. 27514

INTRODUCTION

The theory of dynamical systems has a long and illustrious history. From the time of Kepler and Newton to the present time, each generation of scientists has successively unified and clarified previous approaches, extended the scope of previous methods, even modified the basic problems of consideration as new tools became available. A significant change in approach arose with the studies of Poincaré on celestial mechanics. Previous to that time one expected that each particular problem could be "solved" by finding sufficiently many integrals (constants of motion), special solutions, or by formal manipulations with power series. It was Poincaré who first realized that in general mechanical systems were not completely integrable. He proved that even in the restricted three body problem there did not exist a complete set of "uniform" integrals. This means that in certain cases the restricted three body problem can be regarded as a system of differential equations depending analytically on a parameter μ , and that in those cases it is impossible to find a complete system of integrals depending analytically on μ (see [13,I],[17]). While investigating this problem, Poincaré found a number of interesting and complicated motions to be described. After discovering the complicated homoclinic motions near periodic solutions of the first kind, he wrote, perhaps in exhilaration, perhaps in despair,

> "How does one seek to represent the figure formed by these two curves and their infinitely many intersections each of which corresponds to a doubly asymptotic solution? These intersections form a sort of trellis, of tissue, a network of threads infinitely intertwined. Each of the two curves must never cut through itself but it must fold on itself in a very complex manner in order to cut through all the threads in the network an infinity of times.
>
> One will be struck by the complexity of this figure which I do not even attempt to draw. Nothing more properly gives us an idea of complication of the problem of three bodies and, in general, of all the problems in Dynamics where there is no uniform integral..." ([13,III;p.389]).

Even today we have no definitive description of the closure of the homoclinic motions, but, as we shall see, we have a number of striking results.

Poincaré felt that a knowledge of the location of the periodic motions would be of fundamental importance in Dynamics. He even felt that they would be dense in the totality of solutions to canonical systems $\frac{dx_i}{dt} = \frac{\partial F}{\partial y_i}$, $\frac{dy_i}{dt} = -\frac{\partial F}{\partial x_i}$, $i = 1,\ldots,n$, with $F = F_0 + \mu F_1 + \mu^2 F_2 + \ldots$ where μ is small, $F_0 = F_0(x)$ is independent of y , and F_1 , F_2 ,... are periodic of period 2π in y ([13,I;p.82]). Interpreted with modern precision this is clearly false (take $F_0(x_i,y_i) = \sum_{i=1}^{n} \omega_i x_i$, $F_1 = F_2 = \ldots = 0$, ω_i incommensurate). However, Poincaré was interested in the "general" equation, and , for instance, if each F_i , $i \geq 1$, depends non-trivially on x and y , the result is not known. In the restricted three body problem, Poincaré found a surface of section (in appropriate variables) whose first return map was a map of an annulus rotating boundary components at different rates. Raising this map to an appropriate power gives a so-called "twist" map of an annulus. This is a map having a lift to a strip moving boundary components in opposite directions. In his celebrated "geometric" theorem, Poincaré conjectured that such a map would have at least two fixed points. This theorem was proved by Birkhoff in 1914. It implies that there are infinitely many periodic solutions in the restricted three body problem at certain fixed energy levels. Actually a related theorem of Birkhoff which states that in general an elliptic fixed point of an area preserving map of the plane is a limit of infinitely many periodic orbits suffices to prove the application just mentioned (see [6],[15]).

In the next generation Birkhoff significantly developed and extended Poincaré's ideas. Among other things, he established normal forms near critical points and periodic solutions of Hamiltonian systems, gave a number of examples which could be reduced to surfaces of section, emphasized the importance of geodesic flows, developed various stability problems, and set up a general theory of different kinds of recurrent motions. In this last subject, the pointwise ergodic theorem (in which Birkhoff extended Von Neumann's mean ergodic theorem) is perhaps among the most significant. The collected works of Birkhoff (needless to say of Poincaré) contain many rich ideas which remain to be developed.

In these lectures we discuss some topics close to the interests of both Poincaré and Birkhoff. In lecture 1, we review two topological methods for finding periodic motions - the method of closed geodesics, and the Poincaré-Birkhoff fixed point theorem. Lectures 2,3, and 4 are mainly concerned with conservative systems of two degrees of freedom. In lecture 2, we present the general structure of homoclinic points, and we apply recent developments in smooth ergodic theory (due to Oseledec, Pesin, and Katok) to give a new analytic method for finding periodic and homoclinic motions. Finally, in lectures 3 and 4, we consider some generic properties.

Lecture 1.

This lecture discusses two methods for proving the existence of periodic motions in Hamiltonian systems. The first is the method of "closed geodesics," and the second concerns twist maps of an annulus.

There are many other methods which we will not discuss. We refer the reader to the recent surveys of Rabinowitz [14], and Berestycki [5], and [4].

Let us first recall the modern setting for problems in classical mechanics. General references are Arnold [2] and Abraham-Marsden [1].

In classical mechanics, one studies the motion of a particle in a submanifold Q of Euclidean space subject to various forces. The equations of motion can be written in various forms.

(1) Lagrangian form

Let TQ be the tangent bundle of Q, let $q(t)$ be the position at time t, and let $\dot{q}(t)$ be the velocity at time t. Assume $\alpha : t \longrightarrow (q(t),\dot{q}(t))$ is a C^1 function from the real interval $[t_1,t_2]$ to TQ. Let $L : TQ \longrightarrow \mathbb{R}$ be a smooth function (the Lagrangian). The motion $(q(t),\dot{q}(t))$ must be a critical point of the functional

$$I(\alpha) = \int_{t_1}^{t_2} L(q(t),\dot{q}(t))dt$$

In local coordinates $(q,\dot{q}) = (q_1,\ldots,q_n,\dot{q}_1,\ldots,\dot{q}_n)$ in TQ, such motions $(q(t),\dot{q}(t))$ satisfy the Euler-Lagrange equations

$$\frac{\partial}{\partial t}\left(\frac{\partial L}{\partial \dot{q}_i}(q(t),\dot{q}(t))\right) = \frac{\partial L}{\partial q_i}(q(t),\dot{q}(t)), \quad i = 1,\ldots,n$$

(2) Hamiltonian form

Let T^*Q be the cotangent bundle of Q, let $\pi : T^*Q \longrightarrow Q$ be the projection, and let $T\pi : T(T^*Q) \longrightarrow TQ$ be the derivative of π. There is a canonical one-form θ on T^*Q defined by $\theta(v) = \alpha_q(T\pi(v))$ for $v \in T_{\alpha_q}T^*Q$ and $\alpha_q \in T_q^*Q$ where T_q^*Q is the cotangent space at q. Then, $\omega = -d\theta$ is a symplectic 2-form on T^*Q (i.e. ω is closed and non-degenerate). There are local coordinates $(q,p) = (q_1,\ldots,q_n,p_1,\ldots,p_n)$ about any point in T^*Q such that $\theta = \sum_{i=1}^{n} p_i dq_i$ and $\omega = \sum_{i=1}^{n} dq_i \wedge dp_i$ with $n = \dim Q$. If $H : T^*Q \longrightarrow \mathbb{R}$ is a C^2 function (the Hamiltonian), one defines a vector field X_H on T^*Q by

$$\omega_{\alpha_q}(X_H(\alpha_q),Y) = T_{\alpha_q}H(Y)$$

This is the Hamiltonian vector field associated to H. In the above coordinates (q,p) on T^*Q, $(\dot{q},\dot{p}) = X_H(q,p)$ has the form of Hamilton's equations

$$\dot{q}_i = \frac{\partial H}{\partial p_i}, \qquad \dot{p}_i = -\frac{\partial H}{\partial q_i}, \qquad i = 1,\ldots,n.$$

The equations in (1) and (2) are frequently related. Define the Legendre transformation $FL : TQ \longrightarrow T^*Q$ as follows. Let

$L_q = L \mid T_qQ : T_qQ \longrightarrow \mathbb{R}$ be the restriction of L to the tangent space T_qQ to Q at q. Then T_qQ is a vector space, and $L_q : T_qQ \longrightarrow \mathbb{R}$ is a smooth map. Fixing a point $v_q \in T_qQ$, the derivative $T_{v_q}L_q$ of L_q at v_q may be thought of as a linear map from T_qQ to $\mathbb{R}$; i.e. as an element of T_q^*Q. Set $FL(v_q) = T_{v_q}L_q \in T^*Q$. Following Abraham and Marsden [1], we call L hyperregular if $FL : TQ \longrightarrow T^*Q$ is a diffeomorphism. We set $\omega_L = (FL)^*\omega$ to get a symplectic form on TQ. Define the action function $A : TQ \longrightarrow \mathbb{R}$ by $A(v_q) = FL(v_q)(v_q)$ and the energy function $E : TQ \longrightarrow \mathbb{R}$ by $E = A - L$. Set $H = E \circ (FL)^{-1}$, and consider the Hamiltonian vector field X_E induced by ω_L on TQ. We have $(FL)_*X_E = X_H$, and solutions $(q(t),\dot{q}(t))$ of X_E satisfy the Euler-Lagrange equations for L. Thus, in the hyperregular case, the Lagrangian vector field X_E and the Hamiltonian vector field X_H are equivalent.

In the above situations, one calls Q the configuration space, TQ the velocity phase space, and T^*Q the momentum phase space.

More generally, one can study Hamiltonian systems on arbitrary symplectic manifolds. Such a manifold is a pair (M,ω) where M is a (necessarily even dimensional) smooth manifold and ω is a closed non-degenerate 2-form. Given a C^2 function $H : M \longrightarrow \mathbb{R}$ one defines the Hamiltonian vector field X_H by

$$\omega_x(X_H(x),Y) = T_xH(Y) \text{ for } x \in M \text{ , } Y \in T_xM.$$

The structure of the set of solution curves of X_H for various

(M,ω) and H gives the natural setting for Hamiltonian mechanics.

Examples: 1. Suppose we are given a system of n point masses in the line $\mathbb{R}$. Let m_i denote the mass of the i-th particle, let $q_i(t)$, $\dot{q}_i(t)$, $\ddot{q}_i(t)$ denote its position, velocity, and acceleration at time t, respectively. Suppose we are given an open subset W of $\mathbb{R}^n$ and a C^2 function $U : W \longrightarrow \mathbb{R}$. The system

$$(1) \qquad m_i\ddot{q}_i(t) = -\frac{\partial U}{\partial q_i}(q(t)) \ , \ i = 1,\ldots,n$$

determines the motion of the particles in W with potential energy U. The configuration space is W, the velocity phase space TW may be identified with $W \times \mathbb{R}^n$, and the momentum phase space T^*W may be identified with $W \times \mathbb{R}^n$. Let $p_i = m_i\dot{q}_i$, and set $H(q,p) = \frac{1}{2}\sum_{i=1}^{n}\frac{p_i^2}{m_i} + U(q_1,\ldots,q_n)$. Then, (1) is equivalent to

$$(2) \qquad \begin{aligned} \dot{q}_i &= \frac{\partial H}{\partial p_i} \\ & & i = 1,\ldots,n \\ \dot{p}_i &= -\frac{\partial H}{\partial q_i} \end{aligned}$$

With $\omega = \sum_{i=1}^{n} dq_i \wedge dp_i$,

$$X_H(q_1,\ldots,q_n,p_1,\ldots,p_n) = \left(\frac{\partial H}{\partial p_1},\ldots,\frac{\partial H}{\partial p_n}, -\frac{\partial H}{\partial q_1},\ldots, -\frac{\partial H}{\partial q_n}\right)$$

is the vector field of (2). Letting $T(\dot{q}_1,\ldots,\dot{q}_n) = \frac{1}{2}\sum_{i=1}^{n} m_i\dot{q}_i^2$ be the kinetic energy, we get

$$E(q,\dot{q}) = T(\dot{q}) + U(q), \text{ and } L(q,\dot{q}) = T(\dot{q}) - U(q) .$$

Note that with $n = 3k$, $m_{3i+1} = m_{3i+2} = m_{3i+3}$ for $i = 0,\ldots,k-1$, this example includes the motion of k point masses in an open subset of $\mathbb{R}^3$ with an arbitrary potential energy.

2. <u>The spherical pendulum with arbitrary potential energy.</u> Here $Q = S^2$ is the unit sphere in $\mathbb{R}^3$ a particle of mass m is constrained to move in Q. Let $U : Q \longrightarrow \mathbb{R}$ be the C^2 potential function. Let $\pi : TQ \longrightarrow Q$ be the projection. If $\langle v,w \rangle$ is the usual inner product in $\mathbb{R}^3$, and we think of tangent vectors to S^2 as elements of $\mathbb{R}^3$, we get the kinetic energy function $T : TQ \longrightarrow \mathbb{R}$ defined by $T(q,\dot{q}) = \frac{1}{2}m \langle \dot{q},\dot{q} \rangle$. Then, let $\tilde{U} = U \circ \pi : TQ \longrightarrow \mathbb{R}$, and set $L = T - \tilde{U}$. With the usual gravitational potential energy = mg × height this system is completely integrable (see [17]).

3. <u>Double planar pendulum with arbitrary potential energy</u>

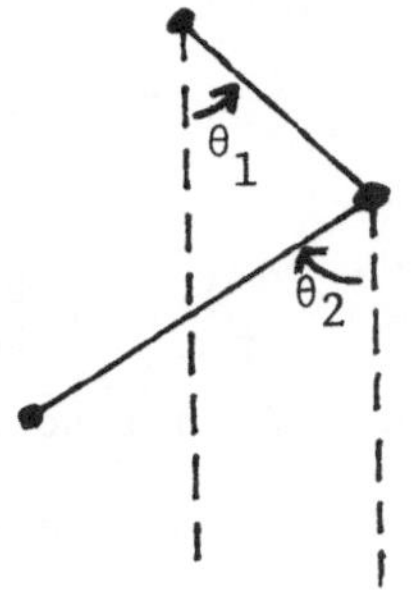

Figure (1.1)

If θ_i is the angular position of the mass m_i, then $\{(\theta_1,\theta_2)\} = Q$ is the 2-torus $T^2 = S^1 \times S^1$ where S^1 is a circle. The kinetic energy is $T(\dot\theta_1,\dot\theta_2) = \frac{1}{2}(m_1\dot\theta_1{}^2 + m_2\dot\theta_2{}^2)$, the potential energy is $U(\theta_1,\theta_2)$ and $L = T - \tilde{U}$ with $\tilde{U} = U \circ \pi$, $\pi : TQ \longrightarrow Q$.

Let us get more general.

4. Let Q be a smooth manifold, and let $<\cdot,\cdot>$ be a Riemannian metric on TQ. Then, g defines a "kinetic energy" function $T : TQ \longrightarrow \mathbb{R}$ by $T(v) = \frac{1}{2} <v,v>$. Given a function $U : Q \longrightarrow \mathbb{R}$, let $\tilde{U} = U \circ \pi$ and let $L : TQ \longrightarrow \mathbb{R}$ be $L = T - \tilde{U}$. Then, L is hyperregular, and we call X_E the Lagrangian vector field induced by T and U.

Concerning periodic motions, one has

Theorem (1.1)(H. Gluck and W. Ziller [10]). With Q,T,U *as above, let* $e \in \mathbb{R}$ *be a regular value of* E*(i.e.* $T_xE \neq 0$ *for all* $x \in E^{-1}(e)$*) such that* $\{q \in Q : U(q) \leq e\}$ *is compact and non-empty. Then,* X_E *has a periodic orbit of energy* e.

5. In the system $m_i\ddot{q}_i = -\frac{\partial U}{\partial q_i}$, any regular value e of U is one of E, so if $\{U \leq e\}$ is compact and non-empty, there is a periodic solution of energy e.

The critical points of U are equilibria for the first order system $\dot{q}_i = v_i$, $m_i \dot{v}_i = -\frac{\partial U}{\partial q_i}$. Thus, if $U^{-1}(e)$ is compact and non-empty, there is a critical point or periodic solution in $E^{-1}(e)$.

6. If Q is compact, then each $e > \max U(q)$ is a regular value of $E = T + \tilde{U}$. Hence, there are periodic motions of all large energy. This holds, for instance, in the pendulum examples 2 and 3.

In Hamiltonian form, Gluck and Ziller prove the following version of theorem 1.

Theorem (1.1) Let $H : T^*Q \longrightarrow \mathbb{R}$ *be convex and even on each co-tangent space* T_q^*Q . *Let* $e \in \mathbb{R}$ *be a regular value of* H *such that* $H^{-1}(e)$ *is compact and non-empty. Then there is a periodic motion of* X_H *in* $H^{-1}(e)$.

To say that H is convex on T_q^*Q means the Hessian of H (second derivative of H) on T_q^*Q is positive definite. To say that H is even means $H(-\alpha_q) = H(\alpha_q)$ for $\alpha_q \in T_q^*Q$.

Theorem (1.2) (Weinstein) Let $\dim Q = n$, *and let* $H : T^*Q \longrightarrow \mathbb{R}$ *have a non-degenerate minimum at* $x \in T^*Q$. *Let* $e = H(x)$. *For* $c > e$ *and near* e , $H^{-1}(c)$ *has at least* n *distinct periodic orbits.*

Moser [12] has a simple proof and generalization of theorem 2. The proof of theorem (1.1) makes use of "geodesic" techniques.

To begin to understand the technique, first consider the case in which $U(q) = 0$. Then we seek to extremize $\int_{t_1}^{t_2} T(\dot{q}(t))dt$. If one looks for minima among curves $q(t)$ with $T(\dot{q}(t)) = \text{constant}$, $q(t_1) = q_1$, $q(t_2) = q_2$, then one has the same solutions as in the equation

$$\delta \int_{t_1}^{t_2} |(\dot{q}(t)|dt = 0 = \delta \int_{t_1}^{t_2} \langle \dot{q}(t), \dot{q}(t) \rangle^{\frac{1}{2}} dt = 0 .$$

That is, one seeks to minimize the arclength of the curve $t \longrightarrow q(t)$ joining q_1 to q_2. Curves of such minimal length, if they exist, are called geodesics. At least if q_1 and q_2 are close, then they can be joined by a unique geodesic. These geodesics are the projections to Q of integral curves of the Lagrangian vector field X_E for $U \equiv 0$ (i.e. $L = T$). The flow on TQ which X_E generates is called the geodesic flow. Another way to describe this flow ϕ_t at time t is the following. Given $v \in T_qQ$, let $\gamma(t)$ be the unique geodesic with $\gamma(0) = q$, $\dot{\gamma} = v$. Set $\phi_t(v) = \dot{\gamma}(t)$. This is defined for t in an interval about 0 depending on q. If Q is compact, then $\phi_t(v)$ is defined for all t.

Thus, in the absence of a potential (i.e. free motion) one has the geodesic flow, and periodic motions correspond to closed geodesics. For instance, in the spherical pendulum with zero potential, all orbits are periodic. They move along the geodesics in the standard metric on S^2 (i.e. along the great circles). It is a theorem of Jacobi that even with a non-zero potential function U, the Lagrangian motions are geodesics (up to reparametrization) in a conformally equivalent "metric". This metric degenerates to zero along subsets of TQ where $E - \tilde{U} = 0$. More explicitly, let $\langle \cdot,\cdot \rangle_q$ be the Riemannian metric on Q giving the kinetic energy $T(\dot{q},\dot{q}) = \frac{1}{2}\langle \dot{q},\dot{q} \rangle$, and let $U : Q \longrightarrow \mathbb{R}$ be the potential function with lift $\tilde{U} = U \circ \pi$ to TQ. Fix a value e of the energy E. Set $\langle \dot{q},\dot{q} \rangle_q' = (e-U(q)) \langle \dot{q},\dot{q} \rangle_q$. Jacobi's theorem says that Lagrangian motions of $L = T - \tilde{U}$ can be reparametrized to be geodesics of $\langle , \rangle_q'$. The metric $\langle , \rangle_q'$ is called the Jacobi metric (see [1]).

<u>Remarks.</u> 7. If Q is compact, and $e > \max_q U(q)$, then $\langle \dot{q},\dot{q} \rangle_q'$ is a Riemannian metric on TQ. So all orbits at large energy are geodesics.

8. Let $h = \sqrt{e}$ and consider the contraction $\Phi_h : TQ \longrightarrow TQ$ given by $\Phi_h(q,\dot{q}) = (q,h^{-1}\dot{q})$. Let $\langle , \rangle''$ be the induced metric, so that

$$\langle \dot{q},\dot{q} \rangle'' = \langle h^{-1}\dot{q},h^{-1}\dot{q} \rangle' = h^{-2}(e-U(q)) \langle \dot{q},\dot{q} \rangle = (1- \frac{U(q)}{e}) \langle \dot{q},\dot{q} \rangle .$$

Then, Φ_h defines a smooth equivalence between the $\langle \cdot,\cdot \rangle''$ and the $\langle \cdot,\cdot \rangle'$ geodesic flows. For large e, $\langle \cdot,\cdot \rangle''$ is a small perturbation of $\langle \cdot,\cdot \rangle$. So the geodesic flow is the limit as $e \longrightarrow \infty$ of any Lagrangian motion with $L = T - \tilde{U}$. This is a precise way of stating the physically obvious fact that at very large energy the motion is essentially free.

Remark 7 permits us to think geometrically to produce closed geodesics, which in turn give periodic motions for our mechanical system. This is the main ingredient of the proof of theorem (1.1).

Let X be a closed subset of some Euclidean space, let Q be a smooth manifold, and let $I = [0,1]$ be the unit real interval. Two continuous mappings $F : X \longrightarrow Q$ and $G : X \longrightarrow Q$ are called homotopic if there is a continuous mapping $H : X \times I \longrightarrow Q$ such that $H(x,0) = F(x)$ and $H(x,1) = G(x)$ for all $x \in X$. We also say F is deformable to G and write $F \simeq G$. If G is a constant mapping (the image of G is a single point), we say F is deformable to a point or null-homotopic. If $X = S^1$ is the unit circle in the real plane, then a continuous (smooth) mapping $\gamma : X \longrightarrow Q$ is called a closed (smooth) curve in Q.

Theorem (1.3) (Hilbert) Let Q *be any Riemannian manifold, and let* γ *be a closed curve in* Q *not deformable to a point. Then* γ *is deformable to a closed geodesic.*

Corollary (1.4) The double (or multiple) pendulum with arbitrary potential has infinitely many periodic solutions for each large energy value.

The idea of the proof of theorem (1.3) is the following. Imagine a frictionless surface M of genus greater than zero. For instance suppose M is a torus as in the next figure. Suppose M is coated with vaseline or refined oil so as to be slippery. Stretch an elastic closed string γ around M so that it is not deformable to a point and suppose its unstretched length ℓ is small enough that any deformation of γ in M has length longer than ℓ . Let γ move freely on M . Its tension will move it so as to decrease length until it comes to rest in a curve $\tilde{\gamma}$ which locally has minimal length. This is a closed geodesic.

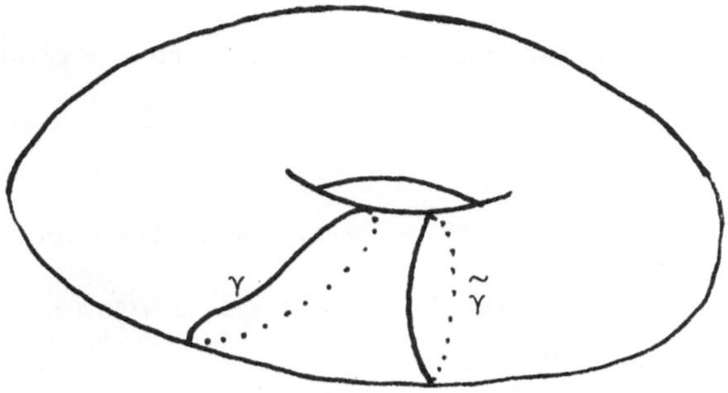

Figure (1.2)

Birkhoff extended this idea introducing a curve shortening process (on any closed manifold) which takes a piecewise smooth closed curve γ with non-vanishing tangent vectors and replaces γ by another such curve $D\gamma$ so that

(a) $\gamma \simeq D\gamma$

(b) length $(D\gamma) \leq$ length (γ) with equality if and only if γ is a closed smooth geodesic.

This is done as follows. Choose a finite set $t_0, t_1, \ldots, t_{k-1}$ of points on S^1 so that $\gamma(t_i)$ is close enough to $\gamma(t_{i+1})$ so that there is a unique geodesic segment α_i joining $\gamma(t_i)$ to $\gamma(t_{i+1})$ for $i \in {}^{Z}/_{kZ}$. Take the midpoint ξ_i of each geodesic segment α_i , let η_i be the geodesic segment joining ξ_i to ξ_{i+1} , and let $D_\gamma = \bigcup_i \eta_i$. Consider $D^2\gamma = D(D\gamma)$ and $D^j\gamma = D(D^{j-1}\gamma)$, $j \geq 1$. If the lengths of the curves $D^j\gamma$ are bounded below by a positive constant, then some subsequence of $\{D^j\gamma\}$ converges to a closed geodesic (in the natural C^1 topology). If the lengths approach zero, then $D^j\gamma$ approaches a constant curve (point). Observing that the operator D is continuous on the set of γ's with the C^1 topology, Birkhoff applied D to continuous families of piecewise smooth curves to prove

Theorem (1.5) (Birkhoff [6]) Any Riemannian metric on the n-dimesional sphere has at least one closed geodesic.

There are the following stronger results

Theorem (1.6) (Fet-Lyusternik [9]) Any Riemannian metric on a compact n-manifold has at least one closed geodesic.

Theorem (1.7) (Lyusternik-Schnirelmann [11]) Any Riemannian metric on the two dimensional sphere has at least three distinct closed geodesics.

Corollary (1.8) The spherical pendulum with arbitrary potential has at least three periodic solutions for each large energy value.

The proof of theorem (1.5) is roughly the following. Consider families of curves covering the n-sphere S^n which are not deformable to a point family; that is, continuous functions $F : S^1 \times D^{n-1} \longrightarrow S^n$ where D^{n-1} is the closed unit ball in $\mathbb{R}^{n-1}$, $F(\cdot,x)$ is a piecewise smooth curve in S^1, and F is not deformable to a point. Given such an F, let L(F) be the longest length of a curve in $\{F(\cdot,x) : x \in D^{n-1}\}$. Then define the family DF so that

(c) $F \simeq DF$

(d) each curve $DF(\cdot,x)$, $x \in D^{n-1}$, is a piecewise smooth geodesic

(e) $L(DF) \leq L(F)$ with equality if and only if the longest curve in $\{DF(\cdot,x)\}$ is a smooth geodesic.

Now, $D^i F \simeq F$ for each i, so no $D^i F$ is null-homotopic. Then, (exercise), there is a positive α such that $L(D^i F) > \alpha$ for all i. There is a subsequence $D^{i_1}F$, $D^{i_2}F, \ldots$ which converges to a family $\tilde{F}$ such that $L(D\tilde{F}) = L(\tilde{F})$, so $\tilde{F}$ contains a closed geodesic.

Theorem (1.6) involves only a slight extension. Let S^k denote the k-dimensional sphere. It follows from the Hurewicz isomorphism theorem in topology that if M^n is any compact n-manifold, then there is an integer k with $1 \leq k \leq n$ and a map $G : S^k \longrightarrow M^n$ which is not null-homotopic. Assume $n > 1$ since theorem (1.6) is trivial for the circle S^1. Letting $F : S^1 \times D^{k-1} \longrightarrow S^k$ be the non-null-homotopic map Birkhoff considered, and letting $F_1 = G \circ F$, we have that $F_1 : S^1 \times D^{k-1} \longrightarrow M^n$ is not null-homotopic. Now the above argument works.

We will not discuss the proof of theorem (1.7) referring the reader to [11] and [3].

We now proceed to discuss the Poincaré-Birkhoff theorem

Recall that in many cases in Hamiltonian dynamics one can

reduce the analysis to symplectic maps of a symplectic manifold. Two prime examples follow.

1. Suppose $\{\phi_t\}$ is the flow of X_H and $\gamma = \{\phi_t(x) : 0 \le t \le \tau\}$ is a periodic solution of period τ. Let $\Sigma_x = H^{-1}(H(x))$ be the constant energy set for x. Suppose also that $T_yH : T_yM \longrightarrow \mathbb{R}$ is not zero for each $y \in \gamma$, so that Σ_x is a codimension one submanifold near γ. Let Γ be a local hypersurface in Σ_x transverse to γ at x. For each y in Γ near x there is a least $t(y) > 0$ such that $\phi_{t(y)}(y) \in \Gamma$. There is also an induced symplectic structure on Γ and the map $f(y) = \phi_{t(y)}(y)$ is symplectic in this structure.

2. Let $\dot{q} = H_p(q,p,t)$, $\dot{p} = -H_q(q,p,t)$ be a time-dependent Hamiltonian system in $\mathbb{R}^{2n}$ with $H(q,p,t + 2\pi) = H(q,p,t)$ for all (q,p,t). Introducing t as a new coordinate we consider the autonomous system on $\mathbb{R}^{2n+1}$

$$(*) \qquad \begin{aligned} \dot{q} &= H_p(q,p,t) \\ \dot{p} &= -H_q(q,p,t) \\ \dot{t} &= 1 \end{aligned}$$

Letting $(q(t),p(t),t + \alpha)$ be a solution to $(*)$ one can show that the map

$$(q(0),p(0),\alpha) \longrightarrow (q(2\pi),p(2\pi),\alpha + 2\pi)$$

is symplectic via the form $\sum_{i=1}^{n} dq_i \wedge dp_i$.
Identifying t with $t + 2\pi$ gives a flow on $\mathbb{R}^{2n} \times S^1$ with $\mathbb{R}^{2n} \times \{s\}$ as a cross-section for any $s \in \mathbb{R}^1$.

The Poincaré-Birkhoff theorem provides an example of the use of this technique to find periodic motions.

Theorem (1.9) Let $T : A \longrightarrow A$ *be an area preserving homeomorphism of a plane annulus* A *rotating boundary components in opposite directions. Then* T *has at least two fixed points.*

This result may be frequently applied to give infinitely many periodic orbits as follows.

Theorem (1.10) Let $T : A \longrightarrow A$ *be an area preserving homeomorphism of a plane annulus rotating the boundary components through angles* α *and* β *, respectively, with* $0 < \alpha < \beta < 2\pi$*. Then, for any rational number* $\frac{p}{q}$ *with* $\alpha < \frac{2\pi p}{q} < \beta$ *and* p,q *relatively prime integers,* T^q *has at least two periodic orbits in* A.

Taking a sequence $\frac{p_n}{q_n}$ with $\alpha < \frac{2\pi p_n}{q_n} < \beta$ and q_{n+1} not divisible by q_n gives infinitely many periodic orbits for T.

Application: Consider a twist map

$$T : \begin{array}{l} x_1 = x + \alpha y \\ y_1 = y \end{array}$$

of the plane where x is an angular variable mod 2π, $y > 0$, and

$\alpha > 0$. The map T preserves the circles $y = \text{const}$ and rotates them at different rates. According to the Moser twist theorem, if $f(x,y)$ and $g(x,y)$ are appropriately small functions of period 2π in x, then

$$T' : \begin{aligned} x_1 &= x + \alpha y + f(x,y) \\ y_1 &= y + g(x,y) \end{aligned}$$

has infinitely many invariant circles on which T' acts like a rotation through an irrational angle. Moreover, the angles are different for different circles and each circle is a limit of others. If C_1 and C_2 are two such circles and A is the annulus they bound, then T' maps A to itself, and after a smooth coordinate change, T' satisfies the assumptions of theorem (1.10). Therefore each circle C_i is a limit of periodic motions.

We proceed to prove theorem (1.9). Consider the annulus A as $S^1 \times I$ where $I = \{r \in \mathbb{R} : r_1 \leq r \leq r_2\}$ and $S^1 = \{z \in C : |z| = 1\}$. We have a map $\Phi : \mathbb{R} \times I \longrightarrow A$ defined by $\Phi(\theta,r) = (e^{i\theta},r)$. We assume $T(z,r\ ) = (e^{i\alpha_j}z,r_j)$ for $j = 1,2$ where $-2\pi < \alpha_1 < 0 < \alpha_2 < 2\pi$. There is a unique map $\bar{T} : \mathbb{R} \times I \longrightarrow \mathbb{R} \times I$ such that $T \circ \Phi = \Phi \circ \bar{T}$ and $\bar{T}(\theta,r_1) = (\theta + \alpha_j,r_j)$. The map $\bar{T}$ is a homeomorphism of the strip $\mathbb{R} \times I$ to itself, and it satisfies $\bar{T}(\theta + 2\pi,r) = \bar{T}(\theta,r) + (2\pi,0)$. $\bar{T}$ translates the line $r = r_1$ to the left $|\alpha_1|$ units and translates the line $r = r_2$ to the right α_2 units. We may assume $\bar{T}$ is defined on the whole plane by setting $\bar{T}(\theta,r) = (\theta + \alpha_1,r)$ for $r \leq r_1$ and $\bar{T}(\theta,r) = (\theta + \alpha_2,r)$ for $r \geq r_2$. Also, $\bar{T}$ preserves the area element $rd\theta dr$. Note that if $\bar{T}(P) = P$, then each horizontal $2\pi n$ translate of P , $n \in Z$, is also a fixed point of $\bar{T}$. We will prove there are at least two such families of fixed points in the strip $\bar{A} = \{(\theta,r) : r_1 \leq r \leq r_2\}$.

Let us first prove there is at least one fixed point in the strip $\bar{A}$. Let L_i be the line $r = r_i$. Assuming there are no fixed points, let γ be an embedded arc in $\bar{A}$ joining L_1 to L_2 . Consider $v_\gamma(z) = \dfrac{\bar{T}(z) - z}{|\bar{T}(z) - z|}$ for $z \in \gamma$. This is a unit vector field along γ which is well-defined since $\bar{T}(z) \neq z$ for all $z \in \gamma$. Since $\bar{A}$ is simply connected there is a continuous function ϕ from the set of unit vectors beginning at points in $\bar{A}$ to $\mathbb{R}$ such that $\phi(v)$ equals the angle between v and the vector $(1,0)$ beginning at the same point as v . Let $\psi(\gamma,\bar{T})$ be the angular variation of $v_\gamma(z)$ as z goes up γ ; i.e., if $\gamma(z_i) \in L_i$, then $\psi(\gamma,\bar{T}) = \phi(v_\gamma(z_2)) - \phi(v_\gamma(z_1))$. We will obtain a contradiction by showing that $\psi(\gamma,\bar{T})$ is both π and $-\pi$.

<u>Proof that</u> $\psi(\gamma,\bar{T}) = -\pi$: Let $\varepsilon > 0$ be small, and let T_ε be the upward translation $T_\varepsilon(\theta,r) = (\theta,\varepsilon+r)$. The map $T_\varepsilon\bar{T}$ preserves $rd\theta dr$ and maps L_1 to $L_1 + (0,\varepsilon)$ with the obvious notation. Consider the half open strip $Q_1 = \{(\theta,r) : r_1 < r \leq r_1 + \varepsilon\}$. Clearly, $(T_\varepsilon\bar{T})^{-1}(Q_1)$ lies below L_1 , so Q_1 lies above $(T_\varepsilon\bar{T})^{-1}(Q_1)$. See figure (1.3)

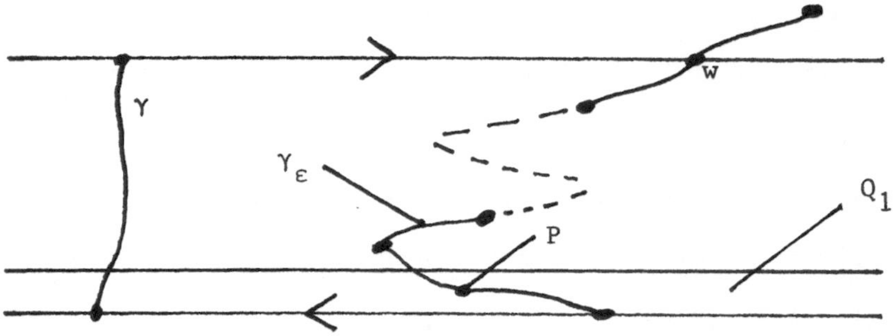

Figure (1.3)

Applying $T_\varepsilon \bar{T}$, we get that $T_\varepsilon \bar{T}(Q_1)$ lies above Q_1. Repeating, we get that $(T_\varepsilon \bar{T})^j Q_1$ lies above Q_1 for each $j \geq 1$. Also, for $j \neq k$, $(T_\varepsilon \bar{T})^j Q_1 \cap (T_\varepsilon \bar{T})^k Q_1 = \phi$. Since the areas of $(T_\varepsilon \bar{T})^j Q_1$ and $(T_\varepsilon \bar{T})^k Q_1$ are equal when projected down to the annulus A by Φ, all the iterates $(T_\varepsilon \bar{T})^j Q_1$, $j > 0$, cannot remain in $\bar{A}$. Thus, there is a $j_1 > 0$ such that $(T_\varepsilon \bar{T})^{j_1}(Q_1)$ contains a point, say P_{j_1} above L_2. Let $P \in Q_1$ be such that $(T_\varepsilon \bar{T})^{j_1}(P) = P_{j_1}$. Let $P_i = (T_\varepsilon \bar{T})^i P$. We may draw an arc η from P to P_1 so that $(T_\varepsilon \bar{T})\eta$ meets η only in P_1. Then taking images of η by $T_\varepsilon \bar{T}$, we get a simple arc $\tilde{\gamma}_\varepsilon$ from P to P_{j_1}. Let w be the first point in $\tilde{\gamma}_\varepsilon \cap L_2$, and add a little curve from L_1 to P in Closure Q_1 to make a curve γ_ε from L_1 to L_2 such that $(T_\varepsilon \bar{T})(\gamma_\varepsilon)$ is further along γ_ε as in figure (1.3).

We claim the angular variation

$$\psi(\gamma_\varepsilon) = \psi(\gamma_\varepsilon, T_\varepsilon \bar{T}) \quad \text{of} \quad \frac{T_\varepsilon \bar{T}(z) - z}{|T_\varepsilon \bar{T}(z) - z|} \quad \text{along } \gamma_\varepsilon .$$

is nearly $-\pi$. Indeed, we can achieve this variation in two steps. First, fix the point z_1 in γ_ε L_1 and move the point z from $T_\varepsilon\bar{T}(z_1)$ along γ_ε until it reaches w . The vector $z - z_1$ always points above L_1 . Then, fix w and move z from z_1 along γ_ε up to $(T_\varepsilon\bar{T})^{-1}w$. The vector $w - z$ still continues to point upward. This two step variation gives the same angular variation as $\psi(\gamma_\varepsilon, T_\varepsilon\bar{T})$, so the latter is nearly $-\pi$. Since γ_ε is deformable to γ with endpoints in $L_1 \cup L_2$, $\psi(\gamma_\varepsilon, T_\varepsilon\bar{T}) = \psi(\gamma, T_\varepsilon\bar{T})$. Letting $\varepsilon \longrightarrow 0$ gives $\psi(\gamma_\varepsilon, T_\varepsilon\bar{T}) \longrightarrow -\pi$ and $\psi(\gamma, T_\varepsilon\bar{T}) \rightarrow \psi(\gamma, \bar{T})$. Hence, $\psi(\gamma, \bar{T}) = -\pi$.

<u>Proof that</u> $\psi(\gamma, \bar{T}) = \pi$: Applying a similar construction to $T_{-\varepsilon}\bar{T}$ ($T_{-\varepsilon}$ is a ε-downward translation) we can construct a curve γ_ε^1 running from L_2 to L_1 so that the angular variation $\psi(\gamma_\varepsilon^1, T_{-\varepsilon}\bar{T})$ is nearly $-\pi$. Again letting $\varepsilon \longrightarrow 0$ gives $\psi(-\gamma, \bar{T}) = -\pi$. Here $-\gamma$ is the reverse curve determined by γ; i.e., if $C(t)$, $0 \le t \le 1$, is a parametrization of γ , then $-\gamma$ is the curve $C(1 - t)$. Since $\psi(\gamma, \bar{T}) = -\psi(-\gamma, \bar{T})$, we get $\psi(\gamma, \bar{T}) = \pi$.

This completes the proof that at least one family $p = \{(\bar{\theta} + 2\pi n , \bar{r}) : n \in Z\}$ of $\bar{T}$-fixed points exists.

Now, following Brown and Neumann [7], we again obtain a contradiction if we assume p is the only family of fixed points. Let C_1 be the curve $t \longrightarrow (\bar{\theta} + \pi, t)$, and C_2 be the curve

$t \longrightarrow (\bar{\theta} - \pi, t)$ with $r_1 \le t \le r_2$. Let $-C_2$ be the curve $t \longrightarrow (\bar{\theta} - \pi, r_1 + r_2 - t)$. Since, C_1 and C_2 differ by a horizontal 2π-translation, we have $\psi(C_1,\bar{T}) = \psi(C_2,\bar{T})$. Also, $\psi(-C_2,\bar{T}) = -\psi(C_2,\bar{T}) = -\psi(C_1,\bar{T})$. Let $D_1 \subset L_1, D_2 \subset L_2$ be such that $\eta \equiv D_1 \cup C_1 \cup D_2 \cup -C_2$ is a simple closed curve. Then, $\psi(\eta,\bar{T}) = 0$. It follows that any closed simple curve η_1 not meeting p has $\psi(\eta_1,\bar{T}) = 0$ using standard index arguments.

For small $\varepsilon > 0$, let

$$T_\varepsilon(\theta,r) = (\theta, r + \varepsilon(|\cos(\bar{\theta} + \theta)| - \cos(\bar{\theta} + \theta))).$$

Then, T_ε is an area preserving homeomorphism of $\mathbb{R}^2$ commuting with 2π-horizontal translation, and T_ε is the identity on the set $W = \{(\theta,r) : r \in \mathbb{R}, |\theta - (\bar{\theta} + 2\pi n)| \le \frac{\pi}{2}, n \in Z\}$. Note that for ε small enough, $T_\varepsilon\bar{T}$ has precisely p as its fixed points also. Reasoning as above we obtain the following facts,

(1) there is a $j > 0$ so that $(T_\varepsilon\bar{T})^j L_1$ contains points above L_2.

(2) there is a curve γ_ε from L_1 to L_2 not meeting p such that $T_\varepsilon\bar{T}$ moves points on γ_ε further along γ_ε.

(3) the angular variation $\psi(\gamma_\varepsilon, T_\varepsilon\bar{T})$ is nearly $-\pi$.

(4) $\psi(\eta_1, T_\varepsilon\bar{T}) = 0$ for any closed simple curve η_1 not meeting p.

Let γ be an embedded curve joining L_1 to L_2 and not meeting p. We assert that $\psi(\gamma, T_\varepsilon\bar{T}) = \psi(\gamma_\varepsilon, T_\varepsilon\bar{T})$ for each small ε. To see this, we construct a family $\{C_n\}$, $n \in Z$, of small closed simple curves such that $(\bar{\theta} + 2\pi n, \bar{r})$ is in the interior of C_n. Then we deform γ into γ_ε in a family $\{\xi_s\}$ with endpoints in $L_1 \cup L_2$ making use of pieces in the C_n's to "cross over" each element of $\mathcal{P}$. Since $\psi(C_n,\bar{T}) = \psi(C_n,T_\varepsilon\bar{T}) = 0$, we have that $\psi(\xi_s,T_\varepsilon\bar{T})$ is independent of s giving our assertion. Now, it follows as above that $\psi(\gamma,\bar{T}) = -\pi$. Working with $T_{-\varepsilon}\bar{T}$ as above would give $\psi(\gamma,\bar{T}) = \pi$ which is our desired contradiction.

Remark : A topological generalization of theorem [1.9] is in Carter [8].

References

1. R. Abraham and J. Marsden, Foundations of Mechanics, Benjamin-Cummings, N.Y., 1978.

2. V. Arnold, Mathematical Methods of Classical Mechanics, English translation.

3. W. Ballman, Der Satz von Lusternik- Schnirelmann, Bonner Math. Schriften 102 (1978).

4. W. Ballmann, G. Thorbergsson, and W. Ziller, On the existence of short closed geodesics and their stability properties, preprint, Math. Dept., Univ. of Penn.

5. H. Berestycki, Solutions Periodiques de Systemes Hamiltoniens, Seminaire Bourbaki, 1982/83, n°603.

6. G. D. Birkhoff, Dynamical Systems, revised edition, 1966., AMS Colloquium Lectures; Collected Papers, Amer. Math. Soc., Providence, R.I., 1950

7. M. Brown and W. Neumann, Proof of the Poincaré-Birkhoff fixed point theorem, Michigan Math. J. 24 (1977), 21-31.

8. P. Carter, An improvement of the Poincaré-Birkhoff fixed point theorem, Trans. Amer. Math. Soc. 269 (1982), 285-299.

9. A. Fet and L. Lyusternik, Variational Problems on closed manifolds, Dokl. Akad. Nauk. SSSR 81 (1951), 17-18.

10. H. Gluck and W. Ziller, Existence of Periodic Motions of Conservative Systems, preprint, Math. Dept., Univ. of Pennsylvania.

11. L. Lyusternik and L. Schnirelmann, Méthodes Topologiques dans les Problèmes Variationels, Hermann, Paris, 1934.

12. J. Moser, Periodic orbits mear an equilibrium and a theorem by Alan Weinstein, Comm. Pure Appl. Math 29 (1976), 727-747.

13. H. Poincaré, Methodes Nouvelles de la Mecanique Celeste, Vols. I,II,III, Gauthier-Villars, Paris, 1892 - 1899.

14. P. Rabinowitz, Periodic solutions of Hamiltonian Systems: A Survey, Siam J. Math. Anal. 13, No. 3, May, 1982, 343-352.

15. J. Moser and C. L. Siegel, Lectures on Celestial Mechanics, Springer-Verlag, N.Y., 1971.

16. S. Sternberg, Celestial Mechanics I,II, Benjamin, 1969.

17. E. Whittaker, A Treatise on the Analytical Dynamics of Particles, and Rigid Bodies, 4th ed., Cambridge, University Press.

Lecture 2.

Here we describe a new method of proving the existence of hyperbolic periodic orbits in systems with two degrees of freedom. Consider a C^r diffeomorphism $f : M^2 \longrightarrow M^2$ of an orientable surface $M = M^2$, $r > 1$. Let ω be a nowhere vanishing 2-form on M. We assume that f is weakly dissipative in the sense that for $x \in M$ and $v, w \in T_xM$, $|\omega_{fx}(T_xf(v), T_xf(w))| \leq |\omega_x(v,w)|$. Thus, T_xf does not increase area. It may be noted that our methods apply to three dimensional flows which weakly decrease volume.

First we recall the structure associated to homoclinic points. Let p be a hyperbolic periodic point of f of least period $\tau > 0$. Thus, $f^\tau(p) = p$, $f^k(p) \neq p$ for $0 < k < \tau$, and T_pf^τ has eigenvalues of absolute value different from 1. Let d be the distance function in M, and set

$$W^s(p) = \{y \in M : d(f^n y, f^n p) \rightarrow 0 \quad \text{as} \quad n \rightarrow \infty\},$$
$$W^u(p) = \{y \in M : d(f^n y, f^n p) \rightarrow 0 \quad \text{as} \quad n \rightarrow -\infty\}.$$

One calls $W^s(p)(W^u(p))$ the stable (unstable) manifold of p. It is known that $W^s(p)$ and $W^u(p)$ are C^r injectively immersed curves in M. Let $O(p)$ denote the orbit of p, and let

$$W^s(O(p)) = \bigcup_{0 \leq j < \tau} f^j W^s(p), \quad W^u(O(p)) = \bigcup_{0 \leq j < \tau} f^j W^u(p)$$

be the stable and unstable manifolds of the orbit of p. A point $q \in W^u(O(p)) \cap W^s(O(p)) - O(p)$ is called a homoclinic point of p (or f). Such a point is doubly asymptotic to the orbit of p.

If q is such a point and $W^u(O(p))$ is not tangent to $W^s(O(p))$ at q, one calls q a transverse homoclinic point. Let us examine the geometric structure produced by a transverse homoclinic point.

Let p be a hyperbolic fixed point for f.

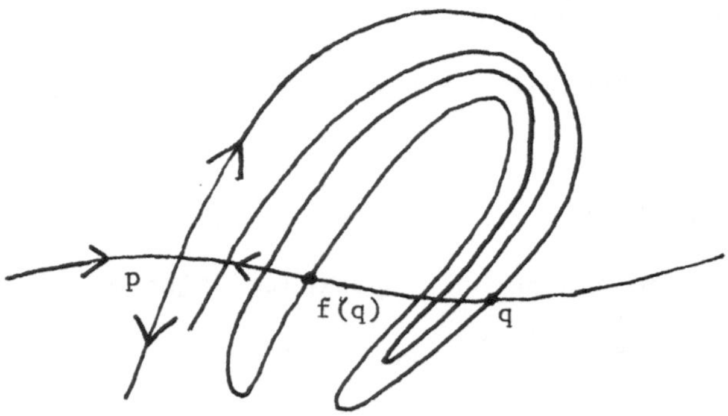

Figure (2.1)

We see that $W^u(p)$ must accumulate on itself oscillating very wildly. This seems to give a situation so complicated as to defy analysis.

Let us now, in addition, assume f is area preserving. Consider pieces I_1, I_2 of $W^u(p)$ and J_1, J_2 of $W^s(p)$ as indicated in figure (2.2).

Let A be the region bounded by $I_1 \cup I_2 \cup J_1 \cup J_2$. Since almost all points are recurrent, there are infinitely many j's with $j > 0$ such that $f^j A \cap A \neq \phi$.

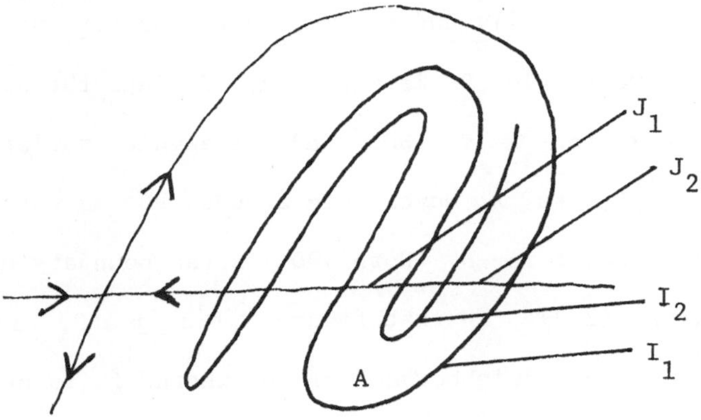

Figure (2.2)

Thus, between any two homoclinic points on a stable manifold there are infinitely many others. On discovering this in the restricted three body problem, Poincaré wrote

> "Nous n'avons pas encore le droit de conclure que les solutions doublement asymptotique sont *überalldicht* sur la surface asymptotique [stable manifold]; mais cela semble probable." (p. 387, 1, vol. III).

In lecture 4 we will show, following Takens, that the above property holds C^1 generically for area preserving diffeomorphisms of surfaces.

The next theorem is due to Birkhoff.

Theorem (2.1). *(Birkhoff* [1]). *Suppose* f *is a diffeomorphism of a surface and* q *is a transverse homoclinic point of* f. *Then in every neighborhood of* q *there are infinitely many periodic orbtis of* f.

We can geometrically understand the reason for this very simply. If a rectangle D is mapped by S into the plane so that $S(x) \neq x$ for x in ∂D, and the angular variation of $v(x) = \frac{x - S(x)}{|x - S(x)|}$ as x moves once around ∂D is not zero, then S has a fixed point in D. Here ∂D is the boundary of D. Consider figure (2.3). In that figure, $f^j q$, $j \geq 0$, and $f^j q$ $j \leq -T$, remain in a neighborhood of p where f is nearly linear. For large n, there is a rectangle D_n so that $f^n D_n$ and $f^{n+T} D_n$ are as indicated. Let $S = f^{n+T}$. It is easily seen that the angular variation of $v(x)$ for $x \in \partial D_n$ is ± 1. So f^{n+T} has a fixed point, say z_n, in D_n.

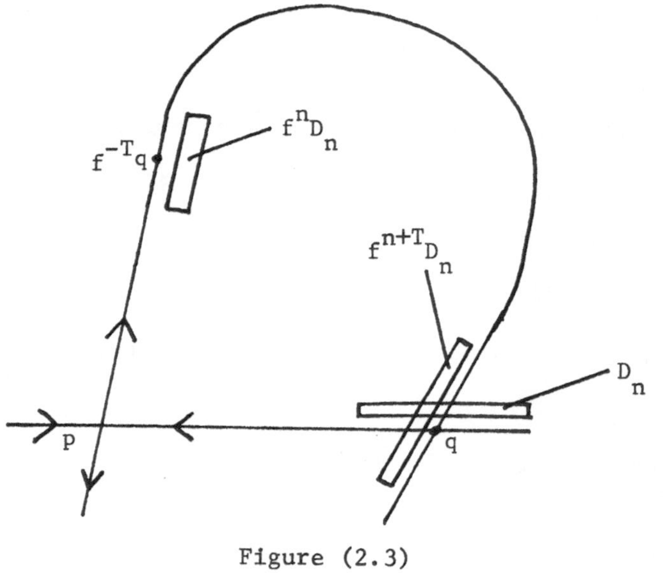

Figure (2.3)

Clearly, $z_n \neq z_m$ for $n \neq m$ since $D_n \cap D_m = \phi$. Thus, there are infinitely many homoclinic points and infinitely many periodic points accumulating on p. Is there some way of putting all these orbits together and getting some understandable structure?

Let p be a hyperbolic periodic point having a transverse homoclinic point. The h-closure of p is the closure of the set of transverse homoclinic points of f. Also, given a homeomorphism f from a compact metric space Λ to itself, we say that f is topologically transitive if f has a dense orbit.

Proposition (2.2) *Each* h-*closure is a closed invariant set on which* f *is topologically transitive.*

Lemma. Suppose $f : \Lambda \longrightarrow \Lambda$ *is a homeomorphism of the compact metric space* Λ *such that for every open set* $U \subset \Lambda$, $\bigcup_{n \geq 0} f^{-n}U$ *is dense in* Λ. *Then* f *has a dense orbit.*

Proof. Let $\{U_i\}$ be a countable basis for the topology of Λ, i.e. for every $x \in \Lambda$ and open set U containing x there is a U_i with $x \in U_i \subset U$. Then, $\bigcup_{n \geq 0} f^{-n}U_i$ is dense and open in Λ. Thus, $\bigcap_i (\bigcup_{n \geq 0} f^{-n}U_i)$ is non-empty (in fact dense). Any point in this latter set has a dense orbit.

Proof of Proposition (2.2) We consider the case when p is fixed. Take transverse homoclinic points q_1, q_2. We need to produce transverse homoclinic points near q_1 whose orbits get near q_2. Consider a small interval I in $W^u(p)$ near q_1 as in figure (2.4). Pieces of the orbit of I accumulate on $W^u(p)$. Hence there is a $j > 0$ such that f^jI has a point q of transverse intersection with $W^s(p)$ near q_2. Then, $f^{-j}q$ is the transverse homoclinic point near q_1 whose orbit gets near q_2.

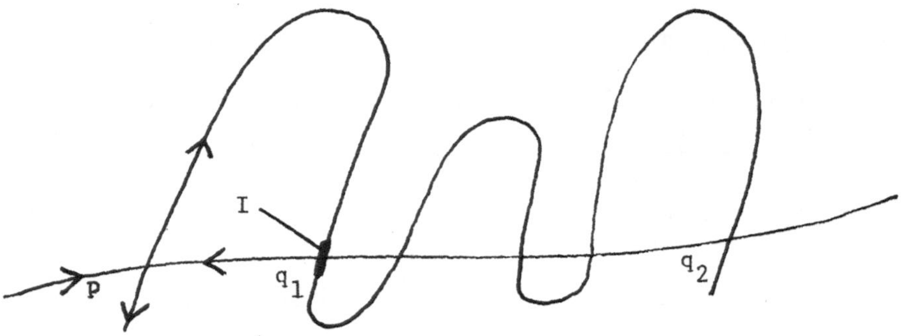

Figure (2.4)

Proposition (2.2) tells us that not only are there infinitely many periodic points, but there is a single point x whose orbit gets arbitrarily close to each of these periodic points. One could think of an h-closure of a hyperbolic periodic point p as a "sphere of influence" of p.

We wish to return to the question of density of the homoclinic points in a stable manifold. Let us first recall the notion of (uniform) hyperbolicity.

Let $f : M \longrightarrow M$ be a diffeomorphism of a compact manifold M, and let Λ be a compact invariant set. One says that Λ is (uniformly) hyperbolic if there are a splitting $T_xM = E_x^s \oplus E_x^u$ for $x \in \Lambda$ and constants $C > 0$, $0 < \lambda < 1$, such that

(a) $T_xf(E_x^s) = E_{fx}^s, T_xf(E_x^u) = E_{fx}^u, x \in \Lambda$

(b) $|T_xf^n(v)| \leq C\lambda^n|v|, n \geq 0, v \in E_x^s$

(c) $|T_xf^{-n}(v)| \leq C\lambda^n|v|, n \geq 0, v \in E_x^u$

The reason for the word "uniformly" is that the number C is independent of x in Λ. Other notions of hyperbolicity (related to our later discussions) allow $C = C(x)$ to be a Borel measurable function of x, and Λ to be a measurable non-compact subset of the Lyapunov regular points with non-zero exponents. For a discussion of hyperbolic sets we refer to Oscar Lanford's lectures, [10], or [11].

A diffeomorphism is called Anosov if all of M is a hyperbolic set. If, in addition, f has a dense orbit, then it is known that f has a single h-closure which is all of M. Thus, the homoclinic points in the stable manifolds of a periodic orbit are dense in those manifolds. A simple example is given by the map $\bar{A}$ on the two dimensional torus T^2 which is induced by $A = \begin{pmatrix} 2 & 1 \\ 1 & 1 \end{pmatrix}$ mapping $\mathbb{R}^2$ to $\mathbb{R}^2$. Now, if $f : M \longrightarrow M$ is Anosov, and dim $M = 2$, then E^s_x furnishes a non-vanishing line bundle on M which is continuous (the bundles E^s_x and E^u_x on a hyperbolic set vary continuously with x). This implies by elementary topology that M is homeomorphic to T^2. It is also known that f is topologically conjugate to a linear toral automorphism. Thus, it is rare for M to be a hyperbolic set.

Proposition (2.3) *Suppose* $f : M^2 \longrightarrow M^2$ *preserves an area form* ω *and* M^2 *is connected. Suppose an* h-*closure* H *is uniformly hyperbolic. Then,* $H = M^2$ *and* f *is Anosov.*

Thus, h-closures are almost never (uniformly) hyperbolic

when f preserves an area element.

Proof. Assume H is a uniformly hyperbolic h-closure. Then H is a closed set. We will also show H is open which implies that $H = M$ since M is connected. For small $\varepsilon > 0$, and $x \in H$ let $W^u_\varepsilon(x)$ be the interval in $W^u(x)$ of length 2ε centered at x, and define $W^s_\varepsilon(x)$ similarly.

We make two assertions.

(1) If $\varepsilon > 0$ is small and $x,y \in H$ are close, then $W^u_\varepsilon(x) \cap W^s_\varepsilon(y) \subset H$.

(2) If $x \in H$, then $H \cap W^u(x)$ is dense in $W^u(x)$ and $H \cap W^s(x)$ is dense in $W^s(x)$.

Proof of (1): It is standard stable manifold theory that $W^u_\varepsilon(x)$ is C^1 near $W^u_\varepsilon(x_1)$ and $W^s_\varepsilon(x)$ is C^1 near $W^s_\varepsilon(x_1)$ for x near x_1 in H. If x and y are close, then $W^u_\varepsilon(x)$ and $W^s_\varepsilon(y)$ have a unique point z of transverse intersection. Let $q_1, q_2, \ldots$ and $q'_1, q'_2, \ldots$ be sequences of transverse homoclinic points of the same periodic orbit $O(p)$ in H such that $q_i \longrightarrow x$ and $q'_i \longrightarrow y$. Then, for large i, $W^u_\varepsilon(q_i) \cap W^s_\varepsilon(q'_i)$ is a point z_i of transverse intersection of $W^u_\varepsilon(q_i)$ and $W^s_\varepsilon(q'_i)$ and $z_i \longrightarrow z$. Now $W^u_\varepsilon(q_i) \subset W^u(O(p))$ and $W^s_\varepsilon(q'_i) \subset W^s(O(p))$, so z_i is a transverse homoclinic point. Thus $z \in H$, proving (1).

Proof of (2). Let $x \in H$ and consider $W^u(x)$. If $H \cap W^u(x)$

is not dense in $W^u(x)$, let $I \subset W^u(x)$ be a maximal interval in $W^u(x) - H$.

<u>Case 1</u>. I is a finite interval, say bounded by y_1, y_2. So $y_1, y_2 \in H$. Let $z \in H$ be a periodic point near x such that $W^u(O(z)) \cap W^u(O(x)) = \phi$. Consider an open rectangle R bounded by I and pieces of $W^s(y_1) \cup W^s(y_2) \cup W^u(z)$ as in figure (2.5). Let $J_1 = \partial R \cap W^s(y_1)$, $J_2 = \partial R \cap W^s(y_2)$ $I_1 = W^u(z) \cap \partial R$.

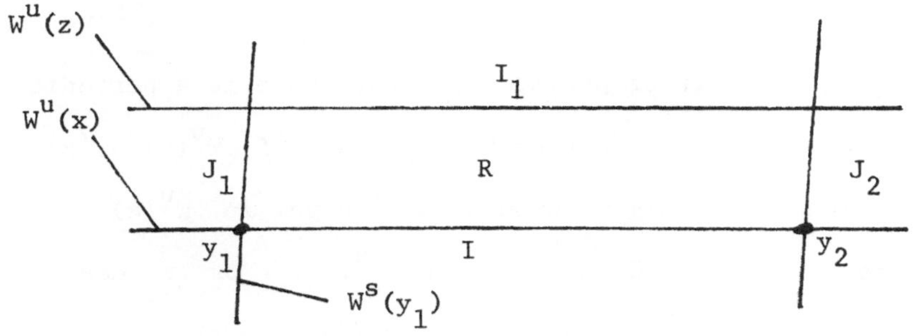

Figure (2.5)

Assume z is close enough to x so that $R \cap H = \phi$. Now, R contains many recurrent points, so $f^n(R) \cap R \neq \phi$ for infinitely many $n > 0$. Choose such an n so that $f^n(J_i)$ is shorter than J_1 and J_2 for $i = 1,2$. Then, $f^n R \neq R$ and, since the areas of $f^n R$ and R are equal, we conclude that $f^n(I \cup I_1) \cap (J_1 \cup J_2) \neq \phi$. But points in the last set are in H, so $(I \cup I_1) \cap H \neq \phi$ which is a contradiction.

<u>Case 2</u>. I is a half infinite interval, say of the form

$(y_1, \infty) \subset W^u(x)$ with $y_1 \in H$. Consider figure (2.6) with I to the right of y_1.

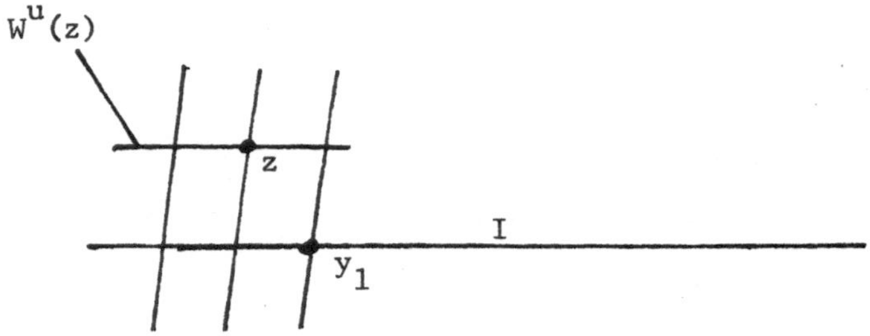

Figure (2.6)

Since y_1 is a limit of periodic orbits, there is a periodic point z near y such that both components of $W^u(z) - \{z\}$ meet H. Then, there is no unbounded interval in $W^u(z) - H$, so by case 1, $W^u(z) \cap H$ is dense in $W^u(z)$. For z near y_1 this gives $I \cap H \neq \phi$, a contradiction, and (2) is proved.

Now let $x \in H$. Then since $W^u(x) \cap H$ and $W^s(x) \cap H$ are dense in $W^u(x)$ and $W^s(x)$, respectively, we have that x is in the interior of H. Thus, H is open.

Even though h-closures are rarely uniformly hyperbolic, they contain many uniformly hyperbolic sets as the next result shows.

Theorem (2.4) (*Smale*) *Near the orbit of a transverse homoclinic point, there are closed invariant zero dimensional uniformly hyperbolic sets* Λ *such that*

(1) *the periodic orbits of* $f|\Lambda$ *are dense in* Λ

(2) $f|\Lambda$ *has a dense orbit*

(3) $f|\Lambda$ *is persistent and locally structurally stable (i.e.* g C^1 *near* f *implies there are a set* $\Lambda(g)$ *near* Λ *and a homeomorphism* $\phi : \Lambda \longrightarrow \Lambda(g)$ *such that* $g \circ \phi = \phi \circ f$*)*

(4) $f|\Lambda$ *is topologically conjugate to a subshift of finite type (see* [11] *for definitions).*

Remark: If H is not uniformly hyperbolic, then one would expect some kind of instability (depending on f). So for this can only be exploited in the C^1 topology.

We proceed to describe our analytic method for producing hyperbolic periodic points and homoclinic points. For $n \in Z^+$, let $J_n = \{1,2,\ldots,n\}$. For $x \in M$ and $y \in M$, we say that $0_+(x,f)$ approaches y with positive frequency if there is a subset F_n of integers in J_n for $n \geq 1$ and a real α with $0 < \alpha \leq 1$ such that

(1) $\limsup\limits_{n\to\infty} \dfrac{\text{cardinality } F_n}{n} = \alpha$

(2) $\text{dist}(\{f^j(x) : j \in F_n\}, y) \longrightarrow 0$ as $n \longrightarrow \infty$.

Now define the upper Lyapunov number $\chi^+(x,f)$ to be $\limsup\limits_{n\to\infty} \frac{1}{n} \log |T_x f^n|$. For $x \in M$, let $\omega(x)$ be the ω-limit set of f. If $0_+(x,f)$ is bounded, then $\omega(x)$ is a closed invariant set.

Theorem (2.5) Suppose $f : M^2 \longrightarrow M^2$ *is weakly dissipative and* $x \in M$ *is such that* $0_+(x,f)$ *is bounded and* $\chi^+(x,f) > 0$.

Then, $\omega(x)$ *contains hyperbolic periodic points. Either* $O_+(x,f)$ *approaches some hyperbolic periodic point with positive frequency or* $\omega(x)$ *contains transverse homoclinic points.*

Remark. The condition $\chi^+(x,f) > 0$ is experimentally verifiable. If $O_+(x,f)$ approaches a hyperbolic periodic point of fairly low period with a frequency α which is not too small then this should also be experimentally verifiable. Thus, if one knew that $O_+(x,f)$ approached no single periodic orbit with positive frequency, then theorem 1 gives the existence of infinitely many periodic orbits.

Before proving theorem (2.5), we discribe some results in smooth ergodic theory.

Let $f : M \longrightarrow M$ be a C^1 diffeomorphism of the compact manifold M (of arbitrary finite dimension). A point $x \in M$ is called a regular point if there are a splitting $T_xM = E_1(x) \oplus \ldots \oplus E_{r(x)}(x)$ and numbers $\lambda_1(x) < \lambda_2(x) < \ldots < \lambda_{r(x)}(x)$ such that

(1) $\lim\limits_{n \to \pm\infty} \frac{1}{n} \log|T_xf^n(v)| = \lambda_i(x)$ for $v \in E_i(x) \setminus \{0\}$.

(2) $\lim\limits_{n \to \pm\infty} \frac{1}{n} \log|\det(T_xf^n)| = \sum\limits_{i=1}^{r(x)} \lambda_i(x) \dim E_i(x)$.

Condition (1) says that for large $|n|$ and $v \in E_i(x) \setminus \{0\}$, $|T_xf^n(v)| \approx e^{n\lambda_i(x)}$ and condition (2) says $|\det(T_xf^n)| \approx e^{n(\sum_{i=1}^{r(x)} \lambda_i(x) \dim E_i(x))}$. It follows from condition (1) that $\lim\limits_{n \to +\infty} \frac{1}{n} \log|T_xf^n(v)|$ equals one of the $\lambda_i(x)$'s and

$\lim_{n\to+\infty} \frac{1}{n} \log|T_x f^{-n}(v)|$ equals the negative of one of the $\lambda_i(x)$'s for each $v \neq 0$ in $T_x M$. Condition (2) guarantees that the angles between $E_i(x)$ and $E_j(x)$ decrease slower than any exponential along orbits. There is good control on the iterates $T_x f^n$ for regular points x. Let $R(f)$ denote the set of regular points of f.

Theorem (2.6) *(Oseledec). Let* f *be a* C^1 *diffeomorphism of a smooth manifold* M *to itself. Then* $R(f)$ *is a Borel set and has full measure for any* f*-invariant probability measure* μ *supported on a compact set in* M. *The functions* $x \longrightarrow E_i(x)$, $x \longrightarrow r(x)$, $x \longrightarrow \lambda_i(x)$ *are Borel measurable functions defined on* $R(f)$. *Moreover,* $r(x)$, $\lambda_i(x)$ *and dim* $E_i(x)$ *are constant along* f*-orbits.*

The numbers $\lambda_i(x)$, $x \in R(f)$, are called the characteristic exponents of x. For a proof of theorem (2.6) we refer to [9] or [4]. One should think of theorem (2.6) as giving the existence of regular points in any compact invariant set F for f. If f has many invariant measures supported on F, then f has many regular points in F. Regular points with non-zero characteristic exponents behave much like points in a hyperbolic set. The following result is stated with proof sketched in [8] (for compact manifolds). The ideas in [8] enable one to improve slightly on differentiability results of Pesin [5].

Theorem (2.7) *Suppose* x *is a regular point for a* C^r *diffeomorphism* $f : M \longrightarrow M$, $r \geq 2$, *with characteristic exponents* $\lambda_1(x) < \lambda_2(x) < \ldots < \lambda_s(x)$ *and assume the orbit of* x *is bounded. Suppose* $\lambda_i(x) < 0$ *for some* $i \in [1,s]$ *and* $\lambda_i(x) < \lambda <$ $\min(\lambda_{i+1}(x),0)$. *Let* $V_\lambda(x) = \{y \in M$ *there is a constant* $K > 0$ *such that* $d(f^n y, f^n x) \leq Ke^{\lambda n}$ *for* $n \geq 0\}$. *Then,* $V_\lambda(x)$ *is a* C^r *manifold tangent at* x *to* $E_1(x)\oplus\ldots\oplus E_i(x)$.

The next result due to Katok gives conditions for the existence of homoclinic points.

Theorem (2.8) (*Katok* [3]) *Suppose* $f : M \longrightarrow M$ *is a* $C^{1+\alpha}$ *diffeomorphism of a manifold,* $\alpha > 0$, *and* μ *is an* f-*invariant probability measure supported on a compact subset* F *of* M. *Suppose there is a subset* $\Gamma \subset F$ *of regular points such that* $\mu(\Gamma) = 1$ *and each* x *in* Γ *has only non-zero characteristic exponents. Then, either* Γ *consists of hyperbolic periodic points or* f *has transverse homoclinic points which accumulate on* Γ.

We will sketch the proof of theorem (2.8) later. Let us now prove Theorem (2.5).

Let $0_+(x,f)$ be bounded and suppose $\chi^+(x,f) > 0$. Then, closure $0_+(x,f)$ is compact. For each $n \in Z^+$ let $\mu_n = \frac{1}{n}\sum_{k=0}^{n-1} \delta_{f^k x}$ where $\delta_{f^k x}$ is the point mass at $f^k x$. Then, there is a subsequence $\{\mu_{n_k}\}$ which converges to a probability measure μ in the weak-* topology (i.e. pointwise on continuous functions $\phi : M \longrightarrow \mathbb{R}$).

It is easy to see that μ is invariant and its support is in closure $0_+(x,f)$. According to the Oseledec theorem, there is a set Γ with $\mu(\Gamma) = 1$ such that for $y \in \Gamma$, $v \in T_yM$, one has $\chi(x,v) = \lim_{n\to\pm\infty} \frac{1}{n} \log|T_y f^n(v)|$. Also, since $\dim M = 2$, for varying v, $\chi(y,v)$ assumes at most two values, say $\chi_1(y,f) \le \chi_2(y,f)$ and $\lim_{n\to\pm\infty} \frac{1}{n} \log|\det T_y f| = \chi_1(y,f) + \chi_2(y,f)$.

<u>Claim</u>: There is a subset $\Gamma' \subset \Gamma$ with $\mu(\Gamma') > 0$ such that $y \in \Gamma'$ implies $\chi_2(y,f) > \frac{1}{2}\chi^+(x,f) > 0$.

Assume the claim for the moment. Since f is weakly dissipative, we have $\chi_1(y,f) + \chi_2(y,f) \le 0$, so $\chi_1(y,f) < -\frac{\chi^+(x,f)}{2}$. Thus, the characteristic exponents of y in Γ' are bounded away from 0. Let $\tilde{\Gamma} = \bigcup_{n\in Z} f^n(\Gamma')$, so that $\tilde{\Gamma}$ is an invariant set with $\mu(\tilde{\Gamma}) > 0$ and $\chi_1(y,f) < -\frac{\chi^+(x,f)}{2}$, $\chi_2(y,f) > \frac{\chi^+(x,f)}{2}$ for $y \in \tilde{\Gamma}$.

Consider the set $A \subset \tilde{\Gamma}$ of periodic orbits in $\tilde{\Gamma}$. Each point in A is hyperbolic, so A is at most countable (any set of hyperbolic periodic orbits of bounded period is finite). If $\mu(A) = \mu(\tilde{\Gamma})$, then some element $p \in A$ has positive μ-measure. Thus, $0_+(x,f)$ approaches p with positive frequency. If $\mu(A) < \mu(\tilde{\Gamma})$, then Katok's theorem applied to μ gives transverse homoclinic points accumulating on the support of μ.

<u>Proof of Claim</u>: Let $\chi^+ = \chi^+(x,f)$. Let n_k be such that $\lim_{k\to\infty} \frac{1}{n_k} \log|T_x f^{n_k}| = \chi^+$. If the claim fails, then $\chi_2(y,f) \le \frac{1}{2}\chi^+$ for all $y \in \Gamma$. Then for $y \in \Gamma$, $\lim_{n\to\infty} \frac{1}{n} \log|T_y f| < \frac{3}{4}\chi^+$. There

is a measurable real-valued function $C(y)$, such that $|T_y f^n| \le C(y)e^{n\frac{3}{4}\chi^+}$ for $n \ge 0$ and $y \in \Gamma$. Let $L > 0$ such that $|T_z f| \le e^L$ for all z. Choose $\delta > 0$ small enough that $3\delta L + \frac{5}{6}\chi^+ < \frac{6}{7}\chi^+$. Choose a compact set $K_\delta \subset \Gamma$ with $\mu(\Gamma - K_\delta) < \delta$ and a constant $C_\delta > 0$ such that $C(y) \le C_\delta$ for $y \in K_\delta$. Thus, $|T_y f^n| \le C_\delta e^{n\frac{3}{4}\chi^+}$ for $n \ge 0$ and $y \in K_\delta$. Choose N large enough so that $C_\delta e^{n\frac{3}{4}\chi^+} < e^{\frac{4}{5}\chi^+ n}$ for $n \ge N$. Then choose a neighborhood U of K_δ so that $y \in U$ implies $|T_y f^N| < e^{\frac{5}{6}\chi^+ N}$. Now consider $\{f^j x : 0 \le j \le n\}$. Let j_1 be the least positive integer such that $f^{j_1}(x) \in U$. Let j_i be the least integer greater than $j_{i-1} + N$ such that $f^{j_i}(x) \in U$.

We have the picture

$$x \ldots, f^{j_1}x, \ldots f^{j_1+N}x, \ldots, f^{j_2}(x), \ldots, f^{j_2+N}(x), \ldots, f^{j_3}(x), \ldots f^{j_3+N}(x),$$

$$\ldots, f^{j_\beta}(x), \ldots\ldots, f^{n_k}(x).$$

where j_β is the largest of the j_i's less than n_k. Let $j_0 = 0$, $m_1 = j_1$, $m_i = j_i - (j_{i-1} + N)$ for $2 \le i \le \beta$, $m_{\beta+1} = n_k - j_\beta$, and let $y_i = f^{j_i}x$.

Then,

$$|T_x f^{n_k}| = |T_{y_\beta} f^{n_k - j_\beta} \circ T_{f^N y_{\beta-1}} f^{m_\beta} \circ T_{y_{\beta-1}} f^N \circ T_{f^N y_{\beta-2}} f^{m_{\beta-1}}$$

$$\circ T_{y_{\beta-2}} f^N \circ \ldots \circ T_{y_1} f^N \circ T_x f^{j_1}|$$

$$\leq e^{\left(\sum_{\ell=1}^{\beta+1} m_\ell\right)L + (\beta - 1)N\frac{5}{6}\chi^+}$$

Now, since $\mu_{n_k} \longrightarrow \mu$ we have $\mu_{n_k}(M - U) < 2\delta$ for large k; hence, $\sum_{\ell=1}^{\beta+1} m_\ell < 3\delta n_k$ for large k, and $n_k = \sum_{\ell=1}^{\beta+1} m_\ell + (\beta-1)N$

So,

$$\log |T_x f^{n_k}| \leq 3\delta L n_k + (n_k)\frac{5}{6}\chi^+ \leq n_k[3\delta L + \frac{5}{6}\chi^+] \leq \frac{6}{7}\chi^+ n_k$$

which contradicts $\lim_{k\to\infty} \frac{1}{n_k} \log |T_x f^{n_k}| = \chi^+$.

<u>Proof of theorem</u> (2.8): Consider the set Γ of regular points in F. For $x \in \Gamma$, let $\lambda_1(x) < \lambda_2(x) < \ldots < \lambda_i(x) < 0 < \lambda_{i+1}(x) < \ldots < \lambda_s(x)$ be the characteristic exponents of x with associated splitting $T_xM = E_1(x)\oplus\ldots\oplus E_s(x)$. Note that i also depends on x. Let $\chi(x) = \frac{1}{10}\min(-\lambda_i(x), \lambda_{i+1}(x))$. Let $E^-(x) = E_1(x)\oplus\ldots\oplus E_i(x)$, and let $E^+(x) = E_{i+1}(x)\oplus\ldots\oplus E_s(x)$. Then $T_xM = E^-(x) \oplus E^+(x)$, and there is a measurable function $C(x) > 0$ such that for $n \geq 0$,

$$|T_x f^n(v)| \leq C(x)e^{-n\chi(x)}|v| \quad \text{for } v \in E^-(x)$$

and $|T_x f^n(v)| \geq C^{-1}(x)e^{n\chi(x)}|v|$ for $v \in E^+(x)$.

This means that the iterates of the derivative of f at x behave hyperbolically with a distortion depending measurably on x.

Let P be the set of points y in Γ such that $\mu(\{y\}) > 0$; i.e., P is the set of atoms in Γ. Then, P is invariant and consists of hyperbolic periodic orbits. Since any set of such orbits with bounded period is finite, P is at most countable. If $\mu(P) = \mu(\Gamma) = 1$, the conclusion of theorem (2.8) holds, so suppose $\mu(\Gamma \setminus P) > 0$. Let $\alpha(x) = \frac{1}{10}\chi(x)$. According to Pesin [5] or Fathi-Herman Yoccoz [2], there is a norm $|\ |'_x$ on T_xM for $x \in \Gamma$, a measurable function $A : \Gamma \longrightarrow \mathbb{R}$, and a constant $\tilde{K} > 0$ such that

(1) $\tilde{K} \le \frac{|v|'_x}{|v|_x} \le A(x)$ for $x \in \Gamma$, $v \in T_xM\backslash\{0\}$

(2) $A(f^n x) \le A(x)e^{\alpha(x)|n|}$ for $n \in Z$

(3) $|T_xf(v)|'_{f(x)} \le e^{-\chi(x)}|v|'_x$ for $v \in E^-(x)$

(4) $|T_xf(v)|'_{f(x)} \ge e^{\chi(x)}|v|'_x$ for $v \in E^+(x)$

The norm $|\ |'_x$ decreases slowly along orbits, and f behaves uniformly hyperbolically in this norm. Denote by $B_\delta(x)$ the open δ-ball about x. There are neighborhoods $B_{\varepsilon(x)}(x)$ where $\varepsilon(x)$ is measurable and depends on $A(x)$ and f in which one has the situation in figure (2.7)

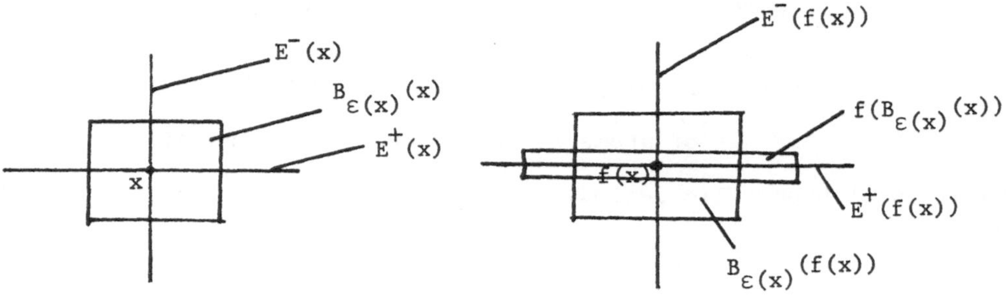

Figure (2.7)

since f contracts along the $E^-(x)$ direction and expands along the $E^+(x)$ direction in the $|\ |'_x$ and $|\ |'_{f(x)}$ norms. To get the $B_{\varepsilon(x)}(x)$ to look like rectangles, we use $|v|' = \max(|v_1|', |v_2|')$ where $v = (v_1,v_2) \in E_1(x) \oplus E_2(x)$. Let us choose a compact set $K \subset \Gamma \setminus P$ with $\mu(K) > 0$ in which the subspaces $E^-(x)$, $E^+(x)$, the functions $\chi(x)$, $\alpha(x)$, $A(x)$, $\varepsilon(x)$, and the norms $|\ |'_x$ depend continuously on x. Suppose $y = f^n x$ with n large, y is near x, and both x and y are in K. We have figure (2.8)

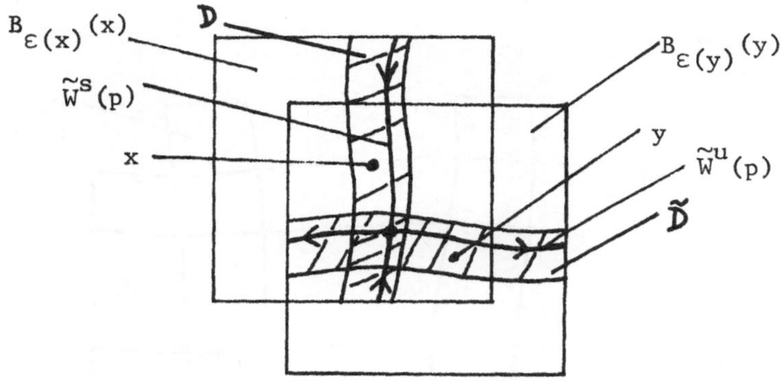

Figure (2.8)

There are disks $D \subset B_{\varepsilon(x)}(x)$ and $\tilde{D} \subset B_{\varepsilon(y)}(y)$ such that $f^n D = \tilde{D}$, f^n expands horizontal directions in D and f^n contracts vertical directions in D, and $D \cap \tilde{D} \neq \phi$.

Let $g = f^n$, and $H = D \cap \tilde{D}$. We may again suppose H is a disk (with piecewise smooth boundary). There will be a hyperbolic fixed point p for g in H, and $\bigcap_{0 \le k \le \infty} g^k H$ will be a piece of its unstable manifold in D. Call this $W^u_{loc}(p)$. Similarly define $W^s_{loc}(p) = \bigcap_{-\infty \le k \le 0} g^k H$ to get a piece of the stable manifold of p in H. Note that $\tilde{W}^u(p) \equiv g(W^u_{loc}(p))$ will stretch completely across $\tilde{D}$ horizontally and $\tilde{W}^s(p) \equiv g^{-1}(W^s_{loc}(p))$ will stretch completely across D vertically as in figure (2.8).

Now μ-almost all points in K recur on themselves in K, so we may choose two points x,y in K and two positive integers n_1,n_2 so that if $z = f^{n_1}x$ and $w = f^{n_2}y$, then z,w are also in K, and x,y,z,w are all close to each other. This will enable us to get disks D_1,D_2 so that $\tilde{D}_1 = f^{n_1}D_1$ and $\tilde{D}_2 = f^{n_2}D_2$ are as in figure (2.9)

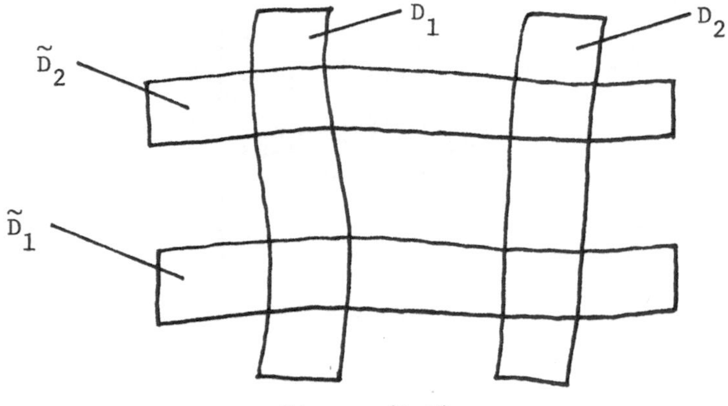

Figure (2.9)

Let $n = n_1 \cdot n_2$. Shrinking D_1 and D_2 horizontally we will be able to keep essentially the same figure for D_1, D_2 and $f^n D_1 = \tilde{D}_1$, $f^n D_2 = \tilde{D}_2$. Then f^n will have hyperbolic fixed points $p \in D_1$ and $q \in D_2$ which are h-related, and p and q will have transverse homoclinic points. The points p and q will be close to the recurrent points x and y, so homoclinic points will accumulate on $K \subset \Gamma$.

References

1. G. D. Birkhoff, Collected Mathematical Papers, Vol. II, p. 350, American Mathematical Society, 1950.

2. A. Fathi, M. Herman, and J. Yoccoz, A proof of Pesin's stable manifold theorem, preprint, Math. Dept., Univ. de Paris, Orsay, France.

3. A. Katok, Lyapunov exponents, entropy, and periodic orbits for diffeomorphisms, Publ. Math. IHES 51 (1980), 137-173.

4. R. Mañe, Lyapunov exponents and stable manifolds for compact transformations, preprint, IMPA, Rio de Janeiro, 1981.

5. Ya Pesin, Families of invariant manifolds corresponding to non-zero characteristic exponents, Math of the USSR-Izvestia 10 (1976), 61 1261-1305.

6. ________, Characteristic Lyapunov exponents and smooth ergodic theory, Russian Math Surveys 32 (1977), 4, 55-114.

7. H. Poincaré, Méthodes Nouvelles de la Mécanique Celeste, Vol. III, Gauthiers-Villars, Pairs, 1899.

8. C. Pugh and M. Shub, Differentiability and Continuity of Invariant Manifolds, Non-linear Dynamics, Ann. of NY Acad of Science (357, (1980), 322-330.

9. D. Ruelle, Ergodic Theory of Differentiable Dynamical Systems Publ. Math. IHES 50 (1979), 27-58.

10. S. Newhouse, Lectures on Dynamical Systems, Progress in Math. 8, Birkhauser, Boston, 1980.

11. M. Shub, Stabilité globale des systemes dynamiques, Asterisque 56.

Lecture 3.

The goal of the theory of dynamical systems is to obtain as much information as possible about the orbit structure of differential equations. Of particular interest are those properties which persist either for slight variation of the initial conditions or for slight variation of the equations themselves. Even for the equations of conservative dynamics, there can be much pathological behavior. For instance any closed set can arise as the set of critical points of some Hamiltonian vector field. So, as a first step toward the above goal one can try to eliminate exceptional cases and describe the "typical" or "general" type of equation. We will study this problem in the context of area preserving diffeomorphisms of a compact orientable two-dimensional manifold M. Let ω be a nowhere zero two-form on M. Let $\mathcal{D}^r_\omega(M)$ be the set of C^r diffeomorphisms of M preserving ω, $r \geq 1$. We allow M to have a boundary ∂M and in this case we consider only those diffeomorphisms mapping ∂M to itself. Of prime importance are the cases when M is a 2-disk, an annulus, or the two-dimensional sphere. The most common way of making the notion of "typical" mapping precise is to introduce a topology in $\mathcal{D}^r_\omega(M)$ and to say that "typical" corresponds to elements in a second category set. Define the C^r topology in $\mathcal{D}^r_\omega(M)$ in the following way.

Let $(U_1,\phi_1),\ldots,(U_m,\phi_m)$ be a collection of (open) coordinate charts in M with $M = \bigcup_{i=1}^{m} U_i$. Fix $f \in \mathcal{D}^r_\omega(M)$. Let $D_1,\ldots,D_s$ be a collection of compact sets such that $M = \bigcup_{i=1}^{s} D_i$ and for each i there are numbers $\alpha(i)$ and $\beta(i)$ such that $D_i \subset U_{\alpha(i)}$ and $f(D_i) \subset U_{\beta(i)}$. We say that a sequence $\{f_j\}$ converges to f in the C^r sense if

(1) for large j, $f_j(D_i) \subset U_{\beta(i)}$ for each i

(2) the mappings $\phi_{\beta(i)} f_j \phi_{\alpha(i)}^{-1}$ and their first r derivatives converge to those of $\phi_{\beta(i)} f \phi_{\alpha(i)}^{-1}$ uniformly on $\phi_{\alpha(i)}(D_i)$ for each i.

This defines the C^r topology in $\mathcal{D}_\omega^r(M)$. A subset $\mathcal{B} \subset \mathcal{D}_\omega^r(M)$ is called residual if it contains a countable intersection of dense open sets, and a countable intersection of residual sets in again residual. A property is "generic" or "typical" if it holds for elements of a residual set.

Recall that a periodic point p with period $\tau > 0$ is called <u>elliptic</u> if the eigenvalues of $T_p f^\tau$ are non-real and of absolute value one.

Theorem (3.1) *Let* $r \geq 1$. *There is a residual set* $\mathcal{B} \subset \mathcal{D}_\omega^r(M)$ *such that if* $f \in \mathcal{B}$, *then the following conditions hold.*

(1) *Each periodic orbit is either elliptic or hyperbolic.*

(2) *Each elliptic periodic orbit is a limit of both elliptic and hyperbolic periodic orbits and transverse homoclinic points.*

(3) *The stable and unstable manifolds of hyperbolic periodic orbits meet transversely (if they meet at all).*

(4) *If* $r \geq 3$, *then each elliptic periodic orbit of period* τ *is a limit of* f^τ*-invariant circles on which* f^τ *is* C^1 *conjugate to a geometric rotation through an angle* α *with* $\alpha/2\pi$ *irrational.*

Let us see the geometry of this theorem near an elliptic fixed point for a typical C^3 f.

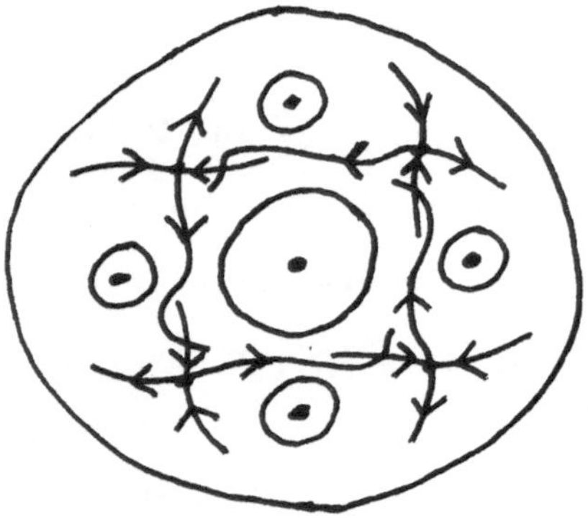

Figure (3.1)

<u>Remarks</u>: 1. Theorem (3.1.1) and (3.1.3) are due to R. C. Robinson [4].

2. Theorem (3.1.4) is the celebrated Kolmogorov-Arnold-Moser (KAM) theorem. The optimum differentiability $r = 3$ has recently been announced by M. Herman.

3. Theorem (3.1.2) is due to J. Moser [1] and E. Zehnder [6]. They actually prove that (3.1.2) is generic for real analytic mappings with a suitable topology. Since the zero set of a non-constant real analytic function $H : \mathbb{R} \longrightarrow \mathbb{R}$ meets each line in a set which is at most countable, an analytic f satisfying (3.1.2) can have no analytic integral (non-constant invariant function).

If the hyperbolic periodic orbits are dense, there are no C^2 integrals. It is known that there are no differentiable integrals of the form $H(x,y) = x^2 + y^2 + o(x^2 + y^2)$. Actually, the question of integrals here is mainly of historical interest. Given what is known about the orbit structure, an integral would be so complicated as to be of little value.

Zehnder and Moser proved theorem (3.1.2) by detailed estimates in local coordinates near an elliptic fixed point. The following result due to Pixton [3] is more topological.

Theorem (3.2) *Suppose* M *can be embedded in the two-dimensional sphere. Then, for* $r \geq 1$, *there is a residual set* $\mathcal{B} \subset \mathcal{D}^r_\omega(M)$ *such that for* $f \in \mathcal{B}$, *every hyperbolic periodic point has transverse homoclinic points.*

The embedding restriction on M arises because the proof requires that M satisfy the Jordan curve theorem: Any closed simple curve γ in $M - \partial M$ is such that $M - (\gamma \cup \partial M)$ consists of two open connected sets with γ as their common boundary in $M - \partial M$. Thus, M may be the whole sphere, or a subset of the sphere bounded by finitely many closed simple curves as in figure (3.1a).

Figure (3.1a)

We proceed with the proof of Pixton's theorem.

For a hyperbolic periodic point p of a diffeomorphism f, let $W^s(O(p),f)$ $(W^u(O(p),f))$ be the stable (unstable) manifold of the orbit $O(p)$ of p. The main step in the proof is the following.

Proposition (3.3). *Suppose* p *is a hyperbolic periodic point of* $f \in \mathcal{D}^r_\omega(M)$ *where* M *is as in* (3.2). *Given any neighborhood* U *of* f *in* $\mathcal{D}^r_\omega(M)$ *there is a* g *in* U *such that* p *is a hyperbolic periodic point for* g *and* $W^u(O(p),g)$ *has a point* q *of transverse intersection with* $W^s(O(p),g) \setminus O(p)$.

Assuming proposition (3.3), one proceeds as follows. Let $\mathcal{B}_{n,m}$ be the set of elements f in $\mathcal{D}^r_\omega(M)$ such that for each hyperbolic periodic point p of f of period less than $n+1$, there are points p_1, $p_2 \in O(p)$ and curves $\gamma_1 \subset W^u(p_1,f)$, $\gamma_2 \subset W^s(p_2,f)$ such that

(1) $p_1 \in \gamma_1$, $p_2 \in \gamma_2$

(2) length $(\gamma_i) < m$

(3) $\gamma_1 \setminus \{p_1\}$ has a non-empty transverse intersection with $\gamma_2 \setminus \{p_2\}$.

Then, for each n, $\bigcup_{m\geq 1} \mathcal{B}_{n,m}$ is open in $\mathcal{D}^r_\omega(M)$, and, by Proposition (3.3), it is dense. Thus $= \bigcap_{n\geq 1}(\bigcup_{m\geq 1} \mathcal{B}_{n,m})$ is the required residual set. For the proof of Proposition (3.3) and subsequent results, it will be necessary to make specific local changes in area preserving maps. The following lemma suffices for these purposes. Let $B_\varepsilon(x)$ denote the open ball of radius ε about x. For a map $\phi : \mathbb{R}^2 \longrightarrow \mathbb{R}^2$, let $||\phi||_r = \sup_{x \in \mathbb{R}^2} \max(|\phi(x)|, |D_x\phi|, |D_x^2\phi|, \ldots, |D_x^r\phi|)$ where $D^j\phi$ is the jth derivative of ϕ at x, $1 \le j \le r$.

Lemma (3.4) *Let* $\varepsilon > 0$ *and* $r \in Z^+$.

(a) *Given* $\delta > 0$ *and* $y \in B_{\frac{\varepsilon}{2}\delta}(0)$, *there is a* C^1 *area preserving diffeomorphism* $\phi : \mathbb{R}^2 \longrightarrow \mathbb{R}^2$ *such that* $\phi(0) = y$, $\phi(x) = x$ *for* $|x| \ge \delta$, *and* $||\phi - \text{id}||_1 < \varepsilon$.

(b) *Given* $\delta > 0$ *and a linear subspace* $H \subset \{(x,y) : |y| \le \frac{\varepsilon}{2}|x|\}$ *there is a* C^1 *area preserving diffeomorphism* $\phi : \mathbb{R}^2 \longrightarrow \mathbb{R}^2$ *such that* $\phi(0) = 0$, $T_0\phi((y = 0)) = H$, $\phi(x) = x$ *for* $|x| \ge \delta$, *and* $||\phi - \text{id}||_1 < \varepsilon$.

(c) *There is a function* $C(r) > 0$ *with the following property. Given* $\delta > 0$ *and* $y \in B_{C(r)\varepsilon\delta^r}(0)$, *there is a* C^r *area preserving diffeomorphism* $\phi : \mathbb{R}^2 \longrightarrow \mathbb{R}^2$ *such that* $\phi(0) = y$, $\phi(x) = x$ *for* $|x| \ge \delta$, *and* $||\phi - \text{id}||_r < \varepsilon$.

We do not prove this here, but remark that it is proved using generating functions. See [2] for a proof.

We proceed to prove Proposition (3.3).

We will use the following notation: $C\ell(E)$ is the closure of E, **int**(E) is the interior of E, $0_+(E) = \bigcup\{f^n(E) : n > 0\}$ is the positive orbit of E, $0_-(E) = \bigcup\{f^n(E) : n < 0\}$ is the negative orbit of E, $0_i^j(E) = \bigcup\{f^n(E) : i \le n \le j\}$, $0_i^\infty(E) = \bigcup\{f^n(E) : i \le n < \infty\}$, id is the identity transformation, supp ϕ

$= \{x : \phi(x) \neq x\}$ is the support of a diffeomorphism ϕ.

First note that it is enough to perturb f to g so that $[W^u(O(p),g) \smallsetminus O(p)] \cap [W^s(O(p),g) \smallsetminus O(p)] \neq \phi$. For then one can make a small subsequent change to produce a transverse intersection.

Let D_1 be a set of intervals in $W^s(O(p)) \smallsetminus O(p)$ such that $W^s(O(p)) \smallsetminus O(p) = \bigcup_{n \in Z} f^n(D_1)$ and let N_1 be a small neighborhood of D_1. Since each point $x \in M$ is non-wandering there is a sequence x_i in N_1, $x_i \longrightarrow x \in D_1$ and $n_i > 0$ such that $f^{n_i} x_i \longrightarrow x$. Let D_2 be a set of intervals in $W^u(O(p)) \smallsetminus O(p)$ such that $W^u(O(p)) \smallsetminus O(p) = \bigcup_{n \in Z} f^n D_2$ and let N_2 be a small neighborhood of D_2. We assume $f^{-2\tau}(\text{int } N_2) \cap \text{int } N_2 = \phi$ and there are integers $m_i < n_i$ such that $f^{m_i} x_i \longrightarrow y \in D_2$. We may pick a neighborhood U of $O(p)$ containing $N_1 \cup N_2$ so that $f^\tau | U$ is nearly linear, (Here τ is the least period of p). We assume N_2 small enough so that any point is N_2 whose negative f orbit leaves U first passes through N_1. Also, we assume $\text{int } f(N_2) \cap U = \phi$. See figure (3.2).

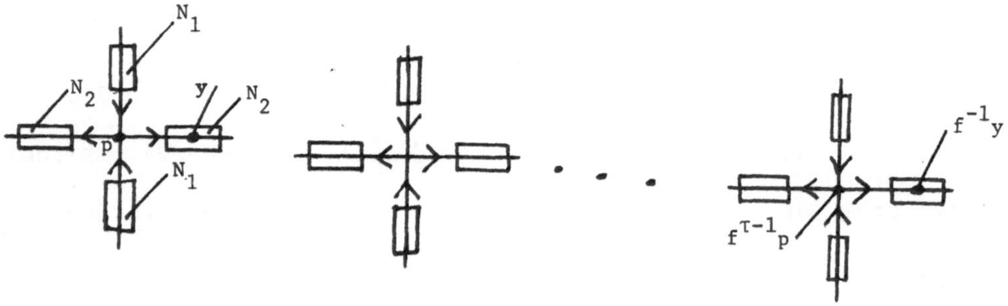

Figure (3.2)

Now let ϕ be C^r near id, supp ϕ near y, and $\phi(y) = f^{m_i} x_i$ for some large i. Consider $\phi \circ f = g$. We claim

$$f^{-1}y \in W^u(p,g) \text{ and } O_+(f^{-1}y,g) \cap N_1 \neq \phi$$

Thus, with a small C^r change we have obtained g with $W^u(O(p),g)$ ε - near $D_1 \subset W^s(O(p),g)$. Doing this with a sequence of $\varepsilon_i \longrightarrow 0$, and replacing f by the limit of the g_i's, we may assume there is a set $D_1 \subset W^s(O(p),f)$ which is a union of intervals such that $f^\tau D_1 \cap D_1 \subset \partial D_1$, $\bigcup_n f^n(D_1) = W^s(O(p),f)$, and $C\ell W^u(O(p),f) \cap \text{int } D_1 \neq \phi$. We also assume $W^u(O(p) \cap W^s(O(p))$ $= O(p)$. The fundamental part of the proof is

Lemma (3.5) *There is a point* $y \in \text{int } D_1$, *a neighborhood* V *of* y *and a sequence of points* $y_q \in W^u(O(p))$ *converging to* y *as* $q \longrightarrow \infty$ *such that* $O_-(y_q,f) \cap \text{int } V = \phi$.

If the lemma holds then one can choose ϕ C^r near id with $\phi = \text{id}$ outside V such that $\phi(y_q) = y$ for some large q. Then $g = \phi \circ f$ is such that $y \in W^u(O(p),g) \cap W^s(O(p),g)$.

To describe the proof of the lemma, let us first assume p is a fixed point and f preserves orientation on $W^s(p)$ and $W^u(p)$, a case considered by R. C. Robinson [5].

Let $y \in W^s(p) \setminus \{p\} \cap C\ell\, W^u(p)$ and suppose the right component of $W^u(p) \setminus \{p\}$ accumulates on y as in Figure (3.3). Call this component $W^u_+(p)$. Let V be a small open disk neighborhood of y such that $f(V) \cap V = \phi$. Let J be the component of $W^s(p) \cap V$ containing y. Let S_ℓ be the component of $V \setminus J$ on the left and S_R be the component of $V \smallsetminus J$ on the right. Let us first assume that $W^u_+(p)$ only accumulates on J from the right, i. e. $W^u_+(p) \cap S_\ell = \phi$. Let z be the first point on $W^u_+(p)$ which meets ∂V. Let γ_s be the interval in $W^s(p)$ joining the bottom of V to p, let L be the interval $[p,z]$ in $W^u(p)$, and let γ be the curve in ∂S_R connecting z to the top of γ_s. See figure (3.3).

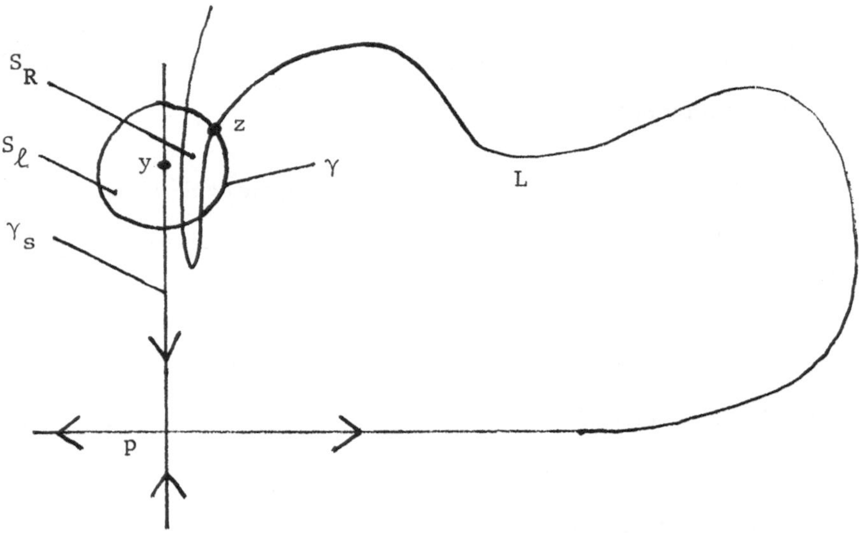

Figure (3.3)

Clearly $C = L \cup \gamma \cup \gamma_s$ is a closed simple curve. Let D be the component of the complement of C not containing y. Note that $f(S_R) \subset D$. Now set $E^q = \bigcup_{0 \le j \le q} f^j(D)$. Note E^q is open.

We have the following facts.

(1) $0_-(L) \cap V = \phi$

(2) $\partial E^q \subset \partial E^0 \cup 0^q_0 L$ for $q \ge 0$.

(3) $y \notin Cl\, E^q$ for $q \geq 0$

(4) $0_1^q(V \cap W_+^u(p)) \subset E^q$

(5) $f(L) \supset L$ and $0_0^\infty L$ accumulates on y.

Proof. (1) is immediate from definition of z since $f^{-1}(L) \subset L$. Let us prove (2). We have

$$\begin{aligned} \partial E^q &\subset \bigcup_{0\le j\le q} \partial f^j D = \bigcup_{0\le j\le q} f^j(\partial D) \\ &= \bigcup_{0\le j\le q} f^j(L \cup \gamma \cup \gamma_s) \\ &= 0_0^q L \cup 0_0^q \gamma \cup 0_0^q \gamma_s \\ &\subset 0_0^q L \cup 0_0^q \gamma \cup \gamma_s \end{aligned}$$

Now, $\gamma \subset \partial E^0$ and $f(\gamma) \subset D$, so $f^j(\gamma) \subset E^q$ for $1 \le j \le q$. Thus, $\partial E^q \cap 0_0^q \gamma \subset \gamma$, so $\partial E^q \subset \gamma \cup \gamma_s \cup 0_0^q L \subset \partial E^0 \cup 0_0^q L$.

We now prove (3). $y \notin Cl\, E^0$ by construction of D. We have $y \notin \partial E^q$ by (2) since $y \notin 0_0^\infty L$. Therefore, if there is a $q \ge 1$ with $y \in Cl\, E^q$ we have $y \in E^q \setminus D$. Thus, $f^{-j}y \in D$ for some $1 \le j \le q$. Let $W_+^s(p)$ be the component of $W^s(p) - \{p\}$ containing γ_s. Then, $W_+^s(p)$ crosses $\partial D - \gamma_s$. Since it cannot meet $0_0^\infty L$, we have $W_+^s(p) \cap \gamma \ne \phi$. Let $w \in W_+^s(p)$ be the first point in $W_+^s(p) \cap \gamma$ gotten by moving along $W_+^s(p)$ from γ_s. The open interval (p,w) in $W_+^s(p)$ is mapped into itself by f, contains y, and does not meet D. This implies for each $j \ge 1$, $f^j w \in (p,w)$, so $f^j w \notin D$, a contradiction. Statement (4) is immediate since $f(S_R) \subset D$. Statement (5) is obvious since $0_0^\infty L = W_+^u(p) \cup \{p\}$.

Now, let $y_q \in C\ell\,(E^q)$ be a point in $C\ell\,(E^q)$ whose distance to y is a minimum. For large q, $y_q \in 0_0^q L$. We assert that $0_-(y_q) \cap V = \phi$. For suppose $f^{-j}(y_q) \in V$ for some $j > 0$. Then $j < q$ because $0_-(L) \cap V = \phi$. So, $y_q = f^j(f^{-j}(y_q)) \in f^j(V \cap W_+^u(p)) \in E^q$. Since E^q is open, there are points in E^q closer to y than y_q, a contradiction.

Now, if $W_+^u(p) \cap S_\ell \neq \phi$, we only know $0_1^q(S_R \cap W_+^u(p)) \subset E^q$. We wish to construct a different E^q satisfying (1) —— (5). Call the E^q just constructed E_R^q, and, label z_R, γ_R, C_R, D_R analogously. Let z_ℓ be the point in $W_+^u(p)$ where $W_+^u(p)$ first meets ∂S_ℓ, and let $L_\ell = [p, z_\ell] \subset W^u(p)$. Let γ_ℓ, C_ℓ, D_ℓ, E_ℓ^q be constructed analogously. Then, (1), (2), (3), (5) hold for E_ℓ^q, L_ℓ, and (4) becomes $0_1^q(S_\ell \cap W^u(p)) \subset E_\ell^q$. Finally, let $L = L_\ell \cup L_R$ and $E^q = E_\ell^q \cup E_R^q$. Shrinking V so that $V \cap L = \phi$, we obtain (1) ——(5), and we may again construct y_q. This proves lemma (3.5) in the case when p is fixed and f preserves orientation on $W^u(p) \cup W^s(p)$.

We now consider the general case. Consider the various components of $W^s(0(p)) \smallsetminus 0(p)$ and $W^u(0(p)) \smallsetminus 0(p)$. Let T be a common period of these components. Since $C\ell\, W^u(0(p)) \cap W^s(0(p)) \smallsetminus 0(p) \neq \phi$, we can find integers $0 \leq j_1 \leq j_2 \leq \tau$ such that a component X_1 of $W^u(f^{j_1}p) \smallsetminus \{f^{j_1}p\}$ accumulates on a component Y_2 of $W^s(f^{j_2}p) \smallsetminus \{f^{j_2}p\}$. Let $X_2 = f^{j_2 - j_1}X_1$, and $Y_3 = f^{j_2 - j_1}Y_2$. Then, $C\ell\, X_2 \cap Y_3 \neq \phi$. Since there are only finitely many periodic points and components, we eventually begin repeating. Thus, there are sequences $p_0, p_1, \ldots, p_\nu$ in $O(p)$ and components $X_i \subset W^u(p_i) \smallsetminus \{p_i\}$, $Y_i \subset W^s(p_i) \smallsetminus \{p_i\}$ such that $C\ell X_i \cap Y_{i+1} \neq \phi$, $0 \le i \le \nu$, and $X_\nu = X_0$, $Y_\nu = Y_0$. Also, $X_i \subset O(X_1)$, $Y_i \subset O(Y_1)$. Call the sequence $X_0 Y_1 X_1 Y_2 X_2 \ldots Y_{\nu-1} X_{\nu-1} Y_\nu$ a <u>cycle</u>. Choose a cycle with a minimum possible number of terms. Say this is $X_0 Y_1 X_1 \ldots Y_{\nu-1} X_{\nu-1} Y_\nu$. Let $y \in Y_0 \cap C\ell X_{\nu-1}$. Suppose $p_i = f^{j_i}(p_0)$. We have a situation like that in figure (3.4) where we assume $X_{\nu-1}$ approaches Y_0 on the left.

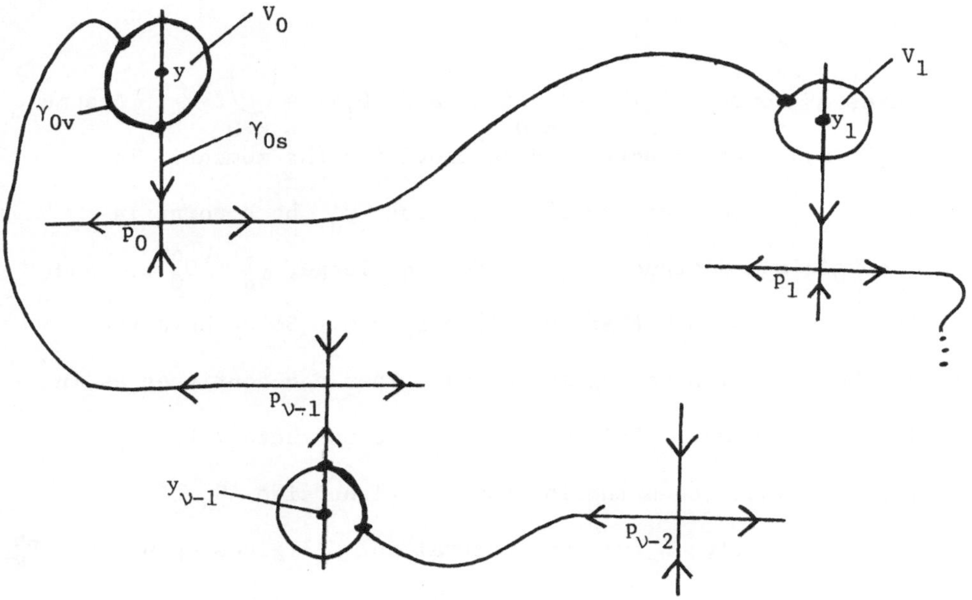

Figure (3.4)

Let V_0 be a small neighborhood of y, and let $C^s(y)$ be the component of y in $W^s(p_0) \cap V_0$ (i.e. $C^s(y)$ is an open interval in $W^s(p_0)$).

Let $V_i = f^{j_i}(V_0)$, $y_i = f^{j_i}y$ and let $S = S_\ell$ be the component of $V_0 \setminus C^s(y)$ to be the left of y. Let $S_\varepsilon = S \cap B_\varepsilon(y)$. We will construct open sets E^q, for $q \geq 1$, a compact curve $L \subset X_{\nu-1}$, and a compact subset $\gamma_s \subset W^s(0(p))$ such that

For $\varepsilon > 0$ small,

(1) $0_-(L) \cap S_\varepsilon = \phi$, $0_+\gamma_s \cap S_\varepsilon = \phi$

(2) $\partial E^q \subset \partial E^1 \cup 0_0^{qT}L$

(3) $y \notin C\ell(E^q)$

(4) For ε small, $0_1^{qT}(S_\varepsilon) \subset E^q$

(5) $\bigcup_{0 \leq k < \infty} f^{kT}L = X_{\nu-1}$

Here, of course, $0_-(L) = \bigcup_{n<0} f^n L$ and $0_+\gamma_s = \bigcup_{n>0} f^n\gamma_s$. Assume E^q, L, γ_s have been constructed as above for the moment. Then as $y \in C\ell\, X_{\nu-1}$, we have $y \in C\ell\, 0_0^\infty L$. Let y_q' be a point in $C\ell\, E^q$ of minimum distance to y. For q large, $y_q' \in 0_0^{qT}L$, and, as before, one can show that $0_-(y_q') \cap S_\varepsilon = \phi$. So we have taken care of the left side of y_0 in $B_\varepsilon(y)$. Then one continues as in the fixed point case. Call the E^q, L constructed, E_ℓ^q, L_ℓ. Repeating the whole construction for the right side $S_{R,\varepsilon}$ of $B_\varepsilon(y) \setminus Y_0$ (actually $B_\varepsilon(y) \setminus$ an interval in Y_0) gives open sets E_R^q and an L_R so that (1) - (5) hold for these with the same γ_s. Then setting $E^q = E_\ell^q \cup E_R^q$, $L = L_\ell \cup L_R$, and shrinking ε we may let $V = B_\varepsilon(y)$ and lemma (3.5) will be proved.

We proceed to construct the E^q, L for $S = S_\ell$. Let γ_{ou} be the shortest curve in $X_{\nu-1}$ joining $p_{\nu-1}$ to V_0. Draw a curve

γ_{ov} in $\partial V_0 \setminus Y_0$ connecting γ_{ou} to Y_0 terminating at the point z_0 closest to p_0. Follow z_0 to p_0 in Y_0 with a curve γ_{os}. Let $\gamma_{iv} = f^{j_i}(\gamma_{ov})$, $\gamma_{is} = f^{j_i}(\gamma_{os})$, $\gamma_{iu} = f^{j_i}(\gamma_{ou})$ for $0 \le i < \nu$. Let $\gamma_v = \bigcup_i \gamma_{iv}$, $\gamma_s = \bigcup_i \gamma_{is}$, $\gamma_u = \bigcup_i \gamma_{iu}$, and let $C = \gamma_v \cup \gamma_s \cup \gamma_u$. Since the cycle has a minimal number of elements we have that C is a simple closed curve; i.e., if C were not simple, then we could find a cycle with fewer elements. Let $L = \gamma_{ou}$. Clearly we have (5), and with ε small, we have (1). We may also assume $y \notin C$, $V_0 \cap [0_+(\gamma_s) \cup 0_1^{T-1}S] = \phi$, $f^T y \notin 0_1^{T-1}V_0$, $f^T\gamma_s \subset \gamma_s$, $f^{-T}\gamma_u \subset \gamma_u$ and $f^T(C\ell\ S) \cup f^{2T}(C\ell\ S) \subset D$ where D is the component of $M \setminus C$ such that $y \notin D$. Set

$$\tilde{D} = \bigcup \{f^j D : 0 \le j < T \text{ and } y \notin f^j D\}$$

and $D^q = \bigcup_{0 \le k \le q} f^{kT}\tilde{D}$, $E^q = 0_1^{T-1} S \cup D^{q-1}$

We have arranged for (1) and (5) to hold . We now check (2), (3), and (4).

<u>Proof of</u> (2): We need two preliminary facts.

(6) For each $0 \le i < \nu$, $f^{T+j_i}(\partial S) \subset D$.

<u>Proof of</u> (6): We orient C so that y lies to the left of $\partial S \cap C$. Thus, D lies to the right of y. Then, $\partial(f^{j_i}S) = f^{j_i}(\partial S)$, and $y_i = f^{j_i}(y)$ lies to the left of $f^{j_i}(\partial S)$ since f^{j_i} preserves orientation on M. Now clearly (from figure (3.4)), y_i and $f^{T+j_i}(S)$ are separated by C, so $f^{T+j_i}(C\ell\ S)$ lies to the right of C i.e. in D.

(7) $$\partial[\bigcup_{0\le k\le q} f^{kT}D] \subset \partial D \cup O_0^{qT}L$$

<u>Proof of (7)</u>: Clear for $q = 0$. Assume true for q inductively. Then,

$$\partial[\bigcup_{0\le k\le q+1} f^{kT}D] = \partial[D \cup f^T[\bigcup_{0\le k\le q} f^{kT}D]]$$

$$\subset \partial D \cup f^T\partial D \cup O_1^{(q+1)T}L \quad \text{(by inductive assumption)}$$

Now,

$$f^T(\partial D) = f^T(\gamma_u) \cup f^T(\gamma_s) \cup f^T(\gamma_v) \subset O_0^{2T}L \cup \gamma_s \cup \bigcup_{0\le i<\nu} f^{T+j_i}(\partial S)$$

$\subset O_0^{2T}L \cup \gamma_s \cup D$ by (6). Since $D \cap \partial(\bigcup_{0\le k\le q+1} f^{kT}D) = \phi$ and $\gamma_s \subset \partial D$, we have (7) for $q + 1$.

We can now prove (2). Let $R = \{r \in [0,T) : y \notin f^r D\}$. Then,

$$\partial E^q = \partial[\bigcup_{0\le k\le q-1} f^{kT}\tilde{D} \cup O_1^{T-1}S] = \partial[\bigcup_{0\le k\le q-1} f^{kT}\tilde{D} \cup E^1]$$

$$= \partial[\bigcup_{r\in R} f^r[\bigcup_{0\le k\le q-1} f^{kT}D] \cup E^1]$$

$$\subset \partial E^1 \cup \bigcup_{r\in R} f^r(\partial(\bigcup_{0\le k\le q-1} f^{kT}D)) \subset \partial E^1 \cup \bigcup_{r\in R} f^r(\partial D) \cup O_0^{qT}L.$$

Since $\partial E^q \cap \bigcup_{r\in R} f^r(\partial D) \subset \partial E^q \cap C\ell E^1 = \partial E^q \cap \partial E^1 \subset \partial E^1$, we get (2).

<u>Proof of (3)</u>: It is true for $q = 1$. Assume it holds for q and fails for $q + 1$. We will get a contradiction. If $y \in C\ell\, E^{q+1}$, then $y \in E^{q+1}$, since $y \notin \partial E^q$ for all q. So

$y \in E^{q+1} - E^q = f^{qT}\tilde{D} - f^{(q-1)T}\tilde{D}$ which implies $f^{-T}y \in f^{(q-1)T}\tilde{D} \subset E^q$. Thus, the open segment $(y, f^{-T}y)$ joining y to $f^{-T}y$ in $W^s(p_0)$ crosses $\partial E^q \subset \partial E^1 \cup 0_0^{qT}(L)$. By construction, $f^T(y, f^{-T}y) = (f^T y, y)$ cannot meet $f^T(\partial E^1 \cup 0_0^{qT}(L))$ which is a contradiction.

<u>Proof of (4)</u>: We first note that for $0 < m < T$, if $y \in f^m D$, then $f^m D$ contains the segment $[f^{2T}y, y] \subset W^s(p_0)$. Otherwise we would have $[f^{2T}y, y] \cap \partial f^m D \neq \phi$. This latter set is contained in $0_1^{2T-1}\gamma_{0s}$. From the geometry of ∂D and f, if $\tilde{z}$ is the point in $\gamma_{0s} \cap 0_1^{2T-1}\gamma_{0s}$ closest to y, then the segment $(\tilde{z}, y)$ must meet $M - f^m D$. Since $y \in f^m D$, there are points in $\partial f^m D \cap [f^{2T}y, y]$ closer to y than $\tilde{z}$, a contradiction.

Next, observe that for $\varepsilon > 0$ small enough, $y \in f^m(D)$ and $0 < m < T$ imply $f^m(D) \dot{\supset} f^T(S_\varepsilon) \cup f^{2T}(S_\varepsilon)$. Now (4) obviously holds for $q = 1$. Assume it holds for q. Then,

$$\begin{aligned} 0_1^{(q+1)T} S_\varepsilon &= 0_1^{T-1} S_\varepsilon \cup 0_T^{2T-1} S_\varepsilon \cup 0_{2T}^{(q+1)T} S_\varepsilon \\ &\subset E^1 \cup 0_T^{2T-1} S_\varepsilon \cup f^T(E^q) \\ &\subset E^1 \cup 0_T^{2T-1} S_\varepsilon \cup E^{q+1}, \end{aligned}$$

so we are done if we show $0_T^{2T-1} S_\varepsilon \subset E^1$. If not, there is a $0 \le t < T$ such that $f^{T+t} S_\varepsilon \not\subset \tilde{D}$. So if $f^{T+t} S_\varepsilon \subset f^r D$ with $0 \le r < T$, then $y \in f^r D$. We obtain a contradiction by showing

(*) $\quad f^{T+t} S_\varepsilon \subset f^{[kt]} D$ for all $k \ge 1$

where $[kt]$ = mod T reduction of kT. For then with $k = T$, we get $y \in D$. Now, (*) is true for $k = 1$. Suppose it is true for k. Let $kt = T\ell + m$, with $0 \leq m < T$, so $[kt] = m$. Then, $[(k+1)t] = m + t$ or $m + t - T$. By inductive assumption $f^{T+t}S_\varepsilon \subset f^m D$, so $y \in f^m D$. This implies $m \neq 0$. But then as we have noted $f^T S_\varepsilon \cup f^{2T} S_\varepsilon \subset f^m D$. Now, $f^T S_\varepsilon \subset f^m D$ implies $f^{t+T} S_\varepsilon \subset f^{m+t} D$, and $f^{2T} S_\varepsilon \subset f^m D$ implies $f^{2T+t-T} S_\varepsilon = f^{t+T} S_\varepsilon \subset f^{m+t-T} D$. Thus $f^{t+T} S_\varepsilon \subset f^{[(k+1)t]} D$ and (4) is proved.

References

1. J. Moser, Non-existence of integrals for canonical systems of differential equations, Comm. Pure and Appl. Math. 8 (1955), 409-436.

2. S. Newhouse, Quasi-elliptic periodic points in conservative dynamical systems, Amer. J. of Math. 99 (1975), 1061-1087.

3. D. Pixton, Planar homoclinic points, J. of Differential Equations 44 (1982), 365-382.

4. R. C. Robinson, Generic properties of conservative systems I, II, Amer. J. Math. 92 (1970), 562-603, 897-906.

5. ______________, Closing stable and unstable manifolds on the two sphere, Proc. AMS 41 (1973), 299-303.

6. E. Zehnder, Homoclinic points near elliptic fixed points, Comm. Pure and Appl. Math. 26 (1973), 131-182.

Lecture 4.

We have seen that in many cases one has periodic and homoclinic motions and that homoclinic motions bring invariant complicated topologically transitive sets. It is natural to ask the following questions

(4.1) Where are the periodic and homoclinic motions?

(4.2) What is the structure of the h-closures? In particular, when do h-closures have positive Lebesque measure, or at least maximal Hausdorff dimension?

Concerning question (4.2), we make two remarks.

(a) Suppose $f \in D_\omega^r(M^2)$, $r \geq 2$, and there is a set E in M^2 with positive Lebesque measure such that for $x \in E$, $\limsup_{n\to\infty} \frac{1}{n}\log|T_x f^n| > 0$. Then, there are h-closures with positive Lebesque measure. One can also show (following Pesin [5]) that there is an invariant set $\tilde{E}$ of positive Lebesque measure on which f is ergodic.

(b) There is a residual set $\mathcal{B} \subset D_\omega^1(M^2)$ such that for $f \in \mathcal{B}$, each h-closure of f has maximal Hausdorff dimension [3].

We will now discuss question (4.1). From a historical perspective, it is interesting to consider the following three properties of an $f \in D_\omega^r(M)$.

(4.3) The periodic points are dense in M .

(4.4) The homoclinic points of a given hyperbolic periodic point are dense in its stable and unstable manifolds.

(4.5) The "stable" periodic points are dense in M .

We have already mentioned that Poincare felt that (4.3) and (4.4) were true in many cases. In a paper in 1941 [1] Birkhoff stated that (4.5) would be true with "certain cases aside" where "stable" periodic points p were defined as those which have complete orbits different from $O(p)$ in every small neighborhood of $O(p)$.

The fact is that "certain cases aside" (4.3), (4.4), (4.5) are true for C^1 generically and not known C^r generically for $r \geq 2$. More precisely,

a. (4.3) holds C^1 generically on all compact manifolds.

b. (4.4) holds C^1 generically on all surfaces. [8]

c. (4.5) holds C^1 generically on all compact manifolds unless f is Anosov. [2]

We proceed to sketch the proofs of (4.3) and (4.4) when $\dim M = 2$. A sketch of the proof of (4.5) is in [4].

Theorem (4.6)(Pugh-Robinson [7]*) There is a residual set* $\mathcal{B} \subset D_\omega^{\,1}(M)$ *such that if* $f \in \mathcal{B}$ *, then the periodic orbits of* f *are dense in* M.

The first step is the following.

(4.7) Given $\varepsilon > 0$, $x \in M$, and a neighborhood U of f there is a g in U such that g has a periodic point $p(g)$ ε-close to x which is persistent.

The persistence means that if g_1 is near g , then there is a periodic point $p(g_1)$ for g_1 near $p(g)$. For this, one just arranges for 1 not to be an eigenvalue of $T_p g^{\tau}$ where τ is the period of $p(g) = p$.

Assuming (4.7) one proceeds as follows. Let $\{x_1, x_2, \ldots\}$ be a countable dense set in M. Let $\mathcal{B}_{n,m}$ be the set of f such that f has a persistent periodic point $p_{i,m}$ within $\frac{1}{m}$ of x_i for $i = 1, \ldots, n$. Then, $\mathcal{B}_{n,m}$ is dense an open in $D_{\omega}^{1}(M)$, and $\bigcap_{n,m} \mathcal{B}_{n,m} = \mathcal{B}$ is such that the periodic points of f are dense in M for $f \in \mathcal{B}$.

Main step:

(4.8) Given $\varepsilon > o$, $x \in M$, and a neighborhood U of f in $D_{\omega}^{1}M$, there is a g in U such that g has a periodic point which is ε-close to x.

First, note that x is non-wandering; i.e., for each neighborhood V of x there is an integer $n > 0$ such that $f^n(V) \cap V \neq \phi$. Otherwise, the sequence $\{f^n V : n \geq 0\}$ would consist of disjoint sets having the same positive measure. (Note : If $f^{n_1}(V) \cap f^{n_2} V \neq \phi$, $n_1 < n_2$, then $f^{n_2 - n_1} V \cap V \neq \phi$). Thus, there are sequences $x_i \longrightarrow x$ and $n_i \longrightarrow \infty$ such that $f^{n_i}(x_i) \longrightarrow x$. Let $\varepsilon > 0$ and $\delta > 0$ be small. Consider a point y such that $y, f^n y \in B_{\varepsilon\delta}(x)$ with n large. Here, as before, we let $B_{\delta}(x)$ denote the open δ - ball about x in M (id is the identity map). The idea is to find $\phi \in D_{\omega}^{1}M$ with $\phi = \text{id}$ off $B_{\delta}(x)$ such that $\phi(f^n y) = y$ and let $g = \phi \circ f$. This does <u>not</u> work in most cases because there may exist $0 < j < n$ such that $f^j y \in B_{\delta}(x)$ and $\phi(f^j y) \neq f^j y$. If this occurs, then $g^n y$ may not equal $\phi(f^n y)$, so we have not made y periodic for g.

We need a way of controlling these intermediate intersections. To understand how to do this, let us first assume that f is an isometry. We work in fixed local coordinates about points in M so we regard our local discussion to take place in the plane $\mathbb{R}^2$. Let $d(z_1,z_2) = |z_1 - z_2|$ denote the distance between z_1 and z_2. Let z be the midpoint of the line segment between y and $f^n y$, and let $d_0 = |y - f^n y|$. Suppose $|z - f^j y| < \frac{2}{3}d_0$ for some $0 < j < n$. Set $y_1 = y$ and $j_1 = j$ if $|y - f^j y| \le |f^j y - f^n y|$ and, set $y_1 = f^j y$ and $j_1 = n - j$ if $|y - f^j y| \ge |f^j y - f^n y|$. Let $d_1 = |y_1 - f^{j_1} y_1| \le \lambda d_0$ where $\lambda = (\frac{5}{6})^{\frac{1}{2}}$. Let z_1 be the midpoint of the line segment between y_1 and $f^{j_1} y_1$. If there is a $0 < j < j_1$ with $|f^j y_1 - z_1| < \frac{2}{3}d_1$, we can similarly define y_2, $0 < j_2 < j_1$ such that $|y_2 - f^{j_2} y_2| \le \lambda d_1$. This process must stop since there are only finitely many j's. If it stops at the kth step, then

$$|y_k - f^{j_k} y_k| \le \lambda d_{k-1} \le \lambda^k d_0,$$

$$\min(|y - y_k|, |y_k - f^n y|) \le d_0 \sum_{i \ge 1} \lambda^i = \frac{d_0 \lambda}{1-\lambda}$$

$$\text{and } \min(|y - f^{j_k} y_k|, |f^n y - f^{j_k} y_k|) \le d_0 \sum_{i \ge 1} \lambda^i = \frac{d_0 \lambda}{1-\lambda}$$

Let $B_i = B_{\frac{2d_0}{3}}(f^i z)$. Relabeling y_k as y, $f^{j_k} y_k$ as $f^n y$, d_k as d_0, we may assume

(1) y, $f^n y$ are close to x and $n > 1 + \frac{12}{\varepsilon}$.

(2) $B_i \cap B_j = \phi$ for $0 \le i < j < n$

(3) if $z = \frac{1}{2}y + \frac{1}{2}f^n y$, then $|f^j y - z| \geq \frac{2}{3}d_0$ for $0 < j < n$. Since f is an isometry, we have $|f^{i+j}y - f^i z| \geq \frac{2}{3}d_0$ for all i and $0 < j < n$. For a diffeomorphism ϕ, recall $\operatorname{supp}\phi = \{u : \phi(u) \neq u\}$. We may find ϕ_0 with $\operatorname{supp}\phi_0 \subset B_0$ such that $||\phi_0 - id||_1 < \varepsilon$ and $|\phi_0(f^n y) - y| \leq |f^n y - y| - \frac{\varepsilon}{12}d_0$ $= d_0(1 - \frac{\varepsilon}{12})$.

Now set $f_0 = \phi_0 \circ f$. Then, we have moved $f^n y$ the fixed proportion $\frac{\varepsilon}{12}$ of d_0 closer to y. That is, $|f_0^n y - y| \leq (1 - \frac{\varepsilon}{12})d_0$. Let $z_1 = \phi_0 \circ (f^n y) = f_0^n(y)$. Choose ϕ_1 with $\operatorname{supp}\phi_1 \subset B_1$, $||\phi_1 - id||_1 < \varepsilon$, and $|\phi_1(f(z_1)) - fy| \leq |f(z_1) - fy| - \frac{\varepsilon}{12}d_0 \leq d_0(1 - 2(\frac{\varepsilon}{12}))$. Let $f_1 = \phi_1 \circ f_0$. Then, $|f_1(y) - f_1^{n+1}y| \leq (1 - 2(\frac{\varepsilon}{12}))d_0$. Continuing in this way, setting $f_i = \phi_i \circ f_{i-1}$ with $\operatorname{supp}\phi_i \subset B_i$ and ϕ_i chosen appropriately, we get f_{n-1} $\varepsilon - C^1$ close to f, and

$$|f_{n-1}^{n-1}(y) - f_{n-1}^{2n-1}(y)| \leq \max(1 - (n-1)\tfrac{\varepsilon}{12})d_0, 0) = 0.$$

Thus, we have made z_1 periodic for f_{n-1}. This proves (4.8) if f is an isometry.

Let us proceed to the general case.

Lemma (4.9) There are a unit vector $v \in T_xM$ *and a sequence* $n_1 < n_2 < \ldots$ *of positive integers such that for any unit vector* $w \in T_xM$

$$\lim_{k\to\infty} \frac{|T_x f^{n_k}(w)|}{|T_x f^{n_k}(v)|} \quad \textit{exists}$$

and is less than or equal to 1.

Proof Using a finite covering by coordinate charts and local representatives of f we may regard each $T_x f^k$ as a linear map F_k from $\mathbb{R}^2$ to itself. Consider the sequence $\phi_k(\cdot) = \frac{F_k(\cdot)}{|F_k|}$.

This is a sequence of linear maps of norm 1, so there is a subsequence ϕ_{n_k} converging uniformly on bounded sets to a linear map T of norm 1. Let v be a unit vector such that $|T(v)| = 1$. Then for any w,

$$\lim_{k\to\infty} \frac{|F_{n_k}(w)|}{|F_{n_k}(v)|} = \lim_{k\to\infty} \frac{|F_{n_k}(w)|}{|F_{n_k}(v)|} \cdot \frac{|F_{n_k}(v)|}{|F_{n_k}|}$$

$$= \lim_{k\to\infty} \frac{|F_{n_k}(w)|}{|F_{n_k}|} = T(w) \quad \text{exists, and}$$

$|T(w)| \le 1$. This proves lemma (4.9).

Now, let $E_2(x)$ be the subspace of T_xM spanned by the vector v, and let $E_1(x)$ be the orthogonal complement of $E_2(x)$. Consider the sequence $F_{n_k} = T_x f^{n_k}$ as in lemma (4.9).

Case 1. If $w \in E_1(x)$ has $|w| = 1$, then $|T(w)| = 1$.

In this case, forgetting the first few terms, we may assume each F_{n_k} maps a small circle about x to an ellipse which is nearly a circle. Thus, F_{n_k} is nearly an isometry. So, f^{n_k} is nearly an isometry also for points near x. We proceed more or less as before except we let $B_i = B_{\frac{2d}{3}0}(f^{n_i}(z))$, $B_0 = B_{\frac{2d}{3}0}(z)$, and make sure that $n > n_N$ with $N > \frac{12}{\varepsilon} + 1$. That is, we do our perturbations near $f^{n_i}(y)$ for each $i = 1,\ldots,N$. For d_0 small, we can insure that $f^j B_0 \cap f^k B_0 = \phi$ for $0 \le j < k < n_N$, so the different ϕ_i's do not interfere with each other.

<u>Case 2</u>. $|T(w)| < |w|$ for $w \in E_1(x) - \{0\}$.

In this case, each F_{n_k} maps circles to ellipses. If $T(w) = 0$, the eccentricities approach 1 so that the ellipses become very flat. We will have to choose y and $f^n y$ differently. We start with a large integer $N_1 > \frac{12}{\varepsilon}$ and a rectangle R whose long side A is along the $E_1(x)$ direction and whose short side B is along the $E_2(x)$ direction and such that $F_{n_k}(A)$ is much longer than $F_{n_k}(B)$ for $1 \le k \le N_1$ and $F_{n_k}(A)$ is much shorter than $F_{n_k}(B)$ for $N_1 < k \le 2N_1$. The integer N_1 is specified in advance, and then finitely many n_k's are specified. Having chosen N_1 and the n_k's we choose a small $\delta > 0$ so that $z \in B_\delta(x)$ implies f^{n_k} on $B_\delta(z)$ is nearly equal to F_{n_k} on $B_\delta(z)$ for $1 \le k \le 2N_1$ and $f^i B_\delta(z) \cap f^j B_\delta(z) = \phi$ for $0 \le i \le j \le n_{2N_1}$.

Now the intermediate intersection property is done not for circles near x but rather for ellipses whose major axis is parallel to A and whose minor axis is parallel to B.

We can work with ellipses as well as circles because they are balls of a fixed radius in a metric linearly deformed from the standard metric. Such a linear deformation brings a constant K into the estimates of the intermediate intersection correction.

This constant is proportional to the maximum ratio of the axes of the ellipses. In the situation at hand, K will depend only on f^j for $0 \le j \le n_{2N_1}$. Thus, we can find $\delta > 0$ small enough, and ellipses

$$\zeta : \frac{u^2}{a^2} + \frac{v^2}{b^2} = 1 \ , \ \tilde{\zeta} : \frac{u^2}{\tilde{a}^2} + \frac{v^2}{\tilde{b}^2} = 1$$

in local coordinates with $u = 0$, $v = 0$ corresponding to a point in $B_\delta(x)$, $(u = 0)$ parallel to $E_1(x)$, $(v = 0)$ parallel to $E_2(x)$ with the following properties

(1) $\tilde{a} = \frac{3}{2}a \quad \tilde{b} = \frac{3}{2}b$

(2) $\tilde{\zeta} \subset B_\delta(x)$

(3) $f^j B_\delta(x) \cap f^k B_\delta(x) = \phi$ for $0 \le j < k \le n_{2N_1}$

(4) there are points y , $f^n y$ in the interior of ζ with $n > n_{2N_1}$ such that $f^j y \notin$ interior $\tilde{\zeta}$ for $0 < j < n$.

Consider the ellipses $\zeta, \tilde{\zeta}$ as in figure (4.1)

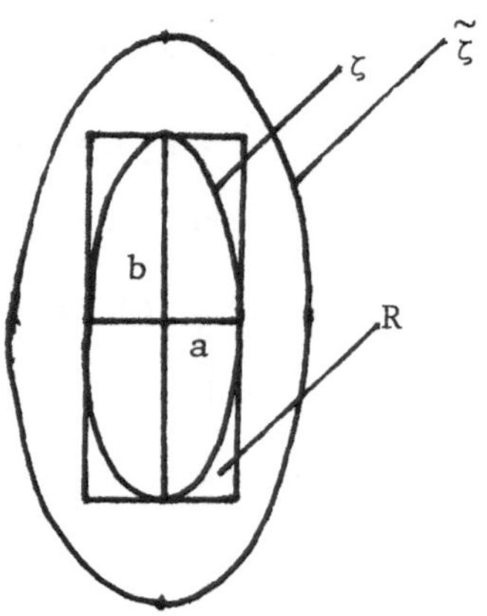

Figure (4.1)

We let R be the smallest rectangle containing ζ with sides parallel to the axes of $\zeta, \tilde{\zeta}$. Let A be the left vertical side of R and B be the bottom side of R . The upper right corner of R has coordinates (a,b) and $\frac{a^2}{\tilde{a}^2} + \frac{b^2}{\tilde{b}^2} = \frac{4}{9} + \frac{4}{9} < 1$, so R is contained in the interior of $\tilde{\zeta}$. In coordinates (u,v) centered at the common center of $\zeta, \tilde{\zeta}$, let $\Phi_\alpha(u,v) = (\alpha u, \alpha v)$. Let $\tilde{R} = \Phi_{\sqrt{\frac{33}{32}}} R$ and $\tilde{\tilde{R}} = \Phi_{\sqrt{\frac{17}{16}}} R$. Thus , $R \subset \tilde{R} \subset \tilde{\tilde{R}} \subset$ interior of $\tilde{\zeta}$ and the ratios of the sides of R , $\tilde{R}$, $\tilde{\tilde{R}}$ are fixed constants. Let $\tilde{A}$, be the left vertical side of $\tilde{R}$, $\tilde{B}$ be the bottom of $\tilde{R}$, and define $\tilde{\tilde{A}}$, $\tilde{\tilde{B}}$ similarly Then ,

(5) y , $f^n y \in R$, and $f^j y \notin \tilde{\tilde{R}}$ for $0 < j < n$

(6) if $f^{n_i} R = R_i$, then the R_i are nearly parallelograms with the length of $f^{n_i} A$ much longer than that of $f^{n_i} B$ for $1 \le i \le N_1$ and the length of $f^{n_i} A$ is much shorter than that of $f^{n_i} B$ for $N_1 < i \le 2N_1$.

Similar statements hold for $\tilde{R}_i = f^{n_i}\tilde{R}$, $\tilde{\tilde{R}}_i = f^{n_i}\tilde{\tilde{R}}$.

Note that the angles between the sides $f^{n_i} A$ and $f^{n_i} B$ may get small.

We draw some possible figures of R_i , $\tilde{R}_i$, $\tilde{\tilde{R}}_i$ in figure (4.2)

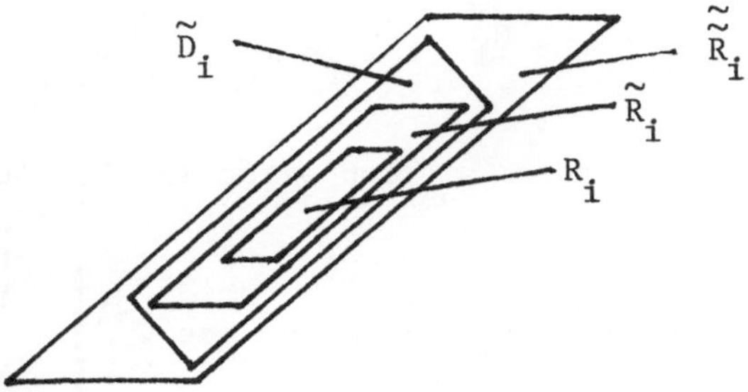

Figure (4.2)

Because of the fixed proportions of side lengths of R_i, $\tilde{R}_i$, $\tilde{\tilde{R}}_i$ there are rectangles D_i, $\tilde{D}_i$ with $R_i \subset D_i \subset \tilde{R}_i \subset \tilde{D}_i \subset \tilde{\tilde{R}}_i$. For $1 \le i \le N_1$, the long side of D_i is nearly parallel to $f^{n_i}A$ and for $N_1 < i \le 2N_1$, the long side of D_i is nearly parallel to $f^{n_i}B$ as in figure (4.3)

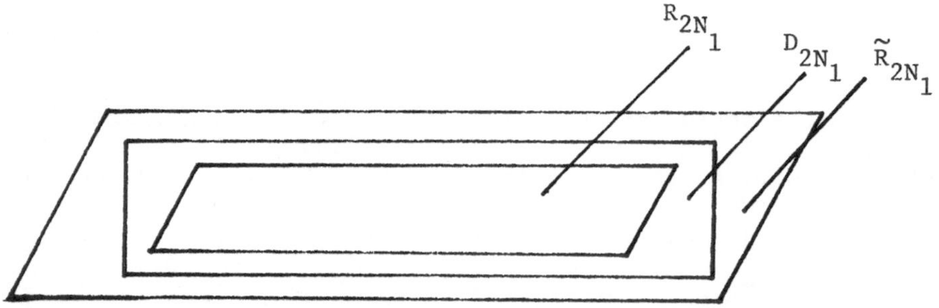

Figure (4.3)

If we push with ϕ_i supported in $\tilde{D}_i$ in the direction of the short altitude of $\tilde{D}_i$ for $1 \le i \le N_1$, we can move $f^n y$ into $f^{N_1}\tilde{A}$. i.e. we can find ϕ_i supported in $\tilde{D}_i$ such that if

$$g(z) = \phi_i \circ f(z) \quad \text{for} \quad z \in f^{-1}\left[\bigcup_{1 \le i \le N_1} \tilde{D}_i\right]$$

$$= f(z) \quad \text{otherwise},$$

$$\text{then } g^{n_{N_1}-n+1}(f^{n-1}y) \in f^{N_1}\tilde{A}.$$

Having done this we can find ψ_i supported in $\tilde{D}_i$ for $N_1 < i < 2N_1$ so that if

$$g_1(z) = \psi_i \circ g(z) \text{ for } z \in g^{-1}\left[\bigcup_{N_1 < i \le 2N_1} \tilde{D}_i\right]$$

$$= g(z) \quad \text{otherwise},$$

then $g_1^{N_{2N_1}-n+1}(f^{n-1}y) \in$ upper boundary of D_{2N_1} .

Using different ψ_i's we can actually attain points on a curve η in D_{2N_1} so that η lies to the left of R_{2N_1} and runs from the top of D_{2N_1} to its bottom. See figure (4.4).

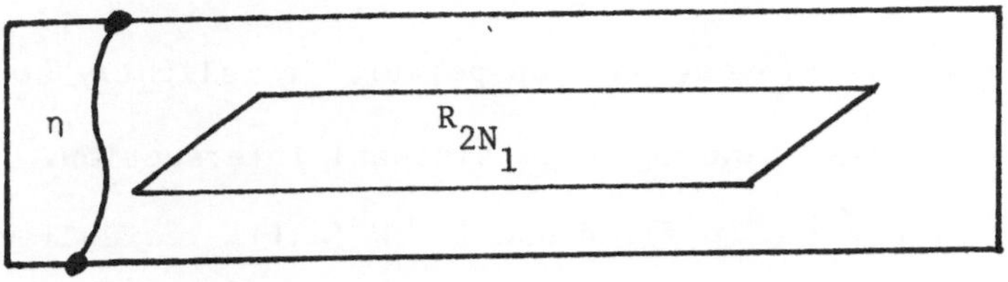

Figure (4.4)

If we choose different ϕ_i's we can make this curve η of attained points for possible g_1's move continuously across D_{2N_1} from left to right covering all of R_{2N} . Thus for some g_1 , we can make $g_1^{n_{2N_1}-n+1}(f^{n-1}y) = f^{n_{2N_1}-n+1}(y)$. This makes $g_1(f^{n-1}y)$ periodic near x for g_1 and completes the proof. For more details and the higher dimensional case see [7].

The next result is

Theorem (4.10)(Takens) There is a residual set $\mathcal{B} \subset D^1_\omega(M)$ *such that if* $f \in \mathcal{B}$, *and* p *is a hyperbolic periodic point of* f , *then* $W^u(p) \cap W^s(p)$ *is dense in* $W^u(p)$ *and* $W^s(p)$.

The main step is

(4.11) Given $f \in D^1_\omega(M)$, p a hyperbolic periodic point of f , $y \in W^u(p)$, $\varepsilon > 0$, and a neighborhood U of f in $D^1_\omega(M)$, there is a $g \in U$ such that p is a hyperbolic periodic point of g and $W^s(p,g)$ has a point z of intersection with $W^u(p,g)$ ε-close to y in $W^u(p,g)$.

Once (4.11) is proved, one can perturb g slightly so that z becomes a transverse (and hence persistent) intersection. Then, one gets $W^u(p,f) \cap W^s(p,f)$ dense in $W^u(p,f)$ residually in a standard way as follows. Let $\{p_i\}_{i=1}^{\infty}$ be the set of hyperbolic periodic points of f , and for each i , let $\{x_{i_j}\}_{j=1}^{\infty}$ be a countable dense set in $W^u(p_i)$. Let $\mathcal{B}_{n,m,s}$ be the set of f's for which each hyperbolic periodic point p_i with $1 \le i \le n$ has a transverse intersection z_{i_j} with $W^s(p_i)$ within $\frac{1}{s}$ of x_{i_j} in $W^u(p_i)$ for $1 \le j \le m$. Then , $\mathcal{B}_{n,m,s}$ is dense and open in $D^1_\omega(M)$, and, hence, $\mathcal{B} = \bigcap_{n,m,s} \mathcal{B}_{n,m,s}$ is residual. For $f \in \mathcal{B}$, $W^u(p) \cap W^s(p)$ is dense in $W^u(p)$ for each p . Repeating the argument for f^{-1} to get $\mathcal{B}^1$, and taking $\mathcal{B} \cap \mathcal{B}^1$ proves theorem (4.10).

Now we proceed to prove (4.11) when p is a fixed point and $f|W^u(p) \cup W^s(p)$ preserves orientation. The proof for arbitrary period. is (unlike in Pixton's theorem) a straightforward generalization.

<u>*Lemma*</u> *Let* m *denote the normalized Lebesque measure induced by the area form* ω . *Suppose* E *and* F *are two measurable sets with* $m(E) = m(F) > 0$ *and suppose for each* $x \in E$ *there is an integer* $n(x) > 0$ *such that* $f^{n(x)}(x) \in F$. *Let* $s(x)$ *be the least such integer, and set* $P(x) = f^{s(x)}(x)$. *Then* $P : E \longrightarrow F$ *preserves* m ; *i.e. for any measurable set* $A \subset E$, $m(PA) = m(A)$.

<u>Proof</u> For each integer $n > 0$, let $E_n = \{x \in E : s(x) = n\}$ Then , $E = \bigcup_{n \geq 1} E_n$, and $E_n \bigcap E_t = \phi$ for $n \neq t$. Also, P is injective , so $P(\bigcap_\alpha A_\alpha) = \bigcap P(A_\alpha)$, $P(\bigcup_\alpha A_\alpha) = \bigcup_\alpha P(A_\alpha)$ for any collection $\{A_\alpha\}$ of subsets of E .

Now, $P|E_n = f^n|E_n$ preserves m . Thus, if $A \subset E$ is measurable, then

$$A = \bigcup_{n=1} (A \bigcap E_n) \ , \ m(A) = \sum_{n=1}^{\infty} m(A \bigcap E_n)$$

$$= \sum_{n=1}^{\infty} m(P(A \bigcap E_n)) = \sum_{n=1}^{\infty} m(P(A) \bigcap P(E_n))$$

$$= m(\bigcup_{n=1}^{\infty} (P(A) \bigcap P(E_n))) = m(P(A) \bigcap P(\bigcup_{n=1}^{\infty} E_n))$$

$$= m(P(A) \bigcap P(E)) = m(P(A \bigcap E)) = m(PA) \ .$$

Now, let $\varepsilon > 0$ and consider a neighborhood U of p in M in which f is nearly linear. Choose local coordinates (u,v) with $(0,0)$ corresponding to p such that $(u=0) \subset W^u(p)$ and $(v=0) \subset W^s(p)$. Let $x = (0,v_0) \in W^u(p)$ with $v_0 > 0$ and let

$(u_1,0)$, $(-u_1,0)$ be points in $W^s(p)$ such that $f(u_1,0) = (\tilde{u}_1,0)$, $f(-u_1,0) = (w_1,0)$.

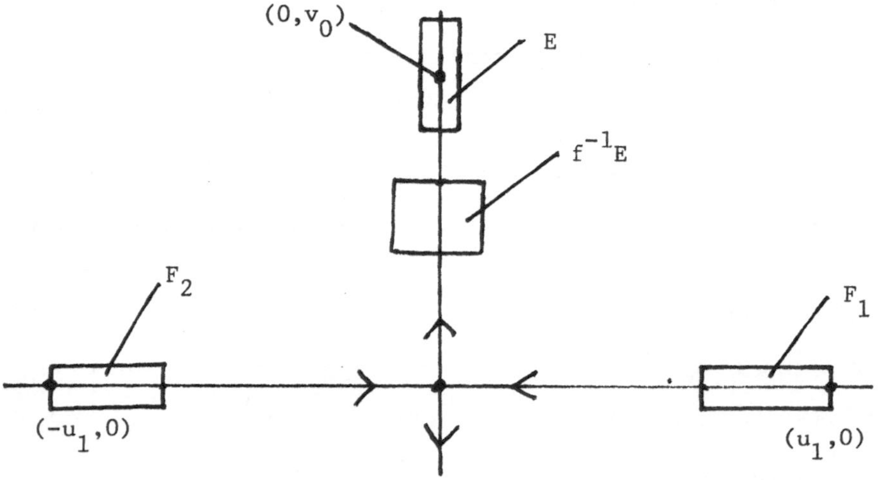

Figure (4.5)

We assume for simplicity that f is linear in U of the form $f(u,v) = (\lambda u, \lambda^{-1}v)$, $0 < \lambda < 1$, so $\tilde{u}_1 = \lambda u_1$, $w_1 = -\lambda u_1$. Let E be a rectangle centered at (o,v_0) as in figure (4.5) with height ε and width $2\lambda^n u_1$, and let $F = F_1 \cup F_2$ be the union of two rectangles symmetric about $[\lambda u_1, u_1]$, $[-u_1, -\lambda u_1]$ in $W^s(p)$ as indicated. Assume $\varepsilon < 1$ and $\varepsilon^2 < 16 v_0(1-\lambda)$. If the height of F_i is $4v_0\lambda^n$, then each point of E which returns to E in the future passes through F first. As almost all points of E are forward recurrent there is a first return map

$P : \tilde{E} \longrightarrow F$ where $\tilde{E} \subset E$ and $m(\tilde{E}) = m(E)$. Assume n large enough so that

$$(1) \quad 2\lambda^n u_1 < \frac{\varepsilon}{4} \ .$$

Consider a box $E_1 \subset E$ centered at $(0,v_0)$ with height $\frac{\varepsilon}{2}$ and width $\varepsilon\lambda^n u_1$. This is depicted in figure (4.6) .

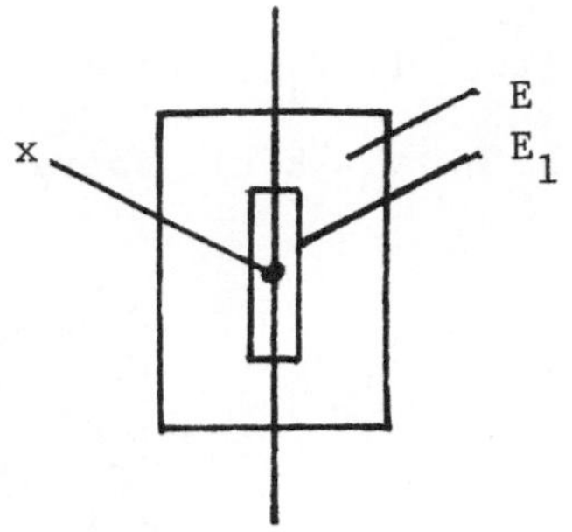

Figure (4.6)

Given any $y \in E_1$ there is a ϕ ε-C^1close to id supported in E such that $\phi^{-1}(y) \in W^u(p)$ because

$$\frac{\frac{1}{2}\text{ width }(E_1)}{\frac{1}{2}\text{ width }(E)} = \frac{\frac{\varepsilon}{2}\lambda^n u_1}{\lambda^n u_1} = \frac{\varepsilon}{2}$$

Now $m(E_1) = \frac{\varepsilon^2}{2}\lambda^n u_1$, and $m(F) = 2[4v_0\lambda^n(1-\lambda)u_1]$.

Thus, $P(\tilde{E} \cap E_1) \subset F$ has measure $m(E_1)$ and, so, for at least one of $i = 1,2$, $m(P(\tilde{E} \cap E_1) \cap F_i) \geq \frac{1}{2}m(E_1)$. Assume $i = 1$, the other case being similar. Then,

$$\frac{m(P(\tilde{E} \cap E_1) \cap F_1)}{m(F_1)} \geq \frac{\frac{1}{2}m(E_1)}{m(F_1)} = \frac{\frac{1}{2}\frac{\varepsilon^2}{2}\lambda^n u_1}{4v_0\lambda^n(1-\lambda)u_1} = C\varepsilon^2 \; , \; C\varepsilon^2 < 1.$$

So $P(\tilde{E} \cap E_1)$ covers a fixed proportion of F_1 independent of n. Let $\tilde{F}_1$ be a subrectangle of F_1 centered at the center of F_1 homothetically related to F_1 so that $m(\tilde{F}_1) = (1 - \frac{C\varepsilon^2}{2})\, m(F_1)$. Then, if $h_1 = \text{dist}(\text{sides of } \tilde{F}_1, \text{sides of } F_1)$ and $h_2 = \text{dist}(\text{top of } \tilde{F}_1, \text{top of } F_1)$, we have $h_1 = \frac{u_1}{2}(1-\lambda)k(\varepsilon)$ and $h_2 = 2v_0\lambda^n k(\varepsilon)$ where $k(\varepsilon) = 1 - (1 - \frac{C\varepsilon^2}{2})^{\frac{1}{2}}$. See figure (4.7).

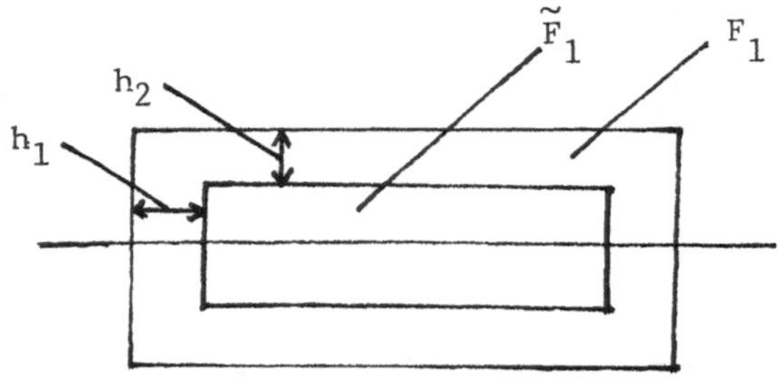

Figure (4.7)

Note that $P(\tilde{E} \cap E_1) \cap \tilde{F}_1 \neq \phi$.

<u>Claim:</u> For n large depending on $\varepsilon, \lambda, u_1, v_0$, and any $z \in \tilde{F}_1 = \tilde{F}_1(n)$, there is a g ε-C^1 close to f such that

(1) $g(w) = f(w)$ off $0_0^n(F_1) = \bigcup_{0 \leq j \leq n} f^j(F_1)$

(2) $z \in W^s(p,g)$.

Once the claim is proved, we fix such an n and choose $z_1 \in \tilde{E} \cap E_1$ such that $P(z_1) \in \tilde{F}_1$. We then use the claim to find g such that $P(z_1) \in W^s(p,g)$. Then we find g_1 with $g_1 = g$ off E such that $z_1 \in W^u(p,g)$ and theorem (4.10) is proved.

<u>Proof of claim:</u> Let $d(p,G)$ denote the distance from the point p to the set G, and let $d(G_1,G_2)$ denote the distance from the set G_1 to the set G_2. Let $h_{1,i} = d(\text{sides of } f^i\tilde{F}_1, \text{sides of } f^iF_1)$, and let $h_{2,i} = d(\text{top of } f^i\tilde{F}_1, \text{top of } f^iF_1)$. Let n_ε be the smallest integer greater than $\frac{2}{\varepsilon}$ and let N_1 be such that $n > N_1$ implies $h_{1,i} > \text{height } f^iF_1$ for $1 \le i \le n_\varepsilon$. For instance, choose N_1 such that $\lambda^{n_\varepsilon} \frac{u_1}{2}(1-\lambda)k(\varepsilon) > 2v_0\lambda^{n-n_\varepsilon}k(\varepsilon)$ for $n > N_1$. Let us use the notation $\phi < G$ to mean that ϕ is ε-C^1 close to id, and supp $\phi \subset G$. Fix $n > N_1$. Now, we can take $w \in f^iF_1$ for $1 \le i \le n_\varepsilon$ and find $\phi < f^iF_1$ such that $d(\phi(w), W^s(p)) \le \max(0, d(\phi(w), W^s(p)) - \frac{\varepsilon}{2}h_{2,i})$. Apply this first to $w = f(z) \in f\tilde{F}_1 = f\tilde{F}_1(n)$. Let $\phi_1 < f\tilde{F}_1(n)$ be such that

$$d(\phi_1(z), W^s(p)) \le \max(0, d(f(z), W^s(p)) - \frac{\varepsilon}{2} h_{2,1}).$$

Then, $g_1 = \phi_1 \circ f$ satisfies

$$d(g_1(z), W^s(p)) \le \max(0, d(f(z), W^s(p)) - \frac{\varepsilon}{2} h_{2,1}),$$

and

$$\begin{aligned} d(g_1^2(z), W^s(p)) &\le \max(0, d(f^2(z), W^s(p)) - \frac{\varepsilon}{2} h_{2,2}) \\ &\le \max(0, (1 - \frac{\varepsilon}{2}) h_{2,2}). \end{aligned}$$

Next, take $\phi_2 < f^2\tilde{F}_1$ such that

$d(\phi_2 \circ g_1^2(z) , W^s(p)) \leq \max (0, d(g_1^2(z) , W^s(p)) - \frac{\varepsilon}{2} h_{2,2})$.

Letting $g_2 = \phi_2 \circ g_1$, note that $g_2^2(z) = \phi_2 \circ g_1^2(z)$, so

$$d(g_2^2(z) , W^s(p)) \leq \max (0, d(f^2(z) , W^s(p)) - 2 \cdot \frac{\varepsilon}{2} h_{2,2}) ,$$

and

$$d(g_2^3(z) , W^s(p)) \leq \max (0, d(f^3(z) , W^s(p)) - 2 \cdot \frac{\varepsilon}{2} h_{2,3})$$
$$\leq \max (0, (1 - 2 \cdot \frac{\varepsilon}{2} h_{2,3}) .$$

Similarly, letting $g_i = \phi_i \circ g_{i-1}$ with $\phi_i < f^i\tilde{F}_1(n)$,

we can get

$$d(g_i^{i+1}(z) , W^s(p)) \leq \max (0, (1 - i \frac{\varepsilon}{2}) h_{2,i+1}).$$

Setting $g = g_{n_\varepsilon}$ proves the claim.

References

1. G. D. Birkhoff, Some unsolved problems of theoretical dynamics, Collected Mathematical Papers of George David Birkhoff, Vol. II, Amer. Math. Soc. 1951, 710-712.

2. S. Newhouse, Quasi-elliptic periodic points in conservative dynamical systems, Amer. J. of Math. 99 (1975), 1061-1087.

3. ___________, Topological entropy and Hausdorff dimension for area preserving diffeomorphisms of surfaces, Asterisque 51 (1978), 323-335.

4. ___________, Generic properties of conservative systems, Chaotic Behavior of Determinisitc Systems, Les Houches, Session XXXVI (1981), G. Iooss, R. H. G. Helleman, R. Stora, eds., North-Holland Publishing Co.,1983, 443-451.

5. Ja. Pesin, Characteristic Lyapunov exponents and smooth ergodic theory, Russ. Math. Surveys 32 (1977), 55-114.

6. H. Poincaré, Methodes Nouvelles de la Mécanique Célèste Vols. I,II.

7. C. C. Pugh and R. C. Robinson, The C^1 closing lemma including Hamiltonians, J. Ergodic Theory and Dynamical Systems, to appear.

8. F. Takens, Homoclinic Points in conservative Systems, Inventiones Math. 18 (1972), 267-292.

CLASSICAL MECHANICS AND RENORMALIZATION GROUP

Giovanni Gallavotti

Istituto Matematico
I[a] Università di Roma
00185 Roma, Italia

ABSTRACT

The theory of Kolmogorov-Arnold-Moser (KAM) is discussed in detail from the point of view of the "renormalization group approach". Similarly we discuss some aspects of the problem of the existence of universal structures in the chaotic transition. The quasi-periodic Schroedinger equation in one dimension is discussed as a special case.

§1. Perturbations of integrable systems

A classical hamiltonian system is said to be integrable on a region W of phase space if:

i) the region W can be mapped via a canonical map $\mathcal{C}$ onto a region of the form $V \times T^{\ell}$ where $V \subset R^{\ell}$ is an open set and T^{ℓ} is the ℓ-dimensional torus.

ii) if the map $\mathcal{C}$ is such that

$$(\underline{p},\underline{q}) = \mathcal{C}(\underline{A},\underline{\phi}) \tag{1.1}$$

then the hamiltonian H, thought as a function h of the $(\underline{A},\underline{\phi})$ variables, becomes a function of the A's alone

$$H(\mathcal{C}(\underline{A},\underline{\phi})) = h(\underline{A}) \tag{1.2}$$

A hamiltonian system with hamiltonian H considered on a phase space region W will be denoted (H,W); in what follows I shall only consider hamiltonian systems and canonical maps of class C^{∞} or analytic.

The notion of canonical map used here will be that usually called "action preserving completely canonical maps" [1] (§3.12 and p. 240): namely a map $\mathcal{C}$, (of class C^∞ at least with inverse of the same regularity), between W and W' is such a map if and only if there is a C^∞-function Φ defined on the graph $G(\mathcal{C})$

$$G(\mathcal{C}) = \{\underline{p},\underline{q},\underline{p}',\underline{q}' \mid (\underline{p},\underline{q}) = \mathcal{C}(\underline{p}',\underline{q}'),\ (\underline{p},\underline{q},\underline{p}',\underline{q}') \in W \times W' \tag{1.2}$$

such that

$$\underline{p} \cdot d\underline{q} = \underline{p}' \cdot d\underline{q}' + d\Phi \qquad \text{on} \quad G(\mathcal{C}) \tag{1.3}$$

If $\mathcal{C}$ is canonical in the above sense and if λ is a closed curve in W' then

$$A(\lambda) = \oint_{\mathcal{C}(\lambda)} \underline{p} \cdot d\underline{q} = \oint_{\lambda} \underline{p}' \cdot d\underline{q}' \tag{1.4}$$

More generally if (H,W), (H',W') are two C^∞ hamiltonian systems one defines the notion of relative integrability: H' is H-integrable on W if there exists a canonical map $\mathcal{C}: W \leftrightarrow W'$ and a C^∞-function F such that

$$H'(\mathcal{C}^{-1}(\underline{p},\underline{q})) = F(H(\underline{p},\underline{q}),\ A_2(\underline{p},\underline{q}),\dots,A_s(\underline{p},\underline{q})) \tag{1.5}$$

where $A_2,\dots,A_s$ is a maximal set of prime integrals in involution with each other and F is also supposed to be invertible with respect to its first argument.

The preceding notion of integrability is a particular case of the general relative integrability, just introduced, and it will be referred to as "integrability by quadratures"; in terms of the relative integrability it can be described as follows: a system (H,W) is integrable by quadratures if (H,W) is integrable with respect to a "free rotators' system". In fact, if (1.2) holds, then in the $(\underline{A},\underline{\phi})$ variables the system moves as

$$\begin{aligned} \dot{\underline{A}} &= \underline{0} \\ \dot{\underline{\phi}} &= \frac{\partial h}{\partial A}(A) \equiv \underline{\omega}(\underline{A}) \end{aligned} \tag{1.7}$$

i.e. each angle rotates uniformly.

In this connection it is useful to recall a theorem "of Arnold-Liouville" which, under some technical assumptions of invertibility of certain changes of coordinates, states that a necessary and sufficient condition of integrability by quadratures of (H,W) is the existence of ℓ independent (in the sense of the rank of the Jacobian matrix) prime integrals in involution $I_1 \equiv H, I_2,\dots,I_\ell$ defined on W and such that the surfaces $I_1 = i_1,\dots,I_\ell = i_\ell$ are compact and their union is W itself, (ℓ being the number of degrees of freedom), [2] (see p 269), (see also appendix A).

Since the Greek times it was firmly believed that the most general motion of a system would be described by equations (1.7) ("motion by epicycles") and only recently it has been realized by the physicists that this is not the case (starting late in the 1800's with Boltzmann and Poincaré).

The surprise has been particularly strong in the case of the theory of systems of the form

$$H_\varepsilon(\underline{A},\underline{\phi}) = h(\underline{A}) + \varepsilon\, f(\underline{A},\underline{\phi}) \tag{1.8}$$

with $(A,\phi) \in V \times T^\ell$, V being a sphere (say). This is a system that one would naively believe to move as the imperturbed rotators. In this case one would have expected the existence of an analytic family of canonical maps $\mathscr{C}_\varepsilon$, defined on $W_\varepsilon = V_\varepsilon \times T^\ell$, $V_\varepsilon \subset V$, and of functions h_ε on $V_\varepsilon \subset V$ such that

1) $\mathscr{C}_\varepsilon : W_\varepsilon \to W$; $\mathscr{C}_\varepsilon$-(identity) $\xrightarrow[\varepsilon\to 0]{} 0$

$h_\varepsilon : V_\varepsilon \to R$; $h_\varepsilon - h \xrightarrow[\varepsilon\to 0]{} 0$

with $\mathscr{C}_\varepsilon$,h_ε analytic in $\varepsilon,\underline{A},\underline{\phi}$.

2) $\mathrm{Vol}\ V_\varepsilon/\mathrm{vol}\ V \xrightarrow[\varepsilon\to 0]{} 1$ (1.9)

3) $H_\varepsilon(\mathscr{C}_\varepsilon(\underline{A}',\underline{\phi}')) \equiv h_\varepsilon(\underline{A}')$ on W_ε

4) $\max\limits_{\underline{A}\in\partial V_\varepsilon} d(\underline{A}',\partial V) \xrightarrow[\varepsilon\to 0]{} 0$

the last property means that V and V_ε differ only because of boundary effects (a difference to be obviously expected).

However, there are easy counter examples:

a) Let $h(\underline{A}) = A_1$ and consider

$$A_1 + \varepsilon\, f(A_2,\dots,A_\ell,\phi_2,\dots,\phi_\ell) \tag{1.10}$$

In this case it is obvious that on a time scale ε^{-1} the evolution of the $A_2,\dots\phi_2,\dots$ variables is entirely determined by the perturbation of f so, if one takes f to be such that the motion governed by the hamiltonian f are not quasi-periodic, the perturbed system is not integrable.

This case is a very pathological one and is called a "fully resonant case" because all the motions of the unperturbed system are periodic. More generally the latter situation is realized every time the unperturbed hamiltonian has the form $\underline{\omega} \cdot \underline{A}$ with $\omega \in R^{\ell}$ having components with rational ratios. Then a situation like the one in the above example arises provided the f depends only on suitable linear combinations of the coordinates

Exercises: 1) show that if ω_i/ω_j is rational then by a linear canonical change of coordinates one can change $\underline{\omega} \cdot \underline{A}$ into $\Omega A'_1$ for some Ω.

2) extend the above problem to the case where there exists $\underline{r} \in Z^{\ell}$, $\underline{r} \neq \underline{0}$, such that $\underline{\omega} \cdot \underline{r} = 0$ but $\omega_1 r_1 + \omega_2 r_2 \neq 0$ for all $(r_1, r_2) \in Z^2$.

3) extend the counter-example in (1.10) to the case when the unperturbed system is the linear system in problem 2).

b) Another type of counter-example to (1.9) is obtained by considering

$$\omega_1 A_1 + \omega_2 A_2 + \varepsilon(A_2 + f(\phi_1, \phi_2)) \qquad |A_1|, |A_2| < 1 \tag{1.11}$$

This system is not integrable in the sense (1.9) even when $\underline{\omega}$ is such that for some C, $\alpha > 0$

$$|\omega_1 r_1 + \omega_2 r_2|^{-1} < C(|r_1|+|r_2|)^{\alpha} \qquad \forall \underline{r} \in Z^2, \underline{r} \neq \underline{0} \tag{1.12}$$

(i.e. "$\underline{\omega}$ is (C,α)-non-resonant"), provided the Fourier transform $f_{r_1 r_2}$ of f never vanishes for $\underline{r} \neq 0$. This is seen by solving explicitly the (trivial) equations of motion and checking that not all the motions are quasi-periodic, at least for the set of values of ε, of full measure, for which $\omega/1+\varepsilon$ is rational (exercise).

c) A classic counter-example due to Poincaré [3] is the system

$$H_{\varepsilon}(\underline{A}, \underline{\phi}) = \frac{1}{2}(A_1^2 + A_2^2) + \varepsilon f(A_1, A_2, \phi_1, \phi_2) \qquad |A_j| < 1 \tag{1.13}$$

with $\hat{f}_{r_1 r_2}(A_1, A_2) \neq 0$, $\forall (r_1, r_2) \neq \underline{0}$ ($f_{\underline{r}}(\underline{A})$ denotes the Fourier's transform of $f(\underline{A}, \underline{\phi}) \in C^{\infty}(V \times T^{\ell})$, for $\underline{r} \in Z^{\ell}$).

If one had integrability in the sense (1.9) one could find $\mathcal{C}_\varepsilon$ and call Φ_ε a generating function for it so that if $(\underline{A},\underline{\phi}) = \mathcal{C}_\varepsilon(\underline{A}',\underline{\phi}')$ it is

$$\underline{A} = \underline{A}' + \frac{\partial \Phi_\varepsilon}{\partial \underline{\phi}} (\underline{A}',\underline{\phi})$$

$$\underline{\phi}' = \underline{\phi} + \frac{\partial \Phi_\varepsilon}{\partial \underline{A}'} (\underline{A}',\underline{\phi})$$

and Φ_ε would be defined for ε small and $|A_j| < 1/2$, say, and $\Phi = \varepsilon\Phi + \varepsilon^2\Phi' + \ldots$. Therefore substituting into the "Hamilton-Jacobi equation" $H_\varepsilon(\mathcal{C}_\varepsilon(\underline{A}',\underline{\phi}')) \equiv h_\varepsilon(\underline{A}')$, see (1.9), and developing both sides in powers of ε one would obtain

$$\frac{1}{2}\underline{A}'^2 + \varepsilon(\underline{A}' \cdot \frac{\partial \Phi}{\partial \underline{\phi}} (A',\phi) + f(\underline{A}',\underline{\phi})) + O(\varepsilon^2) = \frac{1}{2}\underline{A}'^2 + \varepsilon h'(\underline{A}') + O(\varepsilon^2)$$

i.e.

$$\underline{A}' \cdot \frac{\partial \Phi}{\partial \underline{\phi}} (\underline{A}',\underline{\phi}) = - f(\underline{A}',\underline{\phi}) + h'(\underline{A}') \qquad (1.15)$$

which, by integrating both sides with respect to $\underline{\phi}$, implies

$$h'(\underline{A}') = \frac{1}{(2\pi)^\ell} \int_{T^\ell} f(\underline{A}',\underline{\phi})d\phi \equiv \hat{f}_{\underline{0}}(A') \qquad (1.16)$$

where $\hat{f}_{\underline{r}}(\underline{A}')$ denotes the Fourier transform of f, for $\underline{r} \in Z^\ell$. Then by taking the Fourier transform of both sides one finds

$$i\, \underline{A}' \cdot \underline{r}\, \hat{\Phi}_r(\underline{A}') = - \hat{f}_{\underline{r}}(\underline{A}') \qquad \underline{r} \neq \underline{0} \qquad (1.17)$$

which is impossible because $\hat{f}_{\underline{r}}(\underline{A}')$ never vanishes while near every $\underline{A}_0'$ there exist values of $\underline{A}'$ such that $\underline{A}' \cdot \underline{r} = 0$ for some $\underline{r} \neq \underline{0}$!

The above counter-examples' ideas can be combined in various ways to provide other strange examples. However, they basically exhaust the set of all the possible pathologies.

The following theorem illustrates the above statement [4,5].

Theorem: If the unperturbed system h associated with $H_\varepsilon(\underline{A},\underline{\phi}) = h(\underline{A}) + \varepsilon f(\underline{A},\underline{\phi})$ verifies suitable "non-resonance conditions" one can define $\mathcal{C}_\varepsilon$, h_ε of class C^∞ on $V \times T^\ell$ or V, respectively, and $C > 0$, so that

1) $\mathcal{C}_\varepsilon$-(identity) $= 0(\varepsilon)$, $h_\varepsilon - h = 0(\varepsilon)$

2) there exists $V_\varepsilon \subset V$ where*

$$H_\varepsilon(\mathcal{C}_\varepsilon(\underline{A}',\underline{\phi}')) \overset{C^\infty}{=} h_\varepsilon(\underline{A}') \qquad \forall(\underline{A}',\underline{\phi}') \in V_\varepsilon \times T^\ell \tag{1.18}$$

3) vol V_ε/vol $V \underset{\varepsilon\to 0}{\longrightarrow} 1$

4) $V_\varepsilon \supset \{\underline{A}' \mid |\frac{\partial h_\varepsilon}{\partial \underline{A}'}(\underline{A}')\cdot\underline{r}|^{-1} < C_\varepsilon|\underline{r}|^\ell , \quad \forall \underline{r} \in Z^\ell, \underline{r} \neq \underline{0}\}$

and $C_\varepsilon \to +\infty$ as $\varepsilon \to 0$.

There is a variety non-resonance condition under which the above theorem holds. One condition ("an isochrony" or "twist" condition)

$$\det \frac{\partial^2 h}{\partial\underline{A}\partial\underline{A}}(\underline{A}) \neq 0 \qquad \text{on } V \tag{1.19}$$

and it is certainly the simplest. Another condition, valid when $h(\underline{A}) = \underline{\omega}\cdot\underline{A}$, $\underline{\omega} \in R^\ell$, is the existence of $C, \alpha > 0$ such that

$$|\underline{\omega}\cdot\underline{r}|^{-1} < C|\underline{r}|^\alpha \qquad \forall \underline{r} \neq \underline{0} \tag{1.20}$$

and

$$\det \frac{\partial^2 \overline{f}}{\partial\underline{A}\partial\underline{A}}(\underline{A}) \neq 0 \qquad \text{on } V \tag{1.21}$$

where $\overline{f}$ is the average of f over the $\underline{\phi}$-variables.

But there are other conditions which one can impose: the cases studied so far all have the common feature of putting conditions allowing manipulations turning the problem into a problem in which (1.19) is verified. The most brilliant example is provided by the deep results on the three-body problem in [6].

The problem solved by the KAM theorem has a lot of connections with other problems, some of which are of a non-mechanical nature.[7]

*If f,g are two C^∞-functions and S is a closed set one says that $f \overset{C^\infty}{=} g$ on S, or f = g on S in the C^∞-sense, if all their derivatives coincide on S.

§2 Integrability, Wick Ordering and Quasi-periodic Schroedinger Equation

Here we provide an example of a problem (and a related application) which arises in the context of the theory of the systems integrable by quadratures but which is of a somewhat different nature compared to the problem studied by the KAM theorem.

In this problem the connection with the theory of renormalization appears in a different light compared to the one related to the KAM theorem and in these notes I would like to discuss both in some detail.

A problem related to the one of finding non-resonance conditions which make the conclusions of the theorem of §1 valid is that of finding sufficient conditions insuring that H_ε is integrable by quadratures, see (1.8).

Typically, given f, one looks for an operator associating with f a new perturbation $\delta_\varepsilon f$ such that

$$H_\varepsilon(\underline{A},\underline{\phi}) = h(\underline{A}) + \varepsilon\, f(\underline{A},\underline{\phi}) + \delta_\varepsilon\, f(\underline{A},\underline{\phi}) \tag{2.1}$$

is integrable with respect to h itself on a large subset W_ε of the phase space (large in the sense of $(1.9)_4$). One has to think of the perturbations integrable in the above sense with respect to the unperturbed system as forming a "manifold" M_1 in the space of the hamiltonians. Then the operation of replacing the original hamiltonian by

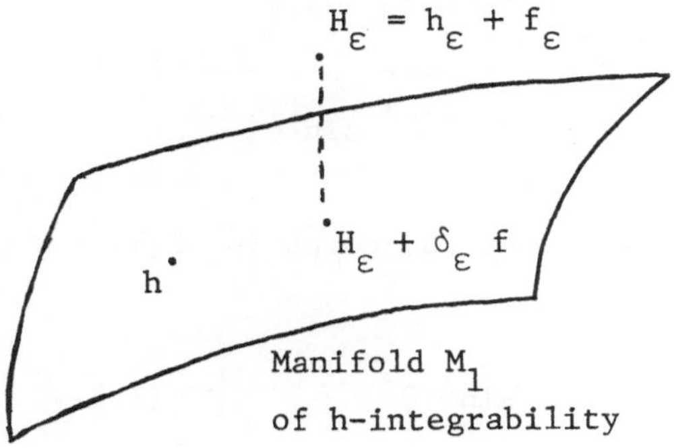

$H_\varepsilon + \delta_\varepsilon f$ is a sort of "projection" of f on M_1 .

Fig. 2

In the applications the "counterterm" $\delta_\varepsilon f$ is not arbitrary (obviously one could always take $\delta_\varepsilon f = -\varepsilon f$!) but its form is restricted by the application itself, e.g. $\delta_\varepsilon f(\underline{A},\underline{\phi})$ can be restricted to be $\underline{\phi}$-independent (in this case the operation of replacing H_ε by $\overline{H_\varepsilon + \delta_\varepsilon f}$ is called "Wick ordering").

The best example where the above problem arises is in an apparently unrelated problem, i.e. in the theory of the Schroedinger equation in a quasi-periodic potential. Consider the equation

$$-\ddot{q} + \varepsilon V(\underline{\omega}t)q = E\,q \qquad t \in R \tag{2.2}$$

where $\underline{\phi} \to V(\underline{\phi}) \geq 0$ is analytic on T^ℓ and $\underline{\omega} \in R^\ell$ verifies (1.20) ("Diophantine condition"), and $E > 0$ is a parameter. This is the Schroedinger equation with quasi-periodic potential $V(\underline{\omega}t)$ on the line. The abscissa on the line, which is usually called x, is called here t.

One interprets the above system as a mechanical system on $R^{1+\ell} \times (R\times T^\ell)$ with the hamiltonian

$$H(p,\underline{B},q,\underline{\phi}) = \frac{p^2}{2} + (E - \varepsilon V(\underline{\phi}))\frac{q^2}{2} + \underline{B}\cdot\underline{\omega} \tag{2.3}$$

where $(p,\underline{B},q,\underline{\phi}) \in R^{1+\ell} \times (R\times T^\ell)$ are canonical variables.

Any solution of (2.2) describes a motion of (2.3) with $\underline{\phi}(0) = \underline{0}$ and vice versa.

For the coming discussion it is useful to change variables replacing (p,q) with the action angle variables of the harmonic oscillator $p^2+Eq^2/2$, which we call (A,θ): $\underline{B}' \equiv \underline{B}$, $\underline{\phi}' \equiv \underline{\phi}$ and

$$A = \frac{1}{2}\frac{1}{\sqrt{E}}(p^2+Eq^2) \qquad p = \sqrt{2A\sqrt{E}}\cos\theta$$
$$\theta = \operatorname{arccotg}\frac{q\sqrt{E}}{\sqrt{p^2+Eq^2}} \qquad q = \frac{\sqrt{2A\sqrt{E}}}{\sqrt{E}}\sin\theta \tag{2.4}$$

(i.e. $(\sqrt{2A\sqrt{E}},\theta)$ are the polar coordinates of the point (p, Eq)).

Then (2.3) becomes

$$\sqrt{E}\,A + \underline{B}\cdot\underline{\omega} - \frac{\varepsilon}{\sqrt{E}}A\,V(\underline{\phi})\sin^2\theta \tag{2.5}$$

and since the map (2.4) is canonical the solutions of (2.5) with $\underline{\phi}(0) = \underline{0}$ and the solutions of the quasi-periodic Schroedinger equation (2.2) are in one-to-one correspondence (easily explicitly written by using (2.4)).

We now consider the problem of finding the values of E for which the Eq. (2.2) admits a quasi-periodic, hence bounded, solution. It is the problem of finding values of E for which the quasi-periodic Schroedinger equation admits "Block waves". This problem is closely related to that of finding the continuous spectrum of the Schroedinger operator with potential $V(\underline{\omega}t)$.

One can clearly find a subset of such values of E by imposing that (2.5) is integrable by quadratures. A way to impose integrability of (2.5) is the following. Let λ be a fixed real number and let $\sqrt{E} = \lambda+a$, $\eta = \varepsilon/(\lambda+a)$ and rewrite (2.5) as

$$\lambda A + \underline{\omega} \cdot \underline{B} - \eta A V(\underline{\phi}) \sin^2\theta + aA \qquad (2.6)$$

Then we can try to impose integrability by fixing conveniently a as a function of η, λ, if this is possible. So in the above context we are led to the problem of determining whether a counterterm of the form $\delta\eta = Aa$ turns the system $\lambda \underline{A} + \underline{\omega}\cdot\underline{B} - \eta A V(\phi)\sin^2\theta$ into an integrable system on, say, the whole phase space $R^{\ell+1} \times T^{\ell+1}$, for η small at least.

If λ is chosen so that for some $\alpha > 0$, $c > 0$

$$|\lambda\mu + \underline{\omega} \cdot \underline{r}|^{-1} \leq C(|\mu|+|\underline{r}|)^{\alpha} \qquad (2.7)$$

$\forall(\mu,\underline{r}) \in Z^{\ell+1}$, $(\mu,\underline{r}) \neq \underline{0}$, then it turns out, as we shall show, that the above choice of a is possible for η small and $a(\eta,\lambda)$ is analytic in η for η small so that

$$a(\eta,\lambda) = \alpha(\lambda)\eta + \beta(\lambda)\eta^2 + \dots \quad , \ \alpha(\lambda) \neq 0 \qquad (2.8)$$

This means that the original Schroedinger equation has the value

$$E = (\lambda + a_\varepsilon)^2 \qquad (2.9)$$

for which there exist Bloch waves, provided a_ε verifies

$$a_\varepsilon = a\left(\frac{\varepsilon}{\lambda+a_\varepsilon}, \lambda\right) \qquad (2.10)$$

which always has a solution for small ε (because $\alpha(\lambda) \neq 0$: actually $\alpha(\lambda) = \{\text{average value of } V(\phi)\cos^2\theta\} \neq 0$, from $V > 0$).

The above-mentioned result is a consequence of the following theorem of which we shall present a proof in these notes [8,9].

<u>Theorem:</u> Suppose that $\underline{\omega} \in R^\ell$ verifies a diophantine condition, (1.20), and let $\underline{f}$ be an R^ℓ-valued analytic function on T^ℓ. Then if ε is small enough there is an analytic function $\varepsilon \to \underline{a}(\varepsilon)$ such that the hamiltonian system described on $R^\ell \times T^\ell$ by

$$\tilde{H}_\varepsilon(A,\phi) = \underline{\omega}\cdot\underline{A} + \varepsilon\underline{A}\cdot f(\phi) + \underline{A}\cdot\underline{a}(\varepsilon) \qquad (2.11)$$

is integrable by quadratures on the whole phase space.

Actually, the above theorem is a slight generalization of [8,9] and is proved in a somewhat different spirit, along the lines of [10]; the only other known result about counterterms is [10], the following

<u>Theorem:</u> Consider $\underline{\omega} \in R^\ell$ verifying a diophantine condition, (1.20) and let f be analytic on $V_r \times T^\ell$ with $V_r = \{\underline{A}|\underline{A} \in R^\ell, |A_j|<r\} \subset R^\ell$. Consider

$$H_\varepsilon(\underline{A},\underline{\phi}) = \underline{A}\cdot\underline{\omega} + \varepsilon f(\underline{A},\underline{\phi}) \qquad (2.1\ )$$

then there is at most one analytic function $N(\underline{A},\varepsilon)$, analytic in $A \in V_r$ and ε for ε small such that $H_\varepsilon - N$ is canonically conjugate to $\underline{\omega}\cdot\underline{A}$ in $V_{r'} \times T^\ell$, $r' < r$, for ε small enough.

Furthermore, there exists a sequence of functions $N_k(A)$ analytic in $\underline{A} \in V_r$, $k = 1,2,\dots$, such that N exists if and only if the following series

$$\sum_{k=1}^{\infty} \varepsilon^k N_k(\underline{A}) \qquad (2.13)$$

converges for small ε. Its sum is then N.

The above result shows that, sometimes, the operator $\delta_\varepsilon f$ is uniquely defined. (Exercise: Compute explicitly the series $\Sigma_k N_k \varepsilon^k$ in the case $\sqrt{2}\, A_1+A_2 + \varepsilon(A_2+f(\phi_1,\phi_2))$).

§3. Existence of Quasi-periodic motions

The KAM theorem admits many proofs. I shall present a proof formulated in the language of the renormalization group, of a slightly weaker version ("Kolmogorov theorem") in which one only shows the existence of a quasi-periodic motion with a given frequency spectrum.

I shall only consider an analytic hamiltonian $h_0(\underline{A})$ with h_0 holomorphic in

$$S_{\rho_0} = \{\underline{A}|\underline{A} \in C^\ell, \ |A_j| < \rho_0, \forall_j\} \tag{3.1}$$

with an analytic perturbation f_0, analytic on $V_\rho \times T^\ell$ where

$$V_\rho = \{\underline{A} \ |A \in R^\ell , \ \ |A_j| < \rho\} \tag{3.2}$$

The analyticity of f_0 will be imposed by demanding that $(\underline{A},\underline{\phi}) \to f_0(\underline{A},\underline{\phi})$ regarded as a function of $A \in V$ and of $\underline{z} = (z_1,\ldots,z_\ell) \equiv (e^{i\phi_1},\ldots,e^{i\phi_\ell})$ is holomorphic in a set of the form

$$C(\rho_0,\xi_0) = S_{\rho_0} \times C(\xi_0) \subset C^{2\ell} \tag{3.3}$$

where S_{ρ_0} is defined in (3.1) and

$$C(\xi_0) = \{\underline{z}|\underline{z} \in C^\ell, \ \ e^{-\xi_0} < |z_j| < e^{\xi_0}\} \tag{3.4}$$

Let us introduce a few parameters measuring the size of various quantities

$$||\underline{W}|| = \sup_j |W_j| \ , \qquad |\underline{W}| = \sum_{j=1}^{p} |W_j| \qquad \forall \underline{W} \in C^p \tag{3.5}$$

and

$$E_0 = ||\frac{\partial h_0}{\partial \underline{A}}||_{\rho_0} \equiv \sup_{A \in S_{\rho_0}} ||\frac{\partial h_0}{\partial \underline{A}} (A)|| \tag{3.6}$$

and E_0^{-1} "measures the time scale" on which the unperturbed (or "free") system evolves. The quantity (in computing norms we regard a matrix $\ell\times\ell$ as a vector in C^p, $p = \ell^2$ and use (3.5)):

$$\eta_0 = ||(\frac{\partial^2 h_0}{\partial \underline{A}\partial \underline{A}})^{-1}||_{\rho_0} \tag{3.7}$$

will be called "resonance mobility" or as "twist parameter" and η_0^{-1} measures how fast the frequencies change as $\underline{A}$ changes, hence how fast $\underline{\omega}(\underline{A})$ moves away from a resonant value as $\underline{A}$ varies. The quantity

$$\varepsilon_0 = \left\|\frac{\partial f_0}{\partial \underline{A}}\right\|_{\rho_0,\xi_0} + \frac{1}{\rho_0}\left\|\frac{\partial f_0}{\partial \underline{\phi}}\right\|_{\rho_0,\xi_0} \tag{3.8}$$

where $\partial/\partial\phi_j \equiv iz_j\partial/\partial z_j$, is related to the time scale on which the perturbation starts revealing itself, which is clearly ε_0^{-1}. Then the following theorem holds.

Theorem: Let h_0 be holomorphic on S_{ρ_0} and suppose that $\underline{\omega}_0 = \partial h_0/\partial \underline{A}\ (\underline{0})$ verifies a non-resonance condition for some $C_0 > 0$

$$|\underline{\omega}_0 \cdot \underline{r}|^{-1} < C_0|\underline{r}|^{\ell} \qquad \forall \underline{r} \in Z^{\ell}, \ |\underline{r}| \neq 0 \tag{3.9}$$

Let f_0 be a perturbation holomorphic in $S_{\rho_0} \times C(\xi_0)$. Consider the hamiltonian system on $V \times T^{\ell}$ with hamiltonian

$$H_0(\underline{A},\underline{\phi}) = h_0(\underline{A}) + f_0(\underline{A},\underline{\phi}) \tag{3.10}$$

Assume, for simplicity, $\varepsilon_0 C_0$, ε_0/E_0, $\xi_0 < 1$.
There exist two functions $\underline{\alpha}$, $\underline{\beta}$ holomorphic on $C(\xi_0/2)$ such that the torus $\mathcal{T}$ with parametric equations

$$\begin{cases} \underline{A} = \underline{\alpha}(\underline{\phi}') \\ \underline{\phi} = \underline{\phi}' + \underline{\beta}(\underline{\phi}') \end{cases} \tag{3.11}$$

is invariant, for ε_0 small, in the flow generated by (3.10) and, actually, this flow is described on σ by $\underline{\phi}' \to \underline{\phi}' + \underline{\omega}_0 t$.

Furthermore $\underline{\alpha}$, $\underline{\beta}$ tend to zero as $\varepsilon_0 \to 0$ and ε_0 small means that there are suitable constants depending only on ℓ, denoted G, a_1, $a_2,\ldots$, $G < 1/2$, such that ε_0 should verify

$$\varepsilon_0 C_0 < G(E_0 C_0)^{-a_1}(E_0\eta_0\rho_0^{-1})^{-a_2}\xi_0^{a_3} \tag{3.13}$$

In other words, in a non-resonant situation ($C_0 < +\infty$, $\eta_0 < +\infty$) a non-resonant torus is not destroyed by a small perturbation in the sense that in its vicinity there is another torus which is invariant for the perturbed flow and is run with a quasi-periodic motion with the same frequencies (i.e. the "same spectrum") as the unperturbed one.

Proof: Introduce a convenient sequence of parameters δ_k, $k = 0, 1, \ldots$, which will be used in the estimates; for instance (so that $\sum_{k=0}^{\infty} 5\,\delta_k = \xi_0/2 > 0$)

$$\delta_k = \xi_0/5\pi(1+k)^2 \tag{3.13}$$

Define the following "renormalization transformation", acting on the pair (h_0, f_0) analytic on $S_{\rho_0} \times C(\xi_0)$ and mapping it in (h_1, f_1) analytic on $S_{\rho_0} \times C(\xi_0 - \delta_0)$ via the following steps.

1) Truncate the Fourier's series of f_0 in two parts

$$f_0 = f_0^{[\leq N_0]} + f_0^{[>N_0]}$$

$$f^{[>N_0]}(\underline{A},\underline{z}) = \sum_{\substack{|\underline{r}|>N_0 \\ \underline{r}\in Z^\ell}} \hat{f}_{\underline{r}}(\underline{A})\, \underline{z}^{\underline{r}} \tag{3.14}$$

where $\underline{z}^{\underline{r}} \equiv z_1^{r_1} \ldots z_\ell^{r_\ell}$, chosing N_0 so that $f^{[>N_0]}$ is (ε_0^2) i.e. for some B_1

$$\left\|\frac{\partial f_0^{[>N_0]}}{\partial \underline{A}}\right\|_{\rho_0,\xi_0-\delta_0} + \frac{1}{\rho_0}\left\|\frac{\partial f_0^{[>N_0]}}{\partial \underline{\phi}}\right\|_{\rho_0,\xi_0-\delta_0} \leq B_1 C_0 \varepsilon_0^2 \tag{3.15}$$

Since the (3.8) implies bounds on the Fourier coefficients $\partial\hat{f}_{0\underline{r}}/\partial\underline{A}\,(\underline{A})$, $|\underline{r}|\hat{f}_{0\underline{r}}(\underline{A})$, as

$$\left\|\frac{\partial \hat{f}_{0\underline{r}}}{\partial \underline{A}}\right\| \leq \varepsilon_0\, e^{-\xi_0|\underline{r}|}, \quad \|\underline{r}\,\hat{f}_{0\underline{r}}(\underline{A})\|_{\rho_0} \leq \varepsilon_0\rho_0\, e^{-\xi_0|\underline{r}|} \tag{3.16}$$

it is easy to see that one can take

$$N_0 = \delta_0^{-1}\log(C_0\varepsilon_0\delta_0^{-\ell})^{-1} \tag{3.17}$$

which if inserted in the l.h.s. of (3.15) provides the bound appearing in (3.16). The number (3.17) will be called the "ultra violet cut-off".

2) Define for $||\underline{A}'|| < \tilde{\rho} < \rho_0/2$

$$\Phi_0(\underline{A}',\underline{z}) = \sum_{0<|\underline{r}|\le N} \frac{\hat{f}_{\underline{r}}(\underline{A}')}{-i\underline{\omega}_0(\underline{A}')\cdot\underline{r}}\, \underline{z}^{\underline{r}} \tag{3.18}$$

where $\tilde{\rho}$ is chosen so small that $\underline{\omega}(\underline{A}')\cdot\underline{r}$ stays away from zero for all $|\underline{r}| \le N_0$, i.e.

$$|\underline{\omega}_0(\underline{A}')\cdot r|^{-1} \le 2\, C_0 |\underline{r}|^{\ell} \qquad ||\underline{A}'|| < \tilde{\rho} \tag{3.19}$$

Using obvious dimensional estimates* on the derivatives of h_0 we find, using $\underline{\omega}(\underline{0}) = \underline{\omega}_0$ and (3.9)

$$|\underline{\omega}_0(\underline{A}')\cdot\underline{r}|^{-1} \equiv |(\underline{\omega}_0 + (\underline{\omega}(A') - \underline{\omega}_0))\cdot\underline{r}|^{-1} \le$$

$$\le |\underline{\omega}_0\cdot\underline{r}|^{-1}\left|1 - \frac{|(\underline{\omega}_0(\underline{A}')-\underline{\omega}_0(\underline{0})\cdot\underline{r}|}{|\underline{\omega}_0\cdot\underline{r}|}\right|^{-1} \le \tag{3.20}$$

$$\le C_0|\underline{r}|^{\ell}\, |1 - \ell C_0 N_0^{\ell+1}\frac{2E_0}{\rho_0}\tilde{\rho}|^{-1}$$

which shows that one can take

$$\tilde{\rho} = \rho_0/2(2\ell C_0 E_0 N_0^{\ell+1}) \tag{3.21}$$

Then we can easily estimate, using (3.16), (3.19), the function Φ_0

$$||\Phi_0||_{\tilde{\rho},\xi_0-\delta_0} \le B_2 C_0 \xi_0 \rho_0 \delta_0^{-2\ell+1} \tag{3.22}$$

hence by dimensional estimates, and by (3.21)

$$\left|\left|\frac{\partial\Phi_0}{\partial\underline{A}}\right|\right|_{\tilde{\rho}/2,\xi_0-2\delta_0} + \frac{2}{\tilde{\rho}}\left|\left|\frac{\partial\Phi_0}{\partial\phi}\right|\right|_{\tilde{\rho}/2,\xi_0-2\delta_0} \le$$

$$\le B_3'\, C_0\varepsilon_0\delta_0^{-2\ell}\, \rho_0/\tilde{\rho} = B_3 C_0\varepsilon_0 C_0 E_0 N_0^{\ell+1}\delta_0^{-2\ell} \tag{3.23}$$

for a suitably chosen B_3', B_3. (Exercise:find explicit estimates for B_2,B_3).

*By dimensional estimate we mean the estimate of a derivative of a holomorphic function in a point by the supremum of the function in a domain divided by the distance of the point to the boundary i.e. $\left|\left|\frac{\partial^2 h_0}{\partial\underline{A}\partial\underline{A}}\right|\right|_{\tilde{\rho}} \le \frac{1}{\rho_0-\tilde{\rho}}\left|\left|\frac{\partial h_0}{\partial\underline{A}}\right|\right|_{\rho_0} \le 2\, E_0/\rho_0$ if $\tilde{\rho}<\rho_0/2$, for instance.

3) Define on $C(\tilde{\rho}/2,\xi_0-2\delta_0)$ the canonical maps, inverse of each other, generated by Φ_0 via

$$\underline{A} = \underline{A}' + \frac{\partial\Phi_0}{\partial\underline{\phi}}(\underline{A}',\underline{\phi}) \qquad (\underline{A},\underline{\phi}) = \mathcal{C}(\underline{A}',\underline{\phi}')$$
$$\Rightarrow \tag{3.24}$$
$$\underline{\phi}' = \underline{\phi} + \frac{\partial\Phi_0}{\partial\underline{A}'}(\underline{A}',\underline{\phi}) \qquad (\underline{A}',\underline{\phi}') = \mathcal{C}'(\underline{A},\underline{\phi})$$

The possibility of this can be imposed by requiring that ε_0 and, hence, Φ_0 and its derivatives be small. By using a simple implicit functions' theorem one finds that, under the condition that $\partial^2\Phi_0/\partial\underline{A}\partial\underline{A}$ is small compared to 1 and $\partial\Phi_0/\partial\underline{A}$, $\tilde{\rho}^{-1}\partial\Phi_0/\partial\underline{\phi}$ are small compared to δ_0,* the maps $\mathcal{C},\mathcal{C}'$ can be defined with the following properties

$$\begin{aligned} &\text{a)}\quad \mathcal{C},\mathcal{C}': C(\tilde{\rho}/4,\xi_0-3\delta_0) \to C(\tilde{\rho}/2,\xi_0-2\delta_0)\\ &\text{b)}\quad \mathcal{C}\mathcal{C}' = \mathcal{C}'\mathcal{C} = \text{identity on } C(\tilde{\rho}/8,\xi_0-4\delta_0)\\ &\text{c)}\quad \mathcal{C}(C(\tilde{\rho}/32,\ \xi_0-5\delta_0)) \supset C(\tilde{\rho}/64,\xi_0-5\delta_0)\ ^{**}\end{aligned} \tag{3.25}$$

and $(\underline{A},\underline{z}) = \mathcal{C}(\underline{A}',\underline{z}')$, $(\underline{A}',\underline{z}') = \mathcal{C}'(\underline{A},\underline{z})$ are defined as

$$\begin{aligned} A &= \underline{A}' + \underline{\Xi}(\underline{A}',\underline{\phi}'), \quad & \underline{A}' &= \underline{A} + \underline{\Xi}'(\underline{A},\underline{\phi})\\ \underline{\phi} &= \underline{\phi}' + \underline{\Delta}(\underline{A}',\underline{\phi}'), \quad & \underline{\phi}' &= \underline{\phi} + \underline{\Delta}'(\underline{A},\underline{\phi})\end{aligned} \tag{3.26}$$

with

$$\begin{aligned} \underline{\Xi}(\underline{A}',\underline{\phi}') &= \frac{\partial\Phi_0}{\partial\underline{\phi}}(A',\underline{\phi}) = -\underline{\Xi}'(\underline{A},\underline{\phi})\\ \underline{\Delta}(\underline{A}',\underline{\phi}') &= -\frac{\partial\Phi_0}{\partial\underline{A}'}(\underline{A}',\phi) = -\underline{\Delta}'(\underline{A},\underline{\phi})\end{aligned} \tag{3.27}$$

so that, by (3.25)

$$\begin{aligned} \|\underline{\Xi}\|_{\tilde{\rho}/8,\xi_0-4\delta_0},\ \|\underline{\Xi}'\|_{\tilde{\rho}/8,\xi_0-4\delta_0} &\le \left\|\frac{\partial\Phi_0}{\partial\underline{\phi}}\right\|_{\tilde{\rho}/2,\xi_0-2\delta_0}\\ \|\underline{\Delta}\|_{\tilde{\rho}/8,\xi_0-4\delta_0},\ \|\underline{\Delta}'\|_{\tilde{\rho}/8,\xi_0-4\delta_0} &\le \left\|\frac{\partial\Phi_0}{\partial\underline{A}'}\right\|_{\tilde{\rho}/2,\xi_0-2\delta_0}\end{aligned} \tag{3.28}$$

The mentioned smallness condition to be imposed on ε_0 to insure the <u>existence</u> of $\mathcal{C},\mathcal{C}'$ with the above properties can be easily

*The first such condition will insure local invertibility while the second will guarantee global extension of the local invertibility and will allow control of the domains.

**32,64,5,6 are here quite arbitrary and this choice will be useful later.

deduced by estimating $\partial^2\Phi_0/\partial\underline{A}\partial\underline{\phi}$, $\partial\Phi_0/\partial\underline{A}$, $\partial\Phi_0/\partial\underline{\phi}$ dimensionally, starting from (3.22). Not surprisingly the three smallness conditions basically coincide and can be imposed through a single condition like

$$B_4\varepsilon_0 C_0 E_0 C_0 \; N_0^{\ell+1} \; \delta_0^{-2\ell-1} < 1 \tag{3.29}$$

The fact that (3.29) is sufficient, for B_4 large, to guarantee (3.25) ÷ (3.28) is easily proved (see [1], §3.11, p. 490, and appendix N).

4) Define

$$\begin{aligned} \tilde{h}_1(\underline{A}') &= h_0(\underline{A}') + \overline{f}_0(\underline{A}') \\ \tilde{f}_1(\underline{A}',\underline{\phi}') &= H_0(\mathcal{C}(\underline{A}',\underline{\phi}')) - \tilde{h}_1(\underline{A}') \end{aligned} \tag{3.30}$$

and $\tilde{h}_1, \tilde{f}_1$ are holomorphic at least on $C(\tilde{\rho}/8, \xi_0 - 4\delta_0)$, by the properties of $\mathcal{C}$: and the motions starting in the $\mathcal{C}$-image of this set can be described in the $(\underline{A}',\underline{\phi})$ - coordinates as hamiltonian motions with hamiltonian $\tilde{H} = \tilde{h} + \tilde{f}$, at least as long as they do not leave the set $\mathcal{C}(C(\tilde{\rho}/8, \xi_0 - 4\delta_0)$.

Then we look for a point $\underline{A}' \in V_{\tilde{\rho}/8}$ which would move with the right spectrum $\underline{\omega}_0$ if $\tilde{f}_1$ were zero (by construction $\tilde{f}_1$ is, in any case, much smaller than f_0 as we shall see) i.e. we look for a solution of

$$\underline{\omega}_0(\underline{A}') + \frac{\partial\overline{f}_0}{\partial\underline{A}'}(\underline{A}') = \underline{\omega}_0 \tag{3.31}$$

where $\overline{f}_0$ denotes the average of f_0 on T^ℓ .

The solution to this equation can be found by recalling that $\underline{\omega}_0 = \underline{\omega}(\underline{0})$ and rewriting it using the Taylor expansion and calling $\underline{M}(\underline{A}) = \partial^2 h_0/\partial\underline{A}\partial\underline{A} = \partial\underline{\omega}_0/\partial\underline{A}$

$$\frac{1}{2}\, M(\underline{0})\underline{A}' + [\underline{\omega}_0(\underline{A}') - \underline{\omega}_0(\underline{0}) - \frac{1}{2}\, M(\underline{0})\underline{A}' + \frac{\partial\overline{f}_0}{\partial\underline{A}'}(\underline{A}')] = \underline{0}$$

i.e. as

$$\underline{A}' + \underline{n}(\underline{A}') = \underline{0}$$

$$\underline{n}(\underline{A}') = M(\underline{0})^{-1}\,\{\frac{\partial f}{\partial\underline{A}'}(\underline{A}') + \int_0^1 d\tau \int_0^\tau d\theta\, \frac{d^2}{d\theta^2}\,\underline{\omega}_0(\theta\underline{A}') \tag{3.32}$$

so that for a suitable $B_5 > 2$ and $\forall\beta < \rho/2$ it is, by a dimensional estimate

$$\|\underline{n}\|_\beta \le B_5\eta_0(\varepsilon_0 + (\frac{\beta}{\rho_0})^2 E_0) \tag{3.33}$$

Hence, if $\beta = \rho_0(\varepsilon_0/E_0)^{1/2}$ it is

$$\|\underline{n}\|_\beta \leq 2B_5\, \eta_0\, \varepsilon_0 \tag{3.34}$$

By an implicit function theorem it can be easily seen (see [1], p. 490) that if $\|\underline{n}\|_\beta$ is much smaller than β the (3.32) has a solution $\underline{A}'_0$ verifying, of course, $|\underline{A}'_0| \leq \|\underline{n}\|_\beta << \beta$. So if B'_6 is a suitably large constant and $B'_6\, \|\underline{n}\|_\beta\, \beta^{-1} < 1$, i.e. with the above value of β and by (3.34) if

$$B_6(\eta_0 E_0 \rho_0{}^{-1}) \sqrt{\frac{\varepsilon_0}{E_0}} < 1 \tag{3.35}$$

there is a solution $\underline{A}'_0$ to (3.31) verifying

$$\|\underline{A}'_0\| \leq 2B_5\, \varepsilon_0\, \eta_0 \tag{3.36}$$

Since we want that $\underline{A}'_0$ be well inside the region where $\hat{h}_1$ is defined, e.g. $\|\underline{A}'_0\| < \tilde{\rho}/16$, we impose $2B_5\, \varepsilon_0\eta_0 < \tilde{\rho}/16$. Recalling the (3.21) we see that the latter condition as well as the (3.35) and (3.29) can be imposed simultaneously by requiring the single inequality

$$B_7(\eta_0 E_0 \rho_0{}^{-1})^2\, \varepsilon_0 C_0 E_0 C_0 N_0{}^{\ell+1} \delta_0{}^{-2\ell-1} < 1 \tag{3.37}$$

(where we use $\varepsilon_0/E_0 = \varepsilon_0 C_0/E_0 C_0 \leq \varepsilon_0 C_0$ because (3.9) implies $C_0 E_0 \geq 1$).

So, if (3.37) holds we can draw around $\underline{A}'_0$ a polidisk of radius $\tilde{\rho}/16$, at least, contained inside $S_{\tilde{\rho}/8}$.

5) Let $\lambda_0{}^{-1} \equiv \tilde{\rho}/32\, \rho_0 << 1$, and let

$$\begin{aligned} \underline{\phi}_1 &\equiv \underline{\phi}' \\ \underline{A}_1 &= (\underline{A}'-\underline{A}'_0)\lambda_0 \qquad \text{for } \|\underline{A}'-\underline{A}'_0\| < \tilde{\rho}/32 \end{aligned} \tag{3.38}$$

Then the variable $\underline{A}_1$ varies again in S_{ρ_0} as $\underline{A}'$ varies. We finally define

$$\begin{aligned} h_1(\underline{A}_1) &= \lambda_0(\tilde{h}_1(\underline{A}'_0 + \lambda_0^{-1}\underline{A}_1) - \tilde{h}_1(\underline{A}'_0)) \\ f_1(\underline{A}_1,\underline{\phi}_1) &= \lambda_0 \tilde{f}_1(\underline{A}'_0 + \lambda_0^{-1}\underline{A}_1, \underline{\phi}_1) \end{aligned} \tag{3.39}$$

for $(\underline{A}_1, \underline{z}_1) \in C(\rho_0, \xi_0 - 5\delta_0)$, and on this set the above functions are holomorphic.

This is the final step in the construction of the renormalization transformation $\mathcal{T}_0$. Its domain consists in the functions h_0, f_0 verifying (3.37).

It is now very easy, although a little long, to check, using dimensional estimates and (3.16), (3.25), that if (3.37) holds, then

$$E_1 \leq E_0 + \varepsilon_0 \quad , \quad \eta_1 \leq \lambda_0 \eta_0 (1 + B_8 \varepsilon_0 \eta_0 \rho_0^{-1})$$

$$\varepsilon_1 \leq B_9 C_0 \varepsilon_0^2 (C_0 E_0)^2 N_0^{\ell+1} \delta_0^{-b_1} \quad , \quad \xi_1 = \xi_0 - 5\delta_0 , \ \rho_1 \equiv \rho_0 \tag{3.40}$$

with B_8, B_9, b_1 suitably chosen. For details see for instance [1], §5.12, pp 510-512.

The above construction can be iterated to build $\mathcal{T}_1, \mathcal{T}_2, \ldots$. The construction is identical to that of $\mathcal{T}_0$ with δ_j replacing δ_0 . The above analysis implies that if we define

$$\begin{cases} \overline{E}_n = (\overline{E}_{n-1} + \overline{\varepsilon}_{n-1}) \\ \overline{\eta}_n = \overline{\lambda}_{n-1} \overline{\eta}_{n-1} \ (1 + B_8 \ \overline{\varepsilon}_{n-1} \eta_{n-1} \rho_0^{-1}) \qquad n \geq 1 \\ \overline{\varepsilon}_n = B_9 C_0 \overline{\varepsilon}_{n-1}^2 \ (C_0 \overline{E}_{n-1})^2 \ \overline{N}_n^{\ell+1} \delta_n^{-b_1} \\ \overline{\xi}_n = \overline{\xi}_{n-1} - 5 \ \delta_{n-1} \ , \ \overline{\rho}_n = \rho_0 \end{cases} \tag{3.41}$$

with $\overline{N}_n = \delta_n^{-1} \log(C_0 \varepsilon_n \delta_n^{-\ell})^{-1}$, $\overline{\lambda}_n = 32 \cdot 4 \ell C_0 \overline{E}_n \overline{N}_n^{\ell+1}$ and for $n = 0$ the barred quantities are identified with the unbarred ones, and if

$$B_7 (\overline{\eta}_n \overline{E}_n \rho_0^{-1})^2 \overline{\varepsilon}_n C_0 \overline{E}_n C_0 \overline{N}_n^{\ell+1} \delta_n^{-2\ell=1} < 1 \tag{3.42}$$

it will result, $\forall \, n \geq 1$

$$E_n \leq \overline{E}_n , \ \varepsilon_n \leq \overline{\varepsilon}_n \ , \ \eta_n \leq \overline{\eta}_n \ , \ \xi_n = \overline{\xi}_n \ , \ \rho_n = \rho_0 \tag{3.43}$$

It is easy to check by induction that if ε_0 / E_0 is small enough the (3.41) imply the (3.42) for all n and

$$(C_0\varepsilon_0)^{2n} \le C_0\overline{\varepsilon}_n \le (C_0\varepsilon_0)^{(3/2)^n}$$

$$\overline{\eta}_n \le \eta_0(\overline{E}_0 C_0)^n(\xi_0^{-1} \log(C_0\varepsilon_0)^{-1})^{2n}\, 2^{2n^2(\ell+1)} \tag{3.44}$$

$$\overline{E}_n \le 2E_0$$

and "small enough" means a condition like the (3.12). For a complete discussion of the implication of (3.44) from (3.41), (3.42) see for instance [1], p 512-515. However, the discussion by induction is so simple that it should be worked out as an exercise. A more careful analysis of (3.41), (3.42) easily leads to the conclusion that in (3.12) $a_1 > 2$, $a_2 > 2$, $a_3 > 2(\ell+1)$ are allowed choices. It is very instructive to try to find an explicit value of G fixed a_1, a_2, a_3 or even to try to optimize all the above bounds. There seems to be no really satisfactory analysis of such bounds.

To summarize, we have found that under the condition (3.12)

$$\mathcal{T}_n\mathcal{T}_{n-1} \ldots \mathcal{T}_0(h_0, f_0) \to (\underline{\omega}_0 \cdot \underline{A}, 0) \tag{3.45}$$

uniformly in the compact sets of $C(\rho_0, \xi_0/2)$ (just use a dimensional estimate on $\partial^2 h_n/\partial\underline{A}\partial\underline{A}$ to see that $h_n \to \underline{\omega}_0 \cdot A$ via Vitali's theorem on sequences of homomorphic functions; $\varepsilon_n \xrightarrow[n\to\infty]{} 0$ implies that $f_n \xrightarrow[n\to\infty]{} 0$. Actually, it only implies that f_n tends to a constant. However, in the proof of (3.41) one finds first a bound on f_n itself (rather than directly a bound on its derivatives) and then by a dimensional estimate the bound ε_n on the derivatives of ε_n is found. It turns out that even the bound on f_n goes to zero. Note, in any case, that for the purposes of the coming discussion replacing 0 by a constant would be irrelevant, see [1], (5.12.69)).

We can read (3.45) as saying that for ε_0 small enough the chain of renormalization transformations drives the system "to a trivial fixed point", i.e. to the harmonic oscillator.

To summarize, we have defined a sequence $\overline{\mathcal{C}}_0, \overline{\mathcal{C}}_1, \ldots$ of maps

$$\overline{\mathcal{C}}_j(\underline{A}_{j+1}, \underline{\phi}_{j+1}) \equiv \mathcal{C}_j(\underline{A}'_j + \lambda_j^{-1}\underline{A}_{j+1}, \underline{\phi}_{j+1}) \tag{3.46}$$

which, although not canonical, map solutions of the hamiltonian equations relative to H_{j+1} on $V_\rho \times T^\ell$ into solutions relative to H_j on $V_\rho \times T^\ell$.

By construction the maps $\mathcal{C}_j$ are close to the identity, see (3.28), (3.44), in their holomorphy domains, within

$$\mathcal{O}((C_0\varepsilon_0)^{(3/2)^n})$$

while the holomorphy domain itself shrinks only by $\overline{\lambda}_j^{-1}$ in the action variables (and almost nothing in the angle variables) which by (3.41) behaves $\simeq \eta_j^{-1}$ i.e. goes to zero at a rate which, even if raised to an arbitrary power, is much slower than

$$\mathcal{O}((C_0\varepsilon_0)^{(3/2)^n}).$$

This immediately implies, by a dimensional argument, that

$$\overline{\mathcal{C}} = \lim_{n\to\infty} \overline{\mathcal{C}}_0 \dots \overline{\mathcal{C}}_n \tag{3.47}$$

exists and, by Vitali's theorem on converging sequences of holomorphic functions, is holomorphic on $S_{\rho_0} \times C(\xi_\infty)$, $\xi_\infty = \xi_0/2$.

Furthermore, $\overline{\mathcal{C}}(\underline{A}',\underline{\phi}') = (\underline{A},\underline{\phi})$ is close to the identity in the $\underline{\phi}'$-variable. However, on $S_{\rho_0} \times C(\xi_\infty)$

$$\frac{\partial\overline{\mathcal{C}}(\underline{A}',\underline{\phi}')}{\partial\underline{A}'} \equiv \underline{0} \tag{3.48}$$

because by construction this derivative is proportional to

$$\prod_{j=1}^{\infty} \lambda_j^{-1} \equiv 0 .$$

This means that $\overline{\mathcal{C}}$ is $\underline{A}'$-independent.

Finally, $\overline{\mathcal{C}}$ still maps solutions of the hamiltonian equations with hamiltonian $H_\infty(\underline{A}',\underline{\phi}') \equiv \underline{\omega}_0 \cdot \underline{A}'$ on $V_{\rho_0} \times T^\ell$ into solutions relative to the hamiltonian H_0 on $V_{\rho_0} \times T^\ell$.

Since $\overline{\mathcal{C}}$ is independent on $\underline{A}'$ and close to the identity in $\underline{\phi}'$ we may write it as $(\underline{A},\underline{\phi}) = \overline{\mathcal{C}}(\underline{A}',\underline{\phi}')$ with

$$\begin{aligned} \underline{A} &= \underline{\alpha}(\underline{\phi}') \\ \underline{\phi} &= \underline{\phi}' + \underline{\beta}(\underline{\phi}') \end{aligned} \tag{3.49}$$

The functions $\underline{\alpha},\underline{\beta}$ are holomorphic on $C(\xi_0/2)$ and small of $\mathcal{O}(C_0\varepsilon_0)$ and the motion $(\underline{A}',\underline{\phi}') \to (\underline{A}',\underline{\phi}' + \underline{\omega}_0 t)$ becomes in the $(\underline{A},\underline{\phi})$ variables, via (3.49) a solution to the equation of motion with hamiltonian H_0 lying on a torus with parametric equations (3.49) and the proof is complete. For more details on the above simple implications of (3.44) see [1], pp 515 - 517, (where the rescalings of the domains by λ_j are not done but, as it is equivalent , one deals with shrinking domains and essentially constant η_j's).

§4. Comments and Generalizations

A) The preceding discussion is very close to the simplest versions of the renormalization group applications. The annoying feature is that we do not iterate the same map $\mathcal{T}$ but we apply recursively $\mathcal{T}_n \dots \mathcal{T}_0$ and $\mathcal{T}_i$ depends on i via the a priori given constants δ_i.

This can be manifestly avoided if instead of defining δ_j a priori we define them through h_j, f_j ; a possible definition could be

$$\delta_j = \xi_j / [\log(\log(C_0 \varepsilon_j)^{-1})]^2 \tag{4.1}$$

This would lead essentially to the same proof with $\mathcal{T}^{n+1}$ replacing $\mathcal{T}_n \mathcal{T}_{n-1} \dots \mathcal{T}_0$.

B) It is nice to give a picture of the proof of the preceding theorem.

Define the "surfaces of resonance of order n" by considering the surfaces

$$\frac{\partial h_n}{\partial \underline{A}} (\underline{A}) \cdot \underline{r} = 0 , \quad 0 < |\underline{r}| \le N_n \tag{4.2}$$

The position of such surfaces depends on how non-linear h_n is (recall that $h_n(\underline{A}) \to \underline{\omega}_0 \cdot \underline{A}$). They are roughly planes in $\underline{A}$-space going through the point $\underline{x}$:

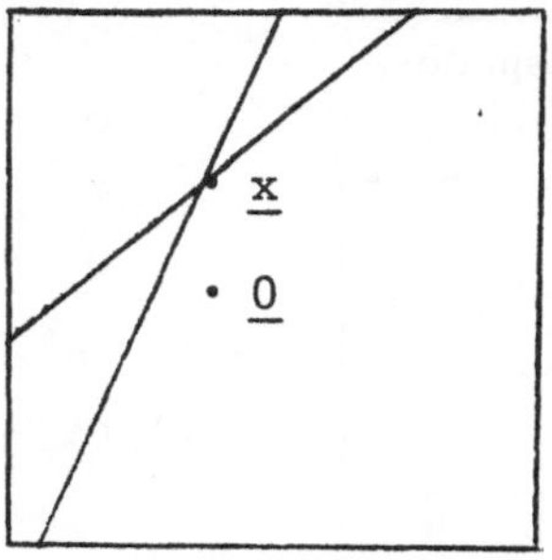

$$\underline{x} = -M_n^{-1} \underline{\omega}_0 \tag{4.3}$$

The larger n is the further from the origin is $\underline{x}$ and the less is the number of such surfaces crossing a given neighborhood of the origin. Note, however, that this is so only because $N_n < \infty$ and $N_n \to \infty$ much slower than M_n^{-1} (as $\approx \varepsilon_n C_0$ compared to $\approx \log(\varepsilon_n C_0)^{-1}$).

Then we can think of the renormalization transformation σ_0 as follows.

1) Draw the resonance surfaces of order 0 and find a non-resonant vicinity $V_{\tilde{\rho}}$ of $\underline{0}$.

2) Map $V_{\tilde{\rho}/8} \times T^{\ell}$ canonically, with a map close to the identity, into $V_{\rho_0} \times T^{\ell}$ over a set containing $V_{\tilde{\rho}/16} \times T^{\ell}$ and locate A'_0 where $\tilde{h}_1$ has spectrum $\underline{\omega}_0$.

3) Take $\underline{A}'_0$ as origin and rescale the $(\underline{A}'-\underline{A}'_0)$ coordinates so that $V_{\tilde{\rho}/32} \times T^{\ell}$ becomes $V\rho_0 \times T^{\ell}$ again and draw the resonances of h_1 of index $\underline{r}$, $0 < |\underline{r}| \leq N_1$, i.e. the "resonances or order 1".

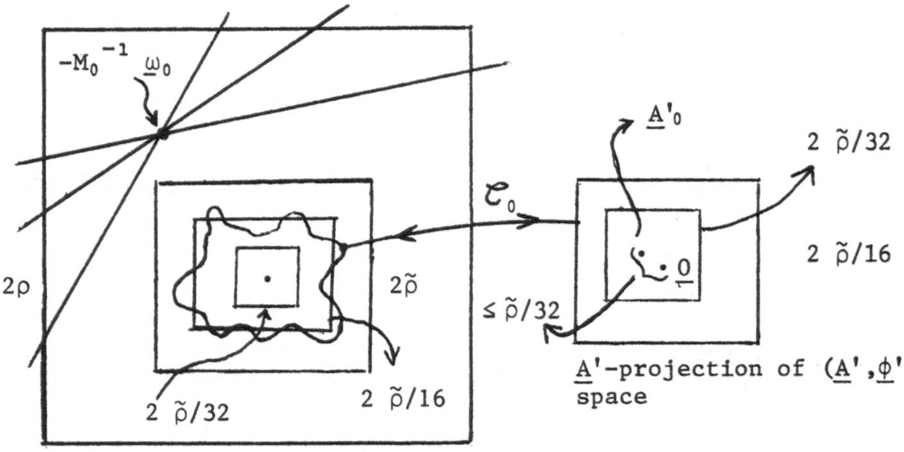

$\underline{A}'$-projection of $(\underline{A}',\underline{\phi}')$-space

$\underline{A}$ projection of $(\underline{A},\underline{\phi})$-space

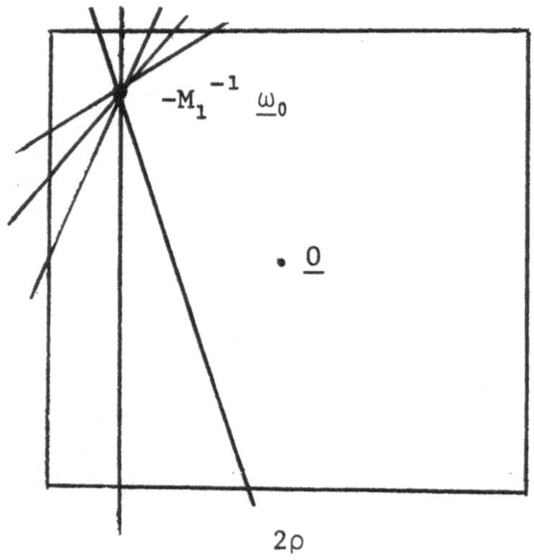

$\lambda_0(\underline{A}'-\underline{A}'_0) \equiv \underline{A}_1$ - projection in $(\lambda_0(\underline{A}'-\underline{A}'_0),\underline{\phi}')$ space

4) Iterate (after a while $-M_n^{-1}\underline{\omega}_0$ gets out of the picture but some resonance lines still may cross it, closer and closer to $\underline{0}$ because $\tilde{\rho} = \rho_0/4\ell C_0 N_n^{\ell+1} E_n \xrightarrow[n\to\infty]{} 0$.

C) Another important comment will be about the approach necessary to prove the full KAM theorem quoted in §2.

In this case we cannot, of course, define a family of renormalization transformations conjugating our system with a given linear oscillator, since every quasiperiodic motion will have its own frequency spectrum.

We shall define our renormalization transformations on triples (h,f,V) where V is an open region. The transformations will still be defined in terms of a sequence of numbers $\delta_n = \xi_0/5\pi(1+n)^2$, $C_n = C_0(1+n)^2$ (this time also C varies with n). We then define $\mathcal{T}_0$ as follows. Draw inside the phase space the resonances of order $\leq N_0$ (the picture deals with the special case $h_0(\underline{A}) = \underline{A}^2/2$, i.e. the surfaces $\partial h_0/\partial \underline{A}\,(\underline{A}) \cdot \underline{r} = 0$, for $0 < |\underline{r}| \leq N_0$. Around them draw corridors of width $\tilde{\rho}_0$, so large that, $\forall\, 0 < |\underline{r}| \leq N_0$

$$|\underline{\omega}_0(\underline{A}) \cdot \underline{r}|^{-1} < C_0 |\underline{r}|^{\ell+1} \tag{4.4}$$

for $\underline{A}$ outside the corridors.

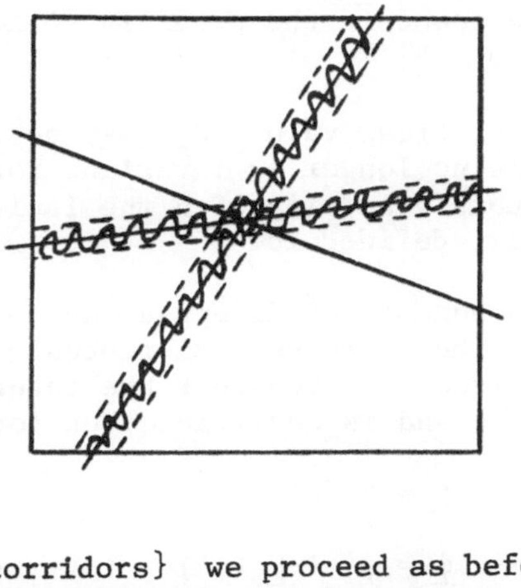

Calling $V_1 = V_0/\{\text{corridors}\}$ we proceed as before to build $(\tilde{h}_1,\tilde{f}_1,V_1)$. This time, however, we do not rescale and we simply set $h_1 \equiv \tilde{h}_1, f_1 \equiv \tilde{f}_1$, and iterate.

At the end one will be left with a set $V_\infty \subset V$ with a dense open complement but with very large measure for C_0 large, because the widths of the corridors are of $O(|\underline{r}|^{-\ell-1})C_0^{-1}$ (as it can be seen from an argument similar to the one following (3.20), (see for details, [11]) hence summable. At the end we define C_0 in terms of ε_0 so that $C_0\varepsilon_0 \xrightarrow[\varepsilon_0\to 0]{} 0$ and $C_0 \xrightarrow[\varepsilon_0\to 0]{} \infty$ to obtain the final result $\text{vol } V_\infty/\text{vol } V_0 \xrightarrow[\varepsilon_0\to 0]{} 1$.

§5. The Chaotic Transition's Problem and a Related Problem

From the discussion of the preceding lectures it emerges clearly that in some sense each invariant torus has to be regarded as sitting in a neighborhood of phase space where the hamiltonian looks like a perturbation of

$$\bar{h}(\underline{A}) = \underline{\omega}_0 \cdot \underline{A} \tag{5.1}$$

with $\underline{\omega}_0 \in R^\ell$ verifying the non-resonance condition (3.9).

We call $\underline{\omega}_0$ the spectral vector of a given torus and we imagine to fix $\underline{\omega}_0$ and to let the perturbation size ε, in the original hamiltonian (1.8), grow. Very close to the selected invariant torus the system will look , in suitable coordinates, as described by a hamiltonian

$$H_\varepsilon(\underline{A},\underline{\phi}) = \underline{\omega}_0 \cdot \underline{A} + \sigma\tilde{f}(\underline{A},\underline{\phi},\sigma) \tag{5.2}$$

where $\sigma \equiv \sigma(\varepsilon) \xrightarrow[\varepsilon\to 0]{} 0$ and $\underline{A},\underline{\phi}$ have to be identified as the rescaled coordinates found in the proof in §3,i.e. with $\underline{A}_n,\underline{\phi}_n$ with n quite large.

As ε grows a critical value ε_c may be reached when for $\varepsilon > \varepsilon_c$ there will be no longer an invariant torus, with spectral vector $\underline{\omega}_0$, continuously merging with the family of tori, with given spectral vector, defined for $\varepsilon < \varepsilon_c$.

To study the phenomenon of disappearance of the torus with spectral vector $\underline{\omega}_0$ there are several routes. A possible one is the following. Fixed $\underline{\omega}_0$ consider the interval where the invariant torus exists and is analytic. Let for $\varepsilon \in [0,\varepsilon_c]$

$$\begin{aligned} \underline{A} &= \underline{\alpha}_\varepsilon(\underline{\phi}') \\ \underline{\phi} &= \underline{\phi}' + \underline{\beta}_\varepsilon(\phi') \end{aligned} \tag{5.3}$$

be its parametric equations, see (3.11). Of course we have only proved in §3 that (5.3) makes sense for ε small. Here, however, we are proceeding in a heuristic, non-rigorous, way.

To analyze a case slightly simpler, but already very non-trivial, suppose $\ell = 2$. Consider the circle

$$\begin{aligned} \underline{A} &= \underline{\alpha}_\varepsilon(\phi_1', 0) \\ \underline{\phi} &= (\phi_1', 0) + \underline{\beta}_\varepsilon(\phi'_1, 0) \end{aligned} \tag{5.4}$$

and the map generated on it by the evolution looked at integer multiples of $2\pi\, \omega_2^{-1}$

$$\begin{aligned} \underline{A} &\to \underline{\alpha}_\varepsilon(\phi'_1 + 2\pi \frac{\omega_1}{\omega_2},\ 0) \\ \underline{\phi} &\to (\phi'_1 + 2\pi \frac{\omega_1}{\omega_2},\ 0) + \beta_\varepsilon(\phi'_1 + 2\pi \frac{\omega_1}{\omega_2},\ 0) \end{aligned} \tag{5.5}$$

We describe this map by introducing the arc length on the circle (5.4), which we call θ and normalize so that the circle's length is 2π, and then rewrite the map (5.5) as the circle's map

$$\theta \to \theta + g_\varepsilon(\theta) \tag{5.6}$$

Of course, by construction, this map is smoothly conjugated to the perfect rotation of the circle, with rotation number $r = \omega_1/\omega_2$, for ε small. As long as the torus exists and is smooth it is clear that the above map of the circle must be smoothly conjugated to a perfect rotation (the conjugation being again given by the map $\theta \leftrightarrow \phi'_1$).

If for some value ε_c such smooth conjugation with the circle's rotation breaks down, it will necessarily mean that the torus (5.3) becomes non-smooth or disappears or more generally the equations (5.3) become singular in $\underline{\phi}'$ and no longer describe a diffeomorphism of T^ℓ into the phase space $V \times T^\ell$.

The smooth conjugability with the circle's rotation can disappear either because g_ε develops a singularity at $\varepsilon = \varepsilon_c$ or because $\theta \to \theta' = g_\varepsilon(\theta) + \theta$ becomes non-invertible as a map of the circle (without any singularity developing in ε), or because of other reasons.

The above discussion shows well the relevance for the theory of the disappearance of the quasi-periodic motions ("chaotic transitions") of the theory of the maps of the circle depending on one parameter ε and having a fixed rotation number.

The study of maps of the form

$$\theta \to \theta + 2\pi\, r(\varepsilon) - \varepsilon \sin\theta \tag{5.7}$$

with $r(\varepsilon)$ so chosen that the above transformation keeps a fixed rotation number $r \equiv r(0)$ ("standard map of the unit circle") has been pursued in considerable detail [12], and it is treated in the parallel lectures. It illustrates what can happen when the conjugacy of the motion with the perfect rotation is broken because $\theta \to \theta + g_\varepsilon(\theta)$ develops at $\varepsilon = \varepsilon_c$ an inflection point becoming non-invertible.

Note also the strong analogy between the problem of finding a function $r(\varepsilon)$ such that (5.7) has a fixed rotation number as ε grows and the problem posed in §1,2 of finding counterterms $\delta_\varepsilon f$ which make (2.1) conjugated canonically with the hamiltonian $\underline{\omega} \cdot \underline{A}$ independently on ε.

This remark leads to the following problem. Imagine that we are given a one-parameter hamiltonian system, with $\ell = 2$, of the form

$$\underline{\omega}_0 \cdot \underline{A} + \varepsilon\, f(\underline{A}, \underline{\phi}, \varepsilon) \tag{5.8}$$

which for ε small is canonically conjugated to $\underline{\omega}_0 \cdot \underline{A}$ on a large phase space region $W_\varepsilon \subset V_\rho \times T^\ell$ (i.e. $\partial W \sim \partial(\overline{V}_\rho \times T^\ell)$).

This problem certainly arises in some applications. In §2 we have explained this in connection with the theory of the quasi-periodic Schroedinger equation, when we fix our attention on a Block wave described by $\underline{\omega}_0 = (\lambda, \underline{\omega})$.

Then it might happen that, as ε grows through ε_c, the integrability is lost in the whole phase space leading to a phenomenon like a phase transition which, by the remarks related to (5.7), could also be relevant to the theory of the chaotic transition. In the next section we discuss in some detail the theory of systems like (5.8) which are canonically conjugated to $\underline{\omega}_0 \cdot \underline{A}$ for all $0 \leq \varepsilon < \varepsilon_c$.

In the final section we point out some problems which in our opinion arise naturally if one tries to study what happens at ε_c in the above systems, using the ideas of the renormalization group.

§6. Renormalization and Isochronous Systems

The purpose of this lecture is to define a renormalization transformation acting on the one parameter hamiltonians of the form

$$H_\varepsilon(\underline{A},\underline{\phi}) = \underline{\omega}_0 \cdot \underline{A} + f(\underline{A},\underline{\phi},\varepsilon) \tag{6.1}$$

with f analytic in ε at $\varepsilon = 0$ and divisible by ε, and $\underline{\omega}_0$ verifying (3.9).

The renormalization transformation $\mathcal{J}$ should be designed so that $\forall\varepsilon \in [0,\bar{\varepsilon}]$ it is $\mathcal{J}^n H_\varepsilon \xrightarrow[n\to\infty]{} \underline{\omega}_0 \cdot \underline{A}$ if H_ε is canonically conjugate to $\underline{\omega}_0 \cdot A$ on a set $W_\varepsilon \supset V_\rho \times T^\ell$ for $\varepsilon \in [0,\bar{\varepsilon})$ where $\rho > 0$ is some given radius.

The renormalization transformation is defined as follows

1) Suppose that f is divisible by ε^k, $2^{n-1} \leq k < 2^n$, $n = 1,2,\ldots$, and define the "truncation" operation on the formal power series as usual*. Then define

$$f^{[<2^n]} \equiv \sum_{k=1}^{2^n-1} f_k \, \varepsilon^k$$
$$\bar{f}(\underline{A}) = \frac{1}{(2\pi)^\ell} \int_{T^\ell} f(\underline{A},\underline{\phi})d\underline{\phi} \tag{6.2}$$

Consider

$$\tilde{H}_\varepsilon(\underline{A},\underline{\phi}) = \underline{\omega}_0 \cdot \underline{A} + f(\underline{A},\underline{\phi},\varepsilon) - \bar{f}^{[<2^n]}(\underline{A},\varepsilon) \tag{6.3}$$

and let Φ be such that

$$\underline{\omega}_0 \cdot \frac{\partial\Phi}{\partial\phi}(\underline{A}',\underline{\phi}) + f^{[<2^n]}(\underline{A},\underline{\phi},\varepsilon) - \bar{f}^{[<2^n]}(\underline{A},\varepsilon) = 0 \tag{6.4}$$

If f is holomorphic in the region

$$C(\rho,\xi,\theta) \equiv S_\rho \times C(\xi) \times S^1_\theta \tag{6.5}$$

---

*i.e. if $g = \sum_{k=0}^{\infty} \varepsilon^k g_k$ one defines $g^{[<n]} \equiv \sum_{k=0}^{n-1} \varepsilon^k g_k$, and $g^{[\geq n]} = g - g^{[<n]}$.

in the variables $(\underline{A},\underline{\phi},\varepsilon)$, where $S^1_\theta = \{\varepsilon | \varepsilon \in C, |\varepsilon| < \theta\}$, then Φ is holomorphic in $C(\rho,\xi,\theta)$ as it follows immediately from the bound

$$n_\delta(\Phi) \equiv ||\Phi||_{\rho e^{-\delta},\xi-\delta,\, e^{-\delta}} \leq B_1 \delta^{-2\ell} C_0 \rho \; ||f||_{\rho,\xi,\theta} \equiv$$

$$\equiv B_1 \; \delta^{-2\ell} C_0 \rho \; n_0(f) \tag{6.6}$$

where $||f||_{\rho,\xi,\theta} = \sup_{C(\rho,\xi,\theta)} |f(\underline{A},\underline{z},\varepsilon)|$ and $n_\beta(f)$ is defined implicitly in (6.6). The proof of (6.6) is the same as that of (3.22) using the estimates analogous to (3.16). This time ρ does not become as small as $\tilde{\rho}$ because in solving (6.4) no small denominators appear because $\underline{\omega}_0$ does not depend on $\underline{A}$ and verifies the diophantine condition (3.9) with a non-resonance constant C_0.

As in the proof of §3 it is possible to use Φ as a generating function of two canonical maps $\mathcal{C}, \mathcal{C}'$ analytic in $\underline{A},\underline{\phi},\varepsilon$

$$\begin{aligned} \mathcal{C}, \mathcal{C}' &: C(\rho e^{-3\delta},\xi-3\delta,\, \theta e^{-\delta}) \to C(\rho e^{-2\delta},\xi-2\delta) \\ \mathcal{C}, \mathcal{C}' &: C(\rho e^{-4\delta},\xi-4\delta,\, \theta e^{-\delta}) \to C(\rho e^{-3\delta},\xi-3\delta) \end{aligned} \tag{6.7}$$

such that $\mathcal{C}\mathcal{C}' = \mathcal{C}'\mathcal{C}$ = identity in $C(\rho e^{-4\delta},\xi-4\delta)$, $\forall \varepsilon$, $|\varepsilon| < \theta e^{-4\delta}$. The maps $\mathcal{C}, \mathcal{C}'$ will have the form (3.26) and verify (3.27) with the consequent bounds of the type (3.28). Of course this will be possible under the condition that Φ is small which, as in §3, will take the form, if $\xi < 1$

$$B_2 \; n(\Phi)\rho^{-1}\delta^{-2} < 1 \tag{6.8}$$

Therefore, if $\lambda = e^{4\delta}$, we define, in analogy with §3, on $C(\rho,\xi-4\delta,\, \theta e^{-2\delta})$

$$H^1_\varepsilon(\underline{A}',\underline{\phi}') = \lambda \tilde{H}_\varepsilon(\mathcal{C}(\lambda^{-1}\underline{A}',\underline{\phi}')) \equiv \underline{\omega}_0 \cdot \underline{A}' + \lambda f_1(\lambda^{-1}\underline{A}',\underline{\phi}'\varepsilon) \tag{6.9}$$

which is holomorphic on $C(\rho,\xi-4\delta,\, \theta e^{-2\delta})$ and f_1 turns out, by construction, to be divisible by ε^{2n}.

Our map $\mathcal{T}_\delta$ will be the map

$$\underline{\omega}_0 \cdot \underline{A} + f(\underline{A},\underline{\phi},\varepsilon) \to \underline{\omega}_0 \cdot \underline{A}' + \lambda f_1(\lambda \underline{A}',\underline{\phi}',\varepsilon) \tag{6.10}$$

with $\lambda = \exp 4\delta$, and its domain of definition will be the set of the f's verifying (6.8).

Let $\delta_k = \xi/5\pi(1+k)^2$, $k = 0,1,\ldots,$ then for all ε, $|\varepsilon| < \overline{\theta} < \theta^k$ for some $\overline{\theta}$ small enough,

$$\mathcal{T}_{\delta_n} \ldots \mathcal{T}_{\delta_0} H \to \underline{\omega}_0 \cdot \underline{A} \tag{6.11}$$

and the maps

$$\overline{\mathcal{C}}_i : (\underline{A}_{i+1}, \underline{\phi}_{i+1}) \to (\underline{A}_i, \underline{\phi}_i) = \mathcal{C}_i(\lambda_i^{-1} \underline{A}_{i+1}, \underline{\phi}_{i+1}) \tag{6.12}$$

where $\mathcal{C}_i$ is the canonical map generated in the construction of $\mathcal{T}_{\delta_i}$, can be decomposed to yield a map

$$\overline{\mathcal{C}} = \lim_n \overline{\mathcal{C}}_n \ldots \overline{\mathcal{C}}_0 \tag{6.13}$$

holomorphic in $C(\rho, \xi/2, \overline{\theta})$.

The map $\overline{\mathcal{C}}$ will not be canonical but, by construction, has a jacobian determinant equal to $(\prod_{i=0}^{\infty} \lambda_i^{-1})^{\ell}$; also by the construction it maps solutions of the hamiltonian equations on $V_\rho \times T^\ell$ with the hamiltonian $\underline{\omega}_0 \cdot \underline{A}$ into solutions of the hamiltonian equations in $\overline{\mathcal{C}}(V_\rho \times T^\ell) \subset V_\rho \times T^\ell$, with hamiltonian

$$\underline{\omega}_0 \cdot \underline{A} + f(\underline{A}, \underline{\phi}, \varepsilon) - \sum_{k=0}^{\infty} \lambda_k \overline{f}_k (\lambda_k^{-1} (\lambda_k^{-1} \underline{A}_k, \varepsilon)^{[<2^{k+1}]} \tag{6.14}$$

where $(\underline{A}_k, \underline{\phi}_k) = \overline{\mathcal{C}}_{k-1}(\underline{A}_{k-1}, \underline{\phi}_{k-1})$, $\underline{A} \equiv \underline{A}_0$, $\underline{\phi} \equiv \underline{\phi}_0$, $\lambda_k = \exp -4\delta_k$

and $\overline{f}_1, \overline{f}_2, \ldots$ are constructed recursively in building the maps $\mathcal{T}_{\delta_0}, \mathcal{T}_{\delta_1}, \ldots$ as the "counterterms" necessary to define the solution Φ of the equation corresponding to (6.4).

In this case the main difference with respect to the case treated in §3 is that $\prod_{k=0}^{\infty} \lambda_k^{-1} > 0$. This means that there is very little loss of analyticity in defining Φ . The phenomenon is due to the systematic subtraction procedure used at each step, which eliminates the resonance problems.

The fact that $\prod_{k=}^{\infty} \lambda_k < +\infty$ also implies that the rescalings performed at each step could be avoided. In fact the maps $\mathcal{C}_i$ can be composed on the set $V_{\rho e^{-\Sigma\delta_i}} \times T^\ell$, for $|\varepsilon| < \overline{\theta}$, and yield a canonical map

$$\mathcal{C}_\infty = \lim_n \mathcal{C}_n \dots \mathcal{C}_0 \tag{6.15}$$

mapping solutions of the hamiltonian equations with hamiltonian $\underline{\omega}_0 \cdot \underline{A}$ into solutions relative to

$$\underline{\omega}_0 \cdot \underline{A} + f(\underline{A},\underline{\phi},\varepsilon) - \sum_{k=0}^{\infty} \overline{f}_k(\underline{A}_k,\varepsilon)^{[<2^{k+1}]} \tag{6.16}$$

if we define recursively $(\underline{A}_k,\underline{\phi}_k)$ such that $(\underline{A}_0,\underline{\phi}_0) \equiv (\underline{A},\underline{\phi})$, $(\underline{A}_k,\underline{\phi}_k) = \mathcal{C}_k(\underline{A}_{k+1},\underline{\phi}_{k+1})$.

Simple arguments on formal power series show that a necessary and sufficient condition for having $\overline{f}_k \equiv 0$, hence for integrability by quadratures of $\omega_0 \cdot A + f(\underline{A},\underline{\phi},\varepsilon)$ and its canonical conjugability to $\underline{\omega}_0 \cdot \underline{A}$ is that the equation

$$\underline{\omega}_0 \cdot (\underline{A}' + \frac{\partial\Phi}{\partial\underline{\phi}}(\underline{A}',\underline{\phi})) + f(\underline{A}' + \frac{\partial\Phi}{\partial\underline{\phi}}(\underline{A}',\underline{\phi}),\underline{\phi},\varepsilon) = \underline{\omega}_0 \cdot \underline{A}' \tag{6.17}$$

with

$$f = \sum_{k=1}^{\infty} \varepsilon^k f^{(k)}(\underline{A},\underline{\phi}) \; ; \; \Phi(\underline{A},\underline{\phi}) = \sum_{k=1}^{\infty} \varepsilon^k \Phi^{(k)}(\underline{A},\underline{\phi}) \, , \qquad \overline{\Phi}^{(k)}(\underline{A}) \equiv 0 \tag{6.18}$$

admits a solution in the sense of the formal power series. It is easy to check that this implies infinitely many conditions on f e.g. for $k = 1$ it implies

$$\overline{f}^{(1)}(\underline{A}) \equiv 0 \tag{6.19}$$

For $k = 2$ it implies that the equation for $\Phi^{(2)}$

$$\underline{\omega}_0 \cdot \frac{\partial\Phi^{(2)}}{\partial\underline{\phi}}(\underline{A}',\underline{\phi}) + [\frac{\partial f^{(1)}}{\partial\underline{A}'}(\underline{A}',\underline{\phi}) \cdot \frac{\partial\Phi^{(1)}}{\partial\underline{\phi}}(\underline{A}',\underline{\phi}) + f^{(2)}(\underline{A}',\underline{\phi})] = 0 \tag{6.20}$$

should have a solution $\Phi^{(2)}$, if

$$\Phi^{(1)}(\underline{A}',\underline{\phi}) = \sum_{\underline{r}\neq 0} \hat{f}^{(1)}_{\underline{r}}(\underline{A}') \, e^{i\underline{r}\cdot\underline{\phi}}(-i\underline{\omega}_0 \cdot \underline{r})^{-1} \tag{6.21}$$

and this means that the average over $\underline{\phi}$ of the function in square brackets in (6.20) vanishes

$$\sum_{\underline{r}\neq\underline{0}} \left(\frac{\partial\hat{f}^{(1)}_{\underline{r}}}{\partial\underline{A}'}(\underline{A}')\cdot\frac{i\underline{r}}{-i\underline{\omega}_0\cdot\underline{r}}\right)\hat{f}^{(1)}_{-\underline{r}}(\underline{A}') = 0 \qquad (6.22)$$

etc.

So we arrive at the following theorem.

<u>Theorem</u>: Consider the hamiltonian system with hamiltonian

$$H_\varepsilon = \underline{\omega}_0\cdot\underline{A} + f(\underline{A},\underline{\phi},\varepsilon) \qquad (6.23)$$

and analytic on $V_\rho \times T^\ell$ and in ε near zero. Suppose that the (6.17) is formally soluble in the sense of the power series. Then it is in fact soluble for ε small enough with a Φ given by a convergent power series in ε .

The above theorem is due to Russmann [13], it was found also in the context of renormalization theory [10]. Its formulation in terms of our analysis is more natural as follows.

<u>Theorem</u>: Under the assumptions above, suppose that the hamiltonian system (6.23) can be conjugated canonically to $\underline{\omega}_0\cdot\underline{A}'$ on large sets of phase space for ε small* . Then the maps $\mathcal{J}_{\delta_i}$ are such that

$$\mathcal{J}_{\delta_n}\cdots\mathcal{J}_{\delta_0}H_\varepsilon \xrightarrow[n\to\infty]{} \underline{\omega}_0\cdot\underline{A}' \qquad (6.24)$$

It would of course be nicer to deal with a single transformation rather than with a sequence of them. This can be easily done by defining a map $\mathcal{J}$ in the space of the analytic one-parameter families of hamiltonians of the form (6.23) as $\mathcal{J}_\delta$ with δ directly defined in terms of f . For instance we could take

$$\delta_f = \xi/(\log[-\log C_0 \,\|f\|_{\rho,\xi,\theta}\,\rho^{-2} + e^2])^2 \qquad (6.25)$$

Let us now derive the following corollary of the above ideas (it is basically equivalent to a theorem of Dinaburg-Sinai on the quasi-periodic Schroedinger equation [8],[9]).

---

*i.e. given $\bar{\rho} < \rho$, $\bar{\xi} < \xi$, $\exists\vartheta > 0$ such that $\underline{\omega}_0\cdot\underline{A}'$ on $V_{\bar{\rho}} \times T^\ell$ is canonically conjugated to (6.23) by a canonical map $\mathcal{C}$ holomorphic on $C(\bar{\rho},\bar{\xi},\bar{\theta})$, close to the identity within $\theta(\varepsilon)$.

Corollary: Let $\underline{f}$ be holomorphic in $C(\xi_0)$ with values in C^ℓ, real on T^ℓ. Consider

$$H_0(\underline{A},\underline{\phi},\varepsilon) = \underline{\omega}_0 \cdot A + \varepsilon \underline{A} \cdot \underline{f}(\underline{\phi}) \tag{6.26}$$

There exist, for ε small enough, a "counterterm" of the form $\underline{A} \cdot \underline{a}(\varepsilon)$ such that

$$\underline{\omega}_0 \cdot \underline{A} + \varepsilon \underline{A} \cdot \underline{f}(\underline{\phi}) - \underline{A} \cdot \underline{a}(\varepsilon) \tag{6.27}$$

is canonically conjugate to $\underline{\omega}_0 \cdot A$ on the whole space $R^\ell \times T^\ell$.

Proof: We apply a recursive algorithm. Imagine labeling its steps by an index $n = 1,2,\ldots$, so that after the n-th step we shall have built a canonical transformation $\mathcal{C}^{(n-1)}$ changing

$$\underline{\omega}_0 \cdot \underline{A} + \varepsilon \underline{A} \cdot \underline{f}(\underline{\phi}) - \underline{A} \cdot \sum_{k=1}^{2^n-1} \varepsilon^k \underline{a}_k \tag{6.28}$$

into

$$H_n(\underline{A}_n,\underline{\phi}_n) = \underline{\omega}_0 \cdot \underline{A}_n + \underline{A}_n \cdot f_n(\underline{\phi}_n,\varepsilon) \tag{6.29}$$

where $(\underline{A},\underline{\phi}) \equiv (\underline{A}_0,\underline{\phi}_0) = \mathcal{C}^{(n-1)}(\underline{A}_n,\underline{\phi}_n)$ and f_n is divisible by ε^{2^n}, holomorphic in $\underline{\phi}_n,\varepsilon$ on $C(\xi_n,\theta_n) \equiv C(\xi_n) \times S^1_{\theta_n}$ and $\underline{a}_1,\ldots,\underline{a}_{2^n-1}$ are also defined inductively together with ξ_n,θ_n; furthermore $\mathcal{C}^{(n-2)^{-1}}\mathcal{C}^{(n-1)} \equiv$ {variation of $\mathcal{C}^{(n-1)}$ with respect to $\mathcal{C}^{(n-2)}$} $\equiv \mathcal{C}_{n-1}$ is holomorphic on $C(\xi_n,\theta_n)$ and maps $C^\ell \times C(\xi_n)$ into $C^\ell \times C(\xi_{n-1})$ as

$$\begin{aligned} \underline{A}_{n-1} &= (1 + J_{n-1}(\underline{\phi}_{n-1}))\underline{A}_n \\ \underline{\phi}_n &= \underline{\phi}_{n-1} + \underline{\Phi}_{n-1}(\underline{\phi}_{n-1}) \end{aligned} \tag{6.30}$$

where $\underline{\Phi}_{n-1}(\underline{\phi}_{n-1}) \cdot \underline{A}_n \equiv \Phi(\underline{A}_n,\underline{\phi}_{n-1})$ is the generating function of $\mathcal{C}_{n-1}$, and $J_{n-1}(\underline{\phi}) \equiv (\partial\underline{\Phi}_{n-1}/\partial\underline{\phi})^T$.

Given $\underline{f}_n$ the function $\underline{\Phi}_n$ and the $\underline{a}_j$, $2^n \leq j < 2^{n+1} - 1$, are determined as follows.

1) Define

$$\bar{a}_n(\varepsilon) \equiv \bar{a}_{2^n} \varepsilon^{2^n} + \ldots + \bar{a}_{2^{n+1}-1} \varepsilon^{2^{n+1}-1} = \tag{6.31}$$

$$= \int \frac{d\,\underline{\phi}}{(2\pi)^\ell} [\underline{f}_n(\underline{A}_n,\underline{\phi}_n,\varepsilon)]^{[2^{n+1}-1]}$$

where $[<p]$ is the series truncation operation considered above, (6.2).

2) Next consider the matrices $J_0(\underline{\phi}_0),\ldots, J_{n-1}(\underline{\phi}_{n-1})$ and note that

$$\underline{A} = (1+J_0(\underline{\phi}_0))^{-1}\ldots(1+J_{n-1}(\underline{\phi}_{n-1}))^{-1}\underline{A}_n \equiv M_n(\underline{\phi}_n)\underline{A}_n \tag{6.32}$$

so that if $\bar{M}_n(\varepsilon)$ denotes the average of $M_n(\underline{\phi})$ over $\underline{\phi}$ and if

$$[(\bar{M}_n(\varepsilon))^{-1}\underline{a}_n(\varepsilon)]^{[<2^{n+1}-1]} = \bar{\underline{a}}_n(\varepsilon) \tag{6.33}$$

defines $\underline{a}_n(\varepsilon) = \underline{a}_{2^n}\varepsilon^{2^n} + \ldots + \underline{a}_{2^{n+1}}\varepsilon^{2^{n+1}-1}$, we see that the function

$$\underline{A}_n \cdot \tilde{\underline{f}}_n(\underline{\phi}_n) = \underline{A}_n \cdot \underline{f}_n(\underline{\phi}_n,\varepsilon) - \underline{A}_0 \cdot \underline{a}_n(\varepsilon) \equiv$$

$$\equiv \underline{A}_n \cdot \underline{f}_n(\underline{\phi}_n,\varepsilon) - \underline{a}_n(\varepsilon) \cdot M_n(\underline{\phi}_n)\underline{A}_n \tag{6.34}$$

has zero average over $\underline{\phi}_n^n$ to order $\varepsilon^{2^{n+1}-1}$.

3) Hence we can define

$$\underline{\Phi}_n(\underline{\phi}_n) = \left[\sum_{\underline{r}\neq 0} \tilde{f}_{n\underline{r}} \frac{e^{i\underline{r}\cdot\underline{\phi}_n}}{-i\underline{\omega}_0\cdot\underline{r}}\right]^{[<2^{n+1}-1]} \tag{6.35}$$

and iterate the procedure.

To discuss the bounds necessary in order that the above construction is really possible let

$$\varepsilon_n = \|\underline{f}\|_{\xi_n,\theta_n}, \quad \eta_n = \|J_{n-1}\|_{\xi_n} \tag{6.36}$$

From (6.31) we find, choosing $\delta_k = \xi_0/5\pi(1+k)^2$, $\rho \geq 0$

$$\|\overline{a}_{n+1}\|_{\theta_n e^{-\delta_n}} \leq B_1\, \varepsilon_n\, \delta_n^{-1} \tag{6.37}$$

using bounds similar to (3.16), and from (6.32)

$$\|M_n\|_{\varepsilon_n-\delta_n,\, \theta_n\, e^{-\delta_n}} \leq \prod_{i=1}^{n-1} (1+B_2\eta_i) \tag{6.38}$$

having also used in (6.37), (6.38) dimensional bounds. So that by (6.33)

$$\|\underline{a}_{n+1}\|_{\theta\, e^{-\delta_n}} \leq B_3\, \delta_n^{-3}\, \varepsilon_n \prod_{j=1}^{n} (1+B_2\eta_j) \tag{6.39}$$

and by (6.34), (6.35)

$$\|\underline{\tilde{f}}_n\|_{\xi_n-\delta_n,\, \theta_n\, e^{-\delta_n}} \leq \varepsilon_n + B_2\, \delta_n^{-3}\, \varepsilon_n \prod_{j=1}^{n} (1+B_2\eta_j)^2 \tag{6.40}$$

$$\|\Phi_n\|_{\xi_n-2\delta_n,\, \theta_n\, e^{-\delta_n}} \leq B_4\, C_0\varepsilon_n\delta_n^{-3} \prod_{j=1}^{n} (1+B_2\eta_j)^2\, \delta_n^{-2\ell-1} \tag{6.41}$$

Therefore, if $\|\Phi\|_n$ is small enough, by the implicit functions' theorems applied several times, the function $\underline{A}'_{n+1}\cdot\underline{\Phi}_n(\underline{\phi}_n)$ generates canonical maps $\mathcal{C}'_n$, $\mathcal{C}_n$ such that

$$\begin{aligned}\mathcal{C}'_n, \mathcal{C}_n:\ & C(\xi_n-3\delta_n,\, \theta_n\, e^{-\delta_n}) \to C(\xi_n-2\delta_n) \\ & C(\xi_n-4\delta_n,\, \theta_n\, e^{-\delta_n}) \to C(\xi_n-3\delta_n)\end{aligned} \tag{6.42}$$

of the form (6.30). The smallness condition is that $\partial\underline{\Phi}_n/\partial\underline{\phi}$ be small compared to δ_n. One easily finds that the precise condition has the form.

$$B_5C_0\varepsilon_n\delta_n^{-4+2\ell} \prod_{j=1}^{n} (1+B_2\eta_j)^2 < 1 \tag{6.43}$$

for a suitable B_5 .

Then from (6.41) it follows

$$\eta_{n+1} \equiv \|J_n\|_{\xi_n-4\delta_n,\ \theta_n e^{-\delta_n}} \leq B_6 C_0 \varepsilon_n \delta_n^{-4-2\ell} \prod_{j=1}^{n} (1+B_2\eta_j)^2 \tag{6.44}$$

Finally, with some patience and using the same methods quoted in §3 to estimate ε_{n+1} in terms of ε_n we find that, if

$$\xi_{n+1} = \xi_n - 5\delta_n, \quad \theta_{n+1} = \theta_n e^{-2\delta_n}$$

$$\|\underline{f}_{n+1}\|_{\xi_{n+1},\ \theta_{n+1}} \equiv \varepsilon_{n+1} \leq B_7 (C_0 \varepsilon_n^{\ 2} + \varepsilon_n e^{-2^{n+1}\delta_n}) \delta_n^{-b} \tag{6.45}$$

where B_7, $b > 0$ are suitably chosen, provided (6.43) holds.

To summarize, if (6.43) holds

$$\begin{aligned} &\xi_{n+1} = \xi_n - 5\delta_n \qquad \theta_{n+1} = \theta_n e^{-2\delta_n} \\ &\varepsilon_{n+1} \leq B_7 (C_0 \varepsilon_n^{\ 2} + \varepsilon_n e^{-2^{n+1}\delta_n}) \delta_n^{-b_1} \\ &\eta_{n+1} \leq C_0 \varepsilon_n \delta_n^{-b_2} \prod_{j=1}^{n} (1+B_2\eta_j)^2 \end{aligned} \tag{6.46}$$

where the B's and the b's are suitable constants (depending only on ℓ).

Proceeding as in §3 one easily deduces from (6.46) that if ε_0 is small enough then η_n, $\varepsilon_n \to 0$ very fast. In turn, similarly in §3, this easily implies the existence of the limit

$$\mathcal{C} = \lim_{n\to\infty} \mathcal{C}^{(n)} \equiv \lim_{n\to\infty} \mathcal{C}_n \cdots \mathcal{C}_0 \tag{6.47}$$

and its holomorphy on $C^{\ell} \times C(\xi_0/2) \times C(\theta_\infty)$ where $\theta_\infty = \theta_0 \exp - 2 \sum_{n=0}^{\infty} \delta_j$. The (6.39) and the rapidity at which ε_n, $\eta_n \to 0$ as $n \to \infty$ also imply that the series for $\underline{a}(\varepsilon)$ converges for $|\varepsilon| < \theta_\infty$ and, by construction,

$$\underline{\omega}_0 \cdot \underline{A} + \varepsilon \underline{A} \cdot \underline{f}(\underline{\phi}) - \underline{A} \cdot \underline{a}(\varepsilon) \tag{6.48}$$

is conjugated by $\mathcal{C}$ to $\underline{\omega}_0 \cdot \underline{A}$.

With some more work one could prove that $a(\varepsilon)$ can be defined over a large set of $\underline{\omega}_0$'s and "depends smoothly on them". The model for such a proof is provided by [14],[12],[5],[11]; precisely.

<u>Theorem</u>: There is a C^∞-function $\underline{\omega}_0 \to \underline{a}(\underline{\omega}_0, \varepsilon)$ such that when $\underline{\omega}_0$ varies in the set of the $\underline{\omega}_0 \in R^\ell$ verifying, for $\underline{r} \in Z^\ell$

$$|\underline{\omega}_0 \, \underline{r}|^{-1} < C_0 |\underline{r}|^\ell \qquad \forall \, 0 < |\underline{r}| \tag{6.49}$$

the hamiltonian

$$\underline{\omega}_0 \cdot \underline{A} + \varepsilon A \cdot \underline{f}(\underline{\phi}) - \underline{A} \cdot \underline{a}(\varepsilon, \underline{\omega}_0) \tag{6.50}$$

is analytically canonically conjugated with $\underline{\omega}_0 \cdot \underline{A}$ on the whole phase space.

See also [15].

A corollary to the last two theorems are the results on the quasi-periodic Schroedinger equation and its Bloch waves, discussed in §2.

§7 Remarks and Problems on the Chaotic Transition in Isochronous Systems

We consider a one-parameter family of systems like (6.1)

$$H_\varepsilon = \underline{\omega}_0 \cdot \underline{A} + f(\underline{A}, \underline{\phi}, \varepsilon) \qquad \text{on } V_\rho \times T^\ell \tag{7.1}$$

which is supposed to be analytically conjugated to the system $\underline{\omega}_0 \cdot \underline{A}$ for ε small enough, in the sense of the footnote to p. 78. The vector $\underline{\omega}_0$ verifies (3.9).

Then the analysis of §6 shows that if we act on H_ε with the transformations $\mathcal{T}$ defined by (6.10), (6.25) and if ε is small enough

$$\mathcal{T}^n H_\varepsilon \to \underline{\omega}_0 \cdot \underline{A} \qquad \text{on } V_{\bar\rho} \times T^\ell \tag{7.2}$$

for $\bar\rho < \rho$.

We now investigate the possibility that for some value of $\varepsilon = \varepsilon_c$ the (7.2) might fail but still $\mathcal{T}^n H_{\varepsilon_c} \to \underline{\omega}_0 \cdot \underline{A} + f^*(\underline{A},\underline{\phi})$ for some f*. It is clear that, if this happens, f* must verify some equation of "fixed point type". The analysis that follows is a heuristic analysis aiming at finding equations for f*.

To understand which could be this equation, recall the definition of $\mathcal{T}$ on $\mathcal{T}^n H_\varepsilon$.

1) Decompose $f_n(\underline{A},\underline{\phi},\varepsilon)$

$$f_n(\underline{A},\underline{\phi},\varepsilon) = f_n^-(\underline{A},\underline{\phi},\varepsilon) + f_n^+(\underline{A},\underline{\phi},\varepsilon) \tag{7.3}$$

where $f^-(\underline{A},\underline{\phi},\varepsilon) = f_n(\underline{A},\underline{\phi},\varepsilon)^{[<2^{n+1}]}$, where $[<p]$ denotes as in §6 the power series truncation operation (with respect to ε).

2) By the integrability hypothesis, see §6, $\bar{f}^- \equiv 0$ and we do not have to perform any subtraction. Define

$$\Phi^-(\underline{A}',\underline{\phi},\varepsilon) = \sum_{\underline{r}\neq\underline{0}} \frac{\hat{f}_{\underline{r}}^-(\underline{A})e^{i\underline{r}\cdot\underline{\phi}}}{-i\underline{\omega}_0\cdot\underline{r}} \tag{7.4}$$

and generate with it a canonical transformation on a suitable $V_{\tilde\rho} \times T^\ell$, via

$$\begin{aligned} \underline{A} &= \underline{A}' + \frac{\partial\Phi^-}{\partial\underline{\phi}}(\underline{A}',\underline{\phi},\varepsilon) \\ \underline{\phi}' &= \underline{\phi} + \frac{\partial\Phi^-}{\partial\underline{A}'}(\underline{A}',\underline{\phi},\varepsilon) \end{aligned} \tag{7.5}$$

3) Then define $\lambda = \rho/\tilde{\rho}$ and

$$\begin{aligned}\mathcal{T}^{n+1}H_\varepsilon &= \lambda\underline{\omega}_0 \cdot (\lambda^{-1}\underline{A}' + \frac{\partial\Phi^-}{\partial\underline{\phi}}(\lambda^{-1}\underline{A}',\underline{\phi},\varepsilon)) + \\ &+ \lambda f^-(\lambda^{-1}\underline{A}' + \frac{\partial\Phi^-}{\partial\underline{\phi}}(\lambda^{-1}\underline{A}',\underline{\phi},\varepsilon),\underline{\phi}) + \\ &+ \lambda f^+(\lambda^{-1}\underline{A}' + \frac{\partial\Phi^-}{\partial\underline{\phi}}(\lambda^{-1}\underline{A}',\underline{\phi},\varepsilon),\underline{\phi},\varepsilon) = \\ &= \underline{\omega}_0 \cdot \underline{A}' + f_{n+1}(\underline{A}',\underline{\phi}',\varepsilon)\end{aligned} \tag{7.6}$$

So we see that the $\mathcal{T}^n H_\varepsilon$ will converge to a limit as $n \to \infty$ for $\varepsilon = \varepsilon_c$ if, for $\varepsilon = \varepsilon_c$, $f_n(\underline{A},\underline{\phi},\varepsilon) \to f^*(\underline{A},\underline{\phi})$. This can be imposed by requiring the (stronger) condition that $f^+ \to f^*_+$, $f^- \to f^*_-$. In this case $f^*_\pm$ have to be related by

$$\lambda f^*_-(\lambda^{-1}\underline{A}' + \frac{\partial\Phi^-}{\partial\underline{\phi}}(\lambda^{-1}\underline{A}',\underline{\phi}),\underline{\phi}) + \lambda f^*_+(\lambda^{-1}\underline{A}' + \frac{\partial\Phi^-}{\partial\underline{\phi}}(\lambda^{-1}\underline{A}',\underline{\phi}),\underline{\phi})$$

$$-\lambda f^*_-(A',\phi,\varepsilon) = f^*_-(\underline{A}',\underline{\phi} + \frac{\partial\Phi^-}{\partial A'}(\lambda^{-1}\underline{A}',\phi)) + f^*_+(\underline{A}',\underline{\phi} + \frac{\partial\Phi^-}{\partial\underline{A}'}(\lambda^{-1}\underline{A}',\underline{\phi}'))$$

$$\Phi^-(\underline{A}',\underline{\phi}) = \sum_{\underline{r}\neq\underline{0}} (\hat{f}^*_-(\underline{A}'))_{\underline{r}}\, e^{i\underline{r}\cdot\underline{\phi}}(-i\underline{\omega}_0 \cdot \underline{r})^{-1} \quad \text{and}$$

$$\overline{f^*_-} \equiv 0 \tag{7.7}$$

for some $\lambda > 1$.

The above equation looks quite difficult to study in general. From the discussions of §6 on renormalization and Wick ordering one gets, however, the idea of inquiring into the existence of special solutions of the form

$$\begin{aligned} f^*_-(\underline{A}',\underline{\phi}) &= \underline{A}' \cdot \underline{f}^*(\underline{\phi}) \\ f^*_+(\underline{A}',\underline{\phi}) &= \underline{A}' \cdot \underline{a}^* \end{aligned} \tag{7.8}$$

Then, if $\underline{\Phi}^*$ is so defined that

$$\underline{\omega}_0 \cdot \frac{\partial(\underline{A}' \cdot \underline{\Phi}^*(\underline{\phi}))}{\partial\underline{\phi}} + \underline{A}' \cdot \underline{f}^*(\underline{\phi}) = 0 \tag{7.9}$$

the above fixed point equation becomes

$$(\underline{\omega}_0+\underline{a}^*) \cdot \frac{\partial(\underline{A}'\cdot\underline{\Phi}^*)}{\partial\underline{\phi}}(\underline{\phi}) + (\underline{A}' + \frac{\partial(\underline{A}'\cdot\Phi^*)}{\partial\underline{\phi}}(\underline{\phi})) \cdot \underline{f}^*(\underline{\phi}) = \\ = \underline{A}'\cdot\underline{f}^*(\underline{\phi}+\underline{\Phi}^*(\underline{\phi})) \tag{7.10}$$

This equation looks better but it is still unclear whether it does admit non-trivial solutions. If one looks for solutions "with separated variables" i.e. of the form

$$\Phi^*(\underline{\phi}) = (\Phi_1^*(\phi_1),\ldots,\Phi_\ell^*(\phi_\ell)) \tag{7.11}$$

One then finds for each component an equation like

$$(\omega + a^*)\frac{\partial\Phi^*}{\partial\phi}(\phi) + (1+\frac{\partial\Phi^*}{\partial\phi}(\phi))\, f^*(\phi) = f^*(\phi+\Phi^*(\phi)) \\ \overline{f^*} = 0\,, \qquad \omega_0\frac{\partial\Phi^*}{\partial\phi}(\phi) + f^*(\phi) = 0 \tag{7.12}$$

which, if we set $g^* = f^*/\omega_0$, $\alpha^* = \alpha^*/\omega_0$, becomes[†]

$$\alpha^* g^*(\phi) + g^*(\phi)^2 + g^*(\phi+\int_0^\phi g^*(\phi')d\phi') = 0 \\ \overline{g^*} = 0 \tag{7.13}$$

It is unclear whether this equation in (α^*,g^*) has a non-trivial solution. This seems quite probable. Even less clear is the existence of a non-trivial solution to (7.10) with non-separated variables (note that (7.10) is a system of ℓ equations because of the arbitrariness of $\underline{A}'$)

The latter question should be relevant for the analysis of the chaotic transition in systems with fixed spectrum. In fact if (7.10) admitted non-trivial solutions, they could provide a key to the theory of the disappearance of the quasi-periodic motions which for $\varepsilon < \varepsilon_c$, fill the phase space, in a way similar to that proposed in [12], [16].

I hope that this matter will be investigated further, possibly trying to find a solution to (7.10) by the numerical methods [17], [18], [19], [16].

---

†Assuming that $\Phi^*(0) = 0$, which is not restrictive.

Appendix A. Arnold-Liouville Theorem

The fact that $I_1,\dots,I_\ell$ have vanishing Poisson brackets has two important consequences.

i) On the surface $\underline{I} = \underline{i}$ one can define ℓ independent vector fields which act on functions via the Poisson bracket with the $I_1,\dots,I_\ell$

$$L_j f = \{I_j, f\} \tag{A.1}$$

The assumption insures that such fields commute. Since they are smooth on the smooth compact surfact $T(\underline{I})$ this implies that $T(\underline{I})$ is a torus.

ii) Consider the relations $I_j(\underline{p},\underline{q}) = I_j$ and imagine that they can be inverted as $\underline{p} = \underline{\alpha}(\underline{I},\underline{q})$. Then $\underline{p}\cdot d\underline{q} \equiv \underline{\alpha}(\underline{I},\underline{q})\cdot d\underline{q}$ is a differential form on $T(\underline{I})$ at fixed $\underline{I}$. Its local integrability conditions coincide, as a simple calculation shows ([1], p. 360), with the condition that $\{I_i, I_j\} = 0$.

So $T(\underline{I})$ are "lagrangian manifolds", i.e. manifolds on which $\underline{p}\cdot d\underline{q}$ is a locally integrable form.

Let $\underline{p} = \underline{\alpha}(\underline{\theta})$, $\underline{q} = \underline{\beta}(\underline{\theta})$ be a parametric representation of the torus $T(\underline{I})$. We consider the ℓ closed curves on $T(\underline{I})$

$$\lambda_j = \{\underline{p},\underline{q} \mid \underline{\theta} = (\theta_1,\dots,\theta_\ell),\ \theta_k = 0 \quad \text{for} \quad k \neq j\} \tag{A.2}$$

Then let

$$A_j = \frac{1}{2\pi} \oint_{\lambda_i} \underline{p}\cdot d\underline{q} \tag{A.3}$$

and suppose that the $\underline{I}$'s can be expressed in terms of the $\underline{A}$'s.

Define the generating function

$$\Phi(\underline{A},\underline{q}) = \int_0^{\underline{q}} \underline{p}\cdot d\underline{q} \tag{A.4}$$

where the integral is along a line on $T(\underline{I})$ from the $(\underline{p},\underline{q})$ corresponding to $\underline{\theta} = \underline{0}$ (assuming that such a point is chosen to be smoothly dependent on $\underline{I}$) and built by segments along the axes. Then let

$$\begin{aligned} \underline{\phi} &= \frac{\partial\Phi}{\partial\underline{A}}(\underline{A},\underline{q}) \\ \underline{p} &= \frac{\partial\Phi}{\partial\underline{q}}(\underline{A},\underline{q}) \end{aligned} \tag{A.5}$$

and (A.4) shows that $\underline{p}$ has the right interpretation.

Suppose that (A.5) can be invented smoothly into a map $(\underline{p},\underline{q}) \leftrightarrow (\underline{A},\underline{\phi})$. It is clear that we have to check, in order to insure this smooth invertability that the $\underline{\phi}$'s admit the interpretation of angles. A necessary condition is that if $\underline{q}$ is thought once as the point $\underline{\theta} = \underline{0}$ and once as the point $\underline{\theta}_{2\pi} = (2\pi, 0, \ldots, 0)$, obtained by moving on the circle λ_1, the variable ϕ_1 varies by 2π. This in fact follows immediately because in the second case

$$\Phi(\underline{A},\underline{q}) = \oint_{\lambda_1} \underline{p} \cdot d\underline{q} = 2\pi A_1$$

So if the map (A.5) can define canonical transformation it will map $T(\underline{I})$ onto $V \times T^{\ell}$ where V is the domain of variability of the $\underline{A}$-functions. Clearly $I_1 \equiv H$ will have the form $H = h(\underline{A})$ and the system will have been integrated by quadratures (actually the quadratures are the ℓ one-dimensional integrals (A.4)).

As an application of the above method of construction of the action angle variables we shall show in Appendx B that any one-dimensional periodic Schroedinger equation gives rise to a hamiltonian system (as in §2) integrable by quadratures.

The above procedure does not necessarily work in the sense that not all the assumptions made in the derivation need to be satisfied even when action-angle variables do exist (and, therefore, have to be built in a different way). The simplest example of such a situation is provided by a one-dimensional system with a hamiltonian $H(p,q)$ such that the curves $H(p,q) = E$ contain a horizontal segment, $\forall\, E > 0$, (Exercise).

Appendix B. Integration by quadratures of the hamiltonian system associated with the periodic Schroedinger equation

We study here the hamiltonian (2.5) which is "equilvalent" to

$$-\ddot{q} = (E - \varepsilon V(\phi_0 + \omega t))q \tag{B.1}$$

in the sense that the map $(p,q,B,\phi) \leftrightarrow (A, \; B,\phi)$

$$A = (p^2 + Eq^2)/2\sqrt{E}$$
$$\cos\theta = p/\sqrt{2\sqrt{E}\,A}\,, \quad \sin\theta = q\,\sqrt{E}/\sqrt{2\sqrt{E}\,A} \tag{B.2}$$

maps solutions of (B.1) into solutions of the Hamiltonian

equations with hamiltonian, $V \in C^{\infty}(T^1)$

$$H = A\sqrt{E} + \omega B - \frac{\varepsilon}{\sqrt{E}} V(\phi)\, A \sin^2\theta \qquad (B.3)$$

Let $k_0 \in [-\frac{\pi}{\omega}, \frac{\pi}{\omega}]$; one defines a Bloch wave with wave number k_0, a solution of (B.1) of the form

$$q(t) = e^{ik_0 t}\, q_0(\omega t+\phi_0) \qquad (B.4)$$

where $q_0 \in C^{\infty}(T^1)$. The equation for q_0 is

$$(k_0^2 - E)q_0 - 2ik_0\,\dot{q}_0 + \varepsilon V(\omega t+\phi_0)q_0 - \omega^2\ddot{q} = 0 \qquad (B.5)$$

The (B.5) is a self-adjoint equation on $L_2(T^1)$ with discrete spectrum. Fixed k_0 we can find a sequence $E_n(k_0)$, $n = 0,1...$, of eigenvalues for (B.5). The parametrization can be made so that $E_n(k_0)$ are, for each n, continuous functions of k_0. So one has the following picture:

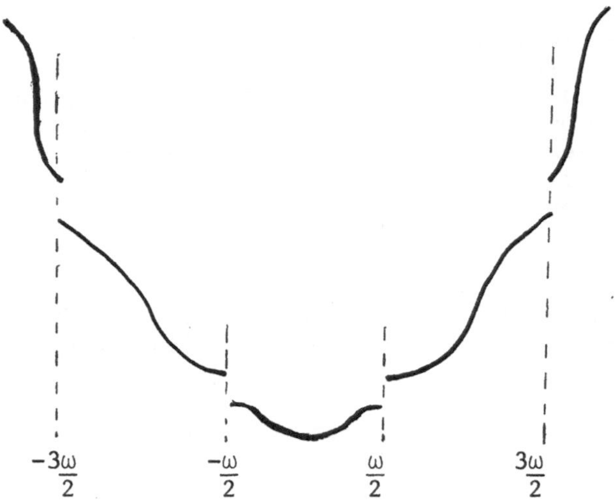

draw $E_0(k_0)$ over the interval $k_0 \in [-\frac{\omega}{2},\frac{\omega}{2}]$. Every value E has double multiplicity because there are two independent Bloch waves with the values of the wave number equal to $\pm k_0$. Then we draw $E_1(k_0)$ over the interval $(\frac{\omega}{2},\frac{3\omega}{2})$ and $E_2(k_0)$ over the interval $(\frac{\omega}{2},\frac{3\omega}{2})$ and $E_2(k_0)$ over the interval $(-\frac{3\omega}{2},-\frac{\omega}{2})$ instead of drawing them again on $(-\omega,\omega)$. In this way the multiplicity is automatically taken into account. Furthermore, $E(k_0)/k_0^2 \xrightarrow[k\to\infty]{}$ constant where $E(k_0)$ is the function whose graph has been constructed in the above picture. It is well known that $E(k_0)$ has discontinuities which can possibly exist only at the points $k_0 = \pm (2k+1)\frac{\omega}{2}$, $k = 0,1\ldots$.

Let $E = E(k_0)$. We claim that with such a value of E the (B.3) is integrable by quadratures if k_0 is different from $\pm (2k+1)\frac{\omega}{2}$, $k = 0,1,\ldots$.

The change of variables (B.2) and the correspondence between solutions of (B.1) and (B.3) suggests the introduction of

$$\begin{aligned} Q(\alpha_1,\alpha_2) &= \mathrm{Re}\, e^{i\alpha_1} q_{k_0}(\alpha_2) \\ P(\alpha_1,\alpha_2) &= \mathrm{Re}(ik_0\, e^{i\alpha_1} q_{k_0}(\alpha_2) + e^{i\alpha_1}\omega q'_{k_0}(\alpha_2)) \\ a(\alpha_1,\alpha_2) &= \frac{P^2 + EQ^2}{2\sqrt{E}} \\ \Gamma(\alpha_1,\alpha_2) &= \frac{\varepsilon}{2} V'(\alpha_2) Q^2(\alpha_1,\alpha_2) \end{aligned} \tag{B.6}$$

where the prime denotes differentiation.

Then, $\forall \rho \in R, \phi_0, \alpha_0 \in T^1$ the following functions

$$\begin{aligned} q(t) &= \rho\, Q(\alpha_0+k_0t, \phi_0+\omega t) \\ \phi(t) &= \phi_0 + \omega t \end{aligned} \tag{B.7}$$

as $(\rho,\alpha_0,\phi_0) \in R \times T^2$ vary we see that we have found a solution to the hamiltonian equations with initial data

$$\begin{aligned} p &= \rho\, \underline{P}(\alpha_0,\phi_0) \qquad & q &= \rho Q(\alpha_0,\phi_0) \\ \phi &= \phi_0 & B &= B_0 \end{aligned} \tag{B.8}$$

provided $B(t)$ is defined so that

$$\begin{aligned} \dot{B} &= \frac{\rho^2}{2}\varepsilon V'(\phi_0+\omega t) Q^2(\alpha_0+k_0t, \phi_0+\omega_0 t) = \\ &= \rho^2\, \Gamma(\alpha_0+k_0t, \phi_0+\omega t) \end{aligned} \tag{B.9}$$

i.e. if $|k_0r_1+\omega r_2|^{-1} \leq C(|r_1|+r_2|)^{\alpha}$, $\forall\ \underline{r} \in z^2$, $\underline{r} \neq \underline{0}$

$$B(t) = B_0 + \rho^2\Gamma_{00}t + \rho^2 \sum_{\underline{r}\neq\underline{0}} \Gamma_{\underline{r}}\, e^{i(r_1\alpha_1+r_2\alpha_2)} \frac{e^{i(r_1k_0+r_2\omega)t}-1}{i(r_1k_0+r_2\omega)} \tag{B.10}$$

where $\Gamma_{\underline{r}}$ are the Fourier's coefficients of Γ .

However, writing $Q(\alpha_1,\alpha_2) = (e^{i\alpha_1}q_0(\alpha_2) + e^{-i\alpha_1}\,\overline{q}_0(\alpha_2))/2$

it is easy to realize that Γ_{00} is proportional to

$$\int_0^{2\pi} V'(\alpha)\,|q_0(\alpha)|^2 d\alpha \tag{B.11}$$

which vanishes as it follows by considering (B.5), multiplying it by $\overline{q}_0'$ and taking the real part of the result and integrating over a period. So $\Gamma_{00} = 0$.

Therefore, the (B.10) with $\Gamma_{00} = 0$ tells us that setting $\gamma_{\underline{r}} = \Gamma_{\underline{r}}/-i(k_0r_1+\omega r_2)$

$$\mathcal{B} = B + \rho^2 \sum_{\underline{r}\neq\underline{0}} \gamma_{\underline{r}}\, e^{i\underline{\alpha}\cdot\underline{r}} \tag{B.12}$$

is a prime integral for H, at least for the values of k_0 such that $\underline{\omega}_0 = (k_0,\omega)$ verifies a diaphantine condition.

For such values we can find action angle variables easily. We have that the surfaces H = constant and ρ = constant (or = constant)* are manifestly tori, see (B.7), of which we know explicit parametric equations

$$p = \rho\, P(\alpha_1,\alpha_2) \qquad q = \rho\, Q(\alpha_1,\alpha_2) \tag{B.13}$$

$$\phi = \alpha_2 \qquad B = \mathcal{B} - \rho^2 \sum_{r\neq 0} \gamma_{\underline{r}}\, e^{i\underline{r}\cdot\underline{\alpha}} \equiv -\rho^2 g(\alpha_1,\alpha_2)$$

So that we can define action angle variables as, (see A.3)

---

* $H, \rho, \mathcal{B}$ are constants of the motion

$$A_1 = \rho^2 \int_0^{2\pi} P(\alpha,0) \frac{\partial Q}{\partial\alpha}(\alpha,0) \frac{d\alpha}{2\pi} + $$

$$+ 2\pi\mathcal{B} - \rho^2 \sum_{r_2\neq 0} \gamma_{0r_2} \tag{B.14}$$

$$A_2 = \rho^2 \int_0^{2\pi} P(0,\beta) \frac{\partial Q}{\partial\beta}(0,\beta) \frac{d\beta}{2\pi} + 2\pi\mathcal{B} - \rho^2 \sum_{r_1\neq 0} \gamma_{r,0}$$

while the conjugate angles can be computed by first finding an expression for $\mathcal{B}$, ρ^2 in terms of A_1, A_2 , which amounts to solving a linear 2 × 2 system, see (B.14), and then using the generating function

$$\Phi(A,q,\phi) = \int_0^{\alpha_1} [\rho^2 P(\alpha, 0) \frac{\partial Q}{\partial\alpha}(\alpha,0) + \mathcal{B} - \rho^2 g(\alpha,0)]d\alpha +$$

$$+ \int_0^{\alpha_2} [\rho^2 P(\alpha_1,\beta) \frac{\partial Q}{\partial\alpha}(\alpha_1,\beta) + \mathcal{B} - \rho^2 g(\alpha_1,\beta)]d\beta \tag{B.14}$$

with ρ^2 and $\mathcal{B}$ expressed in terms of A_1, A_2 .

Another family of solutions can be obtained, likewise, by choosing the imaginary part instead of the real part in (B.6).

It remains to study the values of k_0 such that $\underline{\omega}_0 = (k_0, \omega)$ does not verify a diophantine condition. Clearly the problem here is only that of the convergence of the series in (B.10). From the first and last of (B.6) we see that $\Gamma_{\underline{r}} \equiv 0$ unless $r_1 = \pm 2, 0$. Since $\Gamma_{0,0} = 0$ we see that no small denominators occur in the addends of (B.10) with $r_1 = 0$. The only problem can occur when $-2k_0 + r\omega$, or when $2k_0 + r\omega$, vanish for some integer r. Since we only consider $k_0 \in [-\frac{\omega}{2}, \frac{\omega}{2}]$ this can happen only for $r = 1$ or $r = -1$, i.e. for k_0 at the extremes of the interval ("Brilloin zone").

Clearly this means that (B.3) is integrable for all the values of E in the interior of the "bands", i.e. for $E \in \{E' | E' = E_n(k_0)$ for some n and for $k_0 \in (-\frac{\omega}{2}, \frac{\omega}{2})\}$. This set is a union of intervals which can be easily visualized from the above picture.

Exercise: find explicitly H in terms of A_1, A_2.

References

[1] G. Gallavotti, "The Elements of Mechanics", Springer-Verlag, New York, 1983

[2] V. Arnold, "Méthodes Mathématiques de la Mécanique Classique", MIR, Moscow, 1978

[3] H. Poincaré, "Méthodes Nouvelles de la Mécanique Celiste", Gauthier-Villiers, Paris, 1897

[4] J. Pöshel, Integrability of Hamiltonian Systems on Cantor Sets, Comm. Pure and Appl. Math., 35, 653, 1982

[5] L. Chierchia, G. Gallavotti, Smooth Prime Integrals for Quasi-Integrable Hamiltonian Systems, Nuovo Cimento, 67B, 277, 1982

[6] V. Arnold, Small Denominators and Problems of Stability of Motion in Classical and Celestial Mechanics, Russian Math. Surv., 18(6), 85, 1963

[7] J. Moser, Stable and Random Motions in Dynamical Systems, Princeton University Press, Princeton, 1963

[8] E. Dinaburg, Ja. Sinai, The One-Dimensional Schroedinger Equation with a Quasi-Periodic Potential, Functional Anal. Appl. 9, 279, 1975

[9] H. Rüssmann, On the One Dimensional Schrödinger Equation with a Quasi-Periodic Potential, Ann. New York Acad. Sci., 90, 197

[10] G. Gallavotti, A Criterion of Integrability for Perturbed Nonresonant Harmonic Oscillators. "Wick Ordering" of the Perturbations in Classical Mechanics and Invariance of the Frequency Spectrum, Comm. Math. Phys. 87, 365, 1982

The main result of this reference was actually essentially proved earlier by H. Rüssmann. However, the difference in notation and spirit might make it worthwhile for the reader to consult also my paper after the original reference of Rüssmann, see [13] below.

[11] G. Gallavotti, Lectures on Hamiltonian Systems, to appear in the series, Progress in Physics, ed. J. Fröhlich, Birkhauser, Boston, in 1983.

[12] S. Shenker, L. Kadanoff, Critical Behaviour of a KAM Surface: I, Empirical Results. J. Stat. Phys. 27, 631, 1982; M. Feigenbaum, L. Kadanoff, S. Shenker, Quasi-Periodicity in

Dissipative Systems. A Renormalization Group Analysis, Physica 5D, 370, 1983;
R. McKay, A Renormalization Approach to Invariant Circles in Area Preserving Maps, Princeton Preprint, 1982;
Other papers can be found by looking at the references of those listed above. See also R. McKay, Thesis, Princeton, 1982.

[13] H. Rüssmann, Über die Normalform Analytischer Hamiltonscher Differentialgleichungen in der Nähe Einer Gleichgewichtlösung, Math. Annalen, 169, 55, 1967

[14] V. Arnold, A Proof of a Theorem of A.N. Kolmogorov on the Invariance of Quasi-Periodic Motions under Small Perturbations of the Hamiltonian, Russian Math. Surv., 18(5), 9, 1963

[15] J. Pöshel, Examples of Discrete Schroedinger Operators with Pure Point Spectrum, Commun. Math. Phys. 88, 44,7, 1983

[16] D. Rand, S. Ostlund, J. Sethna, E. Siggia, Universal Transition from Quasi-Periodicity to Chaos in Dissipative Systems, Phys. Rev. Lett., 49, 132, 1982

[17] M. Feigenbaum, Quantitative Universality for a Class of Non-linear Transformations, J. Stat. Phys., 19, 25, 1978

[18] M. Campanino, H. Epstein, D. Ruelle, On Feigenbaum's Functional Equation $g \cdot g(\lambda x) + \lambda g(x) = 0$, Topology, 21, 125, 1982

[19] O. Lanford, A Computer Assisted Proof of Feigenbaum's Conjecture, Preprint, Berkeley, 1981

MEASURES INVARIANT UNDER MAPPINGS OF THE UNIT INTERVAL

Pierre Collet

Centre de Physique Théorique
Ecole Polytechnique
91128 Palaiseau Cedex , France

and

Jean-Pierre Eckmann

Dept. de Physique Théorique
32 B^d d'Yvoy
CH 1211 Genève 4
Suisse

It has been recognized in the last few years that dynamical systems with few degrees of freedom can play an important role in the description of some physical systems which behave in a chaotic way. The rather simple one dimensional dynamical systems can also be used to test some ideas about higher dimensional systems. The question of the statistical description of chaotic motions was already raised several decades ago ([U]) and after the discovery of the ergodic theorems it became obvious that an improtant notion is that of invariant measure. For a given dynamical system, there are in general many invariant measures, and one is led to the problem of choosing the relevant one (if any).

In order to analyze this question we shall, for the sake of simplicity, specialize the discussion to maps of the interval. We denote by f a piecewise C^1 map of the interval $I = [0,1]$. An invariant measure μ for f is a measure on the Borel σ algebra of $[0,1]$ such that for any Borel set A, one has

$$\mu(f^{-1}(A)) = \mu(A).$$

From a physical point of view, one would like to select an invariant measure describing the statistical properties of almost all trajectories. This notion was put into a precise form by Bowen and Ruelle, and the corresponding measures (if they exist) are now called Bowen-Ruelle measures. The definition is as follows. An invariant measure μ is a Bowen-Ruelle measure if for Lebesgue almost any point x in $[0,1]$, and for any continuous function g,

$$\lim_{n \to +\infty} n^{-1} \sum_{\ell=0}^{n-1} g(f^{\ell}(x)) = \int_0^1 g(y)\, d\mu(y) .$$

The problem is now how to characterize Bowen-Ruelle measures. This question was discussed for example in [R1]. By definition, the Bowen-Ruelle measure, if it exists, is unique. We shall restrict our attention to the simpler problem of finding the Bowen-Ruelle measure when it is absolutely continuous with respect to the Lebesgue measure. (This need not necessarily be the case see [CEL] for example). In the sequel, λ will always denote the normalized Lebesgue measure of $[0,1]$. When we shall speak of an absolutely continuous invariant measure (a.c.i.m. for short), we shall always mean absolute continuity with respect to the measure λ.

Assume μ is an invariant measure with a density h with respect to λ. The invariance of μ is equivalent to the relation

$$\mu(f^{-1}([0,x[)) = \mu([0,x[), \ \forall x \in [0,1] ,$$

and it is easy to verify that if f is piecewise C^1, the density h of μ must satisfy the equation

$$Ph = h$$

where

$$Ph(x) = \sum_{y,f(y)=x} h(y)\ |f'(y)|^{-1}.$$

Therefore, the question whether f has an absolutely continuous invariant measure is equivalent to the problem of solving in $L^1([0,1], d\lambda)$ the equation $Ph = h$. The operator P is a positive operator in $L^1([0,1], d\lambda)$ and is sometimes called the Ruelle - Perron - Frobenius operator. It is also very reminiscent of the transfer matrix operator in statistical mechanics and we shall indeed use implicitly this analogy. As a matter of fact, one can formulate a problem of statistical mechanics which is entirely

equivalent to the above problem ([B1] [R1] [L]), but we shall not pursue this idea here. We now give some properties of the operator P which are easy consequences of its definition.
In the sequel, L^1 will always denote the space $L^1([0,1], d\lambda)$. A few useful facts about P are summarized in

<u>Theorem 1</u> : <u>Let</u> $h \in L^1$, <u>then</u>

i) $\|Ph\|_{L^1} \leqslant \|h\|_{L^1}$.

ii) <u>If</u> h <u>is</u> λ a.e. <u>positive then</u>

$$\int (Ph)(x)\, d\lambda(x) = \int h(x)\, d\lambda(x).$$

iii) <u>If</u> $g \in L^\infty([0,1], d\lambda)$ <u>then</u>

$$\int g\circ f^n(x)\, h(x)\, d\lambda(x) = \int g(x)\,(P^n h)(x)\, d\lambda(x).$$

iv) <u>If</u> A <u>is a Borel subset of</u> $[0,1]$, <u>we have</u>

$$\lambda(f^{-n}(A)) = \int_A (P^n 1)(x)\, d\lambda(x).$$

The proof of these facts is a direct consequence of the definition.

There are two different classes of maps of an interval for which results about a.c.i.m. are known : the piecewise expanding maps and the unimodal maps. We now define these two classes.

<u>Definition</u> : A map f is piecewise expanding if one can find a finite sequence $0 = a_0 < a_1 < a_2 < \ldots < a_{p-1} < a_p = 1$ such that for any i with $0 \leqslant i \leqslant p-1$ the restriction of f to $]a_i, a_{i+1}[$ is C^1 and can be extended to a C^1 function on $[a_i, a_{i+1}]$. Moreover, there are two finite numbers σ and ρ, $\sigma > \rho > 1$ such that for any i, $0 \leqslant i \leqslant p-1$

$$\rho \leqslant \left| f'|_{[a_i, a_{i+1}]} \right| \leqslant \sigma .$$

<u>Definition</u> : A map f is unimodal if f belongs to $C^3([0,1])$, is concave, has a (unique) critical point at $x = 1/2$ which satisfies $f(1/2) = 1$ (the general case can be reduced to this case by an affine conjugation). For simplicity, we shall also assume in this case that the Schwarzian derivative of f is negative, i.e.

$$S(f) = \frac{f'''}{f'} - \frac{3}{2}\left(\frac{f''}{f'}\right)^2 \leqslant 0.$$

The case of unimodal maps is more difficult than the piecewise expanding case because f has a critical point. We shall start with an even simpler case : the Markoff case, where the proofs are most easily carried out in the spirit of Renyi's paper [Re] .

II. EXPANDING MAPS, THE MARKOFF CASE

A map f from an interval $I = [0,1]$ into itself is piecewise expansive and Markoff (see $[B_2]$) if one can find a finite sequence $0 = a_o < a_1 < \ldots < a_{p-1} < a_p = 1$, such that if I_k denotes the interval $]a_k, a_{k+1}[$, then

i) $f|_{I_k}$ is C^1 and extends to a C^1 function on $\overline{I_k}$.

ii) There are two finite numbers , ρ and σ , $\sigma > \rho > 1$ such that

$$\sigma > |f'|_{I_k}| > \rho \ , \ \forall k \ 0 \leqslant k \leqslant p-1.$$

iii) If $f(I_k) \cap I_j \neq \phi$, then $I_j \subset f(I_k)$.

It is easy to show that without loss of generality, we can replace iii) by $f(I_k) = I$ for $k = 0,1,\ldots,p-1$ (one replaces f by some iterate of f and considers an invariant interval). From now on, we shall make this assumption . References for this chapter will be given in chapter 3.
In this course, we shall prove in detail some of the more common theorems about invariant measures. While the general ideas are quite standard, we have chosen to consider a relatively large class of functions, which we describe now.

We define, for functions g on an interval J, the quantity Dis (for distortion)

$$K = \text{Dis}\,(g,J,\xi\,)$$

as the smallest number for which

$$\left|\frac{g(x)}{g(y)}\right| \leqslant \exp\ [\,K(\,1 + |\,\text{Log}\,|\,x-y\,||)^{-\xi}\,]\ \forall\, x,y \in J\ .$$

If $J = I$, and whenever there is no confusion possible, we write $\text{Dis}(g)$ for $\text{Dis}(g,I,\xi\,)$.

We denote by $\mathcal{R}_\xi$ the set of continuous functions g for which Dis(g,I, ξ) is finite.

Definition : We say f satisfies the hypothesis H_ξ , $\xi > 0$ if, for some finite C, and every $x,y \in \bar{I}_k$, $k = 0,1,\ldots,p-1$ the inequality

$$|f'(x) - f'(y)| \leq C\ (1 + |\mathrm{Log}\ |x-y||)^{-1-\xi}$$

holds.

This rather curious modulus of continuity is in fact very natural as we shall see. A similar condition was mentioned in [An] .

Theorem 2 : Let f be a piecewise expansive Markoff map satisfying H_ξ. Then f has an a.c.i.m. with density in $\mathcal{R}_\xi$.

Theorem 2 will be an immediate consequence of Proposition 3, below.

We start now a calculation which is at the heart of the proof of existence of a.c.i.m's for f's satisfying condition H_ξ . Denote by int(I) the interior of the interval I. Our task is to control for $x \in \mathrm{int}(I)$,

$$P^n h(x) = \sum_{y,f^n(y)=x} h(y)\ |f^{n'}(y)|^{-1} \quad .$$

We shall bound, for $x,x' \in \mathrm{int}(I)$, the quotient

$$P^n h(x)/P^n h(x') = \frac{\sum_{y:f^n(y)=x} h(y)\ |f^{n'}(y)|^{-1}}{\sum_{y':f^n(y')=x'} h(y')\ |f^{n'}(y')|^{-1}} \quad .$$

We now observe that due to the Markoff character of f, there is a natural bijection between the two sets $\{y\ |f^n(y) = x\}$ and $\{y'|f^n(y') = x'\}$. Namely, for a y in I such that $f^n(y) = x$, there is one and only one y' in I such that $f^n(y') = x'$, and for which $f^j(y) \in I_{k_j}$, $0 \leq j \leq n-1$, (and $0 \leq k \leq p-1$) implies $f^j(y') \in I_{k_j}$.

We shall denote this bijection by b. We have then

$$P^n h(x)/P^n h(x') \leq \sup_{\substack{y, f^n(y)=x \\ y'=b(y)}} |f^{n'}(y')|\ h(y)/|f^{n'}(y)|h(y').$$

We shall estimate separately the quantities $h(y)/h(y')$ and $|f^{n'}(y')|/|f^{n'}(y)|$. The two estimates will rely on a bound for $|f^\ell(y) - f^\ell(y')|$ $0 \leq \ell \leq n-1$, $y'=b(y)$.

Assume for example $f^\ell(y) < f^\ell(y')$ (the other case is similar). Then from $y' = b(y)$ and the Markoff property, we derive that $f^{n-\ell}$ is C^1 on the interval $[f^\ell(y), f^\ell(y')]$, and moreover, its derivative satisfies

$$|f^{n-\ell'}| \geq \rho^{n-\ell} .$$

Therefore

$$|x - x'| = |f^{n-\ell}(f^\ell(y)) - f^{n-\ell}(f^\ell(y'))| \geq \rho^{n-\ell}|f^\ell(y) - f^\ell(y')| ,$$

and hence we get the important inequality

$$|f^\ell(y) - f^\ell(y')| \leq \rho^{-(n-\ell)} |x-x'| . \quad (*)$$

We now make use of the inequality in the definition of H_ξ . This allows us to bound $f^{n'}(y')/f^{n'}(y)$, when $y' = b(y)$. We have

$$|f^{n'}(y')|/|f^{n'}(y)| = \prod_{j=0}^{n-1} (|f'(f^j(y'))/f'(f^j(y))|)$$

$$\leq \prod_{j=0}^{n-1} \left(1 + \frac{|f'(f^j(y')) - f'(f^j(y))|}{|f'(f^j(y))|}\right)$$

$$\leq \exp\left[D \sum_{j=0}^{n-1} K/(1 + |\mathrm{Log}(|f^j(y') - f^j(y)|)|)^{1+\xi}\right]$$

where D is a constant which depends only on ρ and σ. If we set

$$L = DK(1 + \xi^{-1}(\mathrm{Log}\ \rho)^{-1}) ,$$

we obtain

$$|f^{n'}(y') / f^{n'}(y)| \leq \exp[L/(1 + |\mathrm{Log}|x-x'||)^\xi] \qquad (**).$$

Notice that this is the only place where hypothesis (H_ξ) is crucially used.

We have now prepared the calculations for

Proposition 3 : Let f satisfy the assumptions of Theorem 2 and let $h \in \mathcal{R}_\xi$. Then $P^n h \in \mathcal{R}_\xi$ for all n and

$$\mathrm{Dis}(P^n h) \leqslant \mathrm{Dis}(h) + L,$$

for some constant L which only depends on f, but not on h and n.

From the hypothesis on h, we derive, using (*) with $\ell = 0$,

$$h(y)/h(y') \leqslant \exp\left[\mathrm{Dis}(h)/\left(1 + |\mathrm{Log}(|y-y'|)|\right)^{\xi}\right]$$

$$\leqslant \exp\left[\mathrm{Dis}(h)/\left(1 + n\mathrm{Log}\,\rho + |\mathrm{Log}|x-x'||\right)^{\xi}\right]$$

$$\leqslant \exp\left[\mathrm{Dis}(h)/\left(1 + |\mathrm{Log}|\,x-x'||\right)^{\xi}\right]$$

Combining this estimate with (**) yields the result.

QED

The following lemma will be of independent use later on.

Lemma 4 : Let $h \in \mathcal{R}_\xi$, and define the linear operator Q_n by

$$Q_n\, h = n^{-1} \sum_{j=0}^{n-1} P^j\, h\ .$$

Then the sequence $(Q_n h)_{n \in \mathbb{N}}$ is precompact in L^1 .

Proof : Let $m = \int_0^1 h(x)\, d\lambda(x)$, then, as we have seen in

Theorem 1,

$$m = \int_0^1 P^j h(x)\, d\lambda(x) \quad \forall\, j \in \mathbb{N}\ .$$

From Proposition 3, we have easily

$$m\, e^{-(L + \mathrm{Dis}(h))} \leqslant P^j h(x) \leqslant e^{(L + \mathrm{Dis}(h))}\, m, \ \forall\, j \in \mathbb{N}\ .$$

Therefore if x and x' belong to int(I), we have

$$|P^j h(x) - P^j h(x')| = P^j h(x)\ |1 - P^j h(x')/P^j h(x)|$$
$$\leq m\, e^{(L+\mathrm{Dis}(h))}\ |1 - \exp((L+\mathrm{Dis}(h))/(1 + |\mathrm{Log}(|x-x'|)|)^{\xi})|$$

By convex combinations, the same inequality holds for $Q_n h$, and the result follows from Ascoli's compactness theorem.

QED

We now prove Theorem 2.

Proof of Theorem 2 : By Lemma 4, the sequence $Q_n 1$ is precompact in L^1. Let h be an accumulation point of this sequence, and let n_j be a sequence of integers such that

$$Q_{n_j} 1 \xrightarrow[L^1]{} h .$$

It is easy to verify that $h \in \mathcal{R}_\xi$ (in fact $\mathrm{Dis}(h) \leq L$), and

$$PQ_{n_j} 1 = Q_{n_j} 1 + [P^{n_j+1} 1 - 1]/n_j .$$

From

$$\| P^{n_j+1} 1 \|_{L^1} = 1 ,$$

we derive $Ph = h$ by letting j go to infinity.

QED

We now investigate the uniqueness and ergodic properties of the above measure. We shall use extensively the fact that $f(I_k) = I$ for $k = 0,1,\ldots, p-1$. It is not very difficult to extend the previous result to the general case.

Theorem 5 : Under the hypotheses of Theorem 2, f has a unique a.c.i.m. . This measure is ergodic and has the Bowen-Ruelle property.

Proof : We have seen in Lemma 4 that for any g in $\mathcal{R}_\xi$, the sequence $Q_n g$ is precompact in L^1. Let g^* and g^{**} be two accumulation points in L^1, and let n_j and m_j be two infinite sequences such that :

$$Q_{n_i} g \to g^* , \quad Q_{m_j} g \to g^{**} \qquad \text{in } L^1 .$$

From the contractivity of Q_{m_j} in L^1, we derive

$$\| Q_{m_j} Q_{n_i} g - Q_{m_j} g^* \|_{L^1} \leq \| Q_{n_i} g - g^* \|_{L^1} .$$

On the other hand, $Q_{m_j} g^* = g^*$, and $Q_{m_j} Q_{n_i} = Q_{n_i} Q_{m_j}$, therefore

$$\| Q_{n_i} Q_{m_j} g - g^* \|_{L^1} \leq \| Q_{n_i} g - g^* \|_{L^1} .$$

If we let $m_j \to \infty$, we get

$$\| g^{**} - g^* \|_{L^1} \leq \| Q_{n_i} g - g^* \|_{L^1} .$$

Letting $n_i \to +\infty$, we get $g^{**} = g^*$ since both functions belong to $\mathcal{R}_\xi$. The sequence $(Q_n g)_{n \in \mathbb{N}}$ is precompact and has only one accumulation point g^* in L^1 (g^* belongs to $\mathcal{R}_\xi$). Therefore

$$Q_n g \xrightarrow[L^1]{} g^* .$$

We now show the uniqueness of the invariant measure with density in $\mathcal{R}_\xi$. Let h_1 and h_2 be such that $Ph_i = h_i$, $i = 1$ and 2, $h_i \in \mathcal{R}_\xi$ $i = 1,2$ and $h_1 \neq h_2$. Assume moreover $\int_0^1 h_i d\lambda = 1$, $i = 1,2$. Let $\psi = h_1 - h_2$. From $\psi \in C^\circ([0,1])$, and $\int_0^1 \psi d\lambda = 0$, there are two open sets U and V such that $\psi|_U > 0$ and $\psi|_V < 0$. From the Markoff property and the expansivity, it is easy to verify that there is an integer ℓ_0 such that for $\ell > \ell_0$ we have

$$f^\ell(U) = [0,1] \quad \text{and} \quad f^\ell(V) = [0,1] .$$

This implies for $\ell > \ell_0$ and any x in I,

$$|P^\ell \psi(x)| < P^\ell |\psi| (x).$$

However, from $P^\ell \psi = \psi$, we have $|P^\ell \psi| = |\psi|$, hence

$$\int_0^1 |\psi| \, d\lambda = \int_0^1 |P^\ell \psi| \, d\lambda < \int_0^1 P^\ell |\psi| \, d\lambda = \int_0^1 |\psi| \, d\lambda$$

which is a contradiction if $\psi \neq 0$.
Therefore there is a unique $h \in \mathcal{R}_\xi$ such that for any g in $\mathcal{R}_\xi$,

$$Q_n g \xrightarrow[L^1]{} h \int_0^1 g \, d\lambda .$$

We now show that f is ergodic with respect to the invariant measure $h d\lambda$. Let $\varphi \in L^1$ such that $\varphi \circ f = \varphi$, and let $g \in \mathcal{R}_\xi$ we have

$$\int_0^1 \varphi(x) \, g(x) \, dx = n^{-1} \sum_{j=0}^{n-1} \int_0^1 g(x) \, \varphi \circ f^j(x) \, dx = \int_0^1 \varphi(x) \, Q_n g(x) dx$$

$$= \int_0^1 \varphi(x) \, h(x) \, dx \int_0^1 g(x) \, dx$$

by the Lebesgue dominated convergence theorem. Since g is arbitrary in $\mathcal{R}_\xi$, this implies $\varphi(x) = \int_0^1 \varphi(y) \, h(y) \, dy$ a.e, i.e. f is ergodic.

By the ergodic theorem, we have for $h d\lambda$ a.e. x and every continuous function g the identity

$$\lim_{n \to +\infty} n^{-1} \sum_{\ell=0}^{n-1} g(f^\ell(x)) = \int_I g h d\lambda .$$

Since $h \in \mathcal{R}_\xi$, we have $h \neq 0$ a.e. and hence $h d\lambda$ is a Bowen-Ruelle measure.

We next show that f has a unique a.c.i.m. Assume $\tilde{h}$ is a positive function in L^1 such that $\tilde{h} \, d\lambda$ is an a.c.i.m.. Since the linear combinations of elements of $\mathcal{R}_\xi$ are dense in L^1, for any $\varepsilon > 0$, one can find a linear combination g_ε of elements of $\mathcal{R}_\xi$, such that

$$\| \tilde{h} - g_\varepsilon \|_{L^1} < \varepsilon .$$

From $Q_n g_\varepsilon \to h \int_0^1 g_\varepsilon d\lambda$, we derive that for n sufficiently large, we have

$$\| Q_n \tilde{h} - h \int_0^1 \tilde{h} d\lambda \|_{L^1} < 3 \varepsilon .$$

If we let $\varepsilon \to 0$, and use $Q_n \tilde{h} = \tilde{h}$, we obtain that $\tilde{h}$ is a.e. a multiple of h, hence f has a unique a.c.i.m.. Since $h \neq 0$ a.e., we conclude that h is the unique Bowen-Ruelle measure for the map f.

QED

From now on we shall denote by h the density of this unique a.c.i.m.. We now investigate the decay of correlations. If $g \in L^1$ we shall denote by $\langle g \rangle$ the number $\int gh d\lambda$. We shall first prove the following useful lemma.

Lemma 6 : Let $g \in \mathcal{R}_\xi$, and assume $\mathrm{Dis}(g) \leqslant L$. There is a number R and a number $\gamma > 0$, which depend only on ρ, σ and ξ, such that for any integer n,

$$\| P^n g - h \int_0^1 g d\lambda \|_{L^1} \leqslant R\, n^{-\gamma} \int_0^1 g d\lambda$$

Moreover, for ρ and σ fixed, we have $\gamma = \mathcal{O}(\xi)$ for ξ large enough.

Proof : If g is a positive function for which

$$g(x)/g(x') \leqslant \exp(\beta(|x - x'|))$$

then it follows from (*) and (**) that

$$P^n g(x)/ P^n g(x') \leqslant \exp\{\beta(\rho^{-n}|x-y|) + L/(1+|\log(|x-y|)|)^\xi\}.$$

(It is not necessary to have $\beta(t) = \text{const.}/(1 + |\log t|)^\xi$.)
Define now $\tilde{g} = P^n g - \eta h$, where h is the invariant density and

$$\eta = \exp\{- \beta(\rho^{-n}) - 2L - 1\}.$$

Notice that $\tilde{g} > 0$, and that η and $\tilde{g}$ depend on n. We have

$$\tilde{g}(x)/\tilde{g}(y) = [P^n g(x)/P^n g(y)].\frac{[1 - \eta h(x)/P^n g(x)]}{[1 - \eta h(y)/P^n g(y)]}$$

$$\leqslant \exp[\beta(\rho^{-n}|x - y|) + L/(1 + |\mathrm{Log}(|x - y|)|)^\xi] \,.$$

$$. \exp(2\eta\, |h(x)/P^n g(x) - h(y)/P^n g(y)|)$$

since $\eta h(x)/P^n g(x) \leqslant \frac{1}{2}$. We also have

$$2\eta \; |h(x)/P^n g(x) - h(y)/P^n g(y)| = 2[\eta \; h(x)/P^n g(x)] \cdot |1 - h(y) \; P^n g(x)/h(x) \; P^n g(y)|$$
$$\leq |1 - \exp[\beta(\rho^{-n}|x-y|) + L/(1 + |\text{Log} \; |x-y||)^{\xi}]|$$
$$\leq A[\beta(\rho^{-n}|x-y|) + L/(1 + |\text{Log}|x-y||)^{\xi}]$$

where A is independent of n and β provided $\beta(\rho^{-n}) < L$ (notice that A is only a function of L).

Therefore, we find

$$\tilde{g}(x)/\tilde{g}(y) \leq \exp \beta'(|x-y|)$$

where

$$\beta'(t) = (A_n \; \beta)(t)$$
$$= (1+A)\{\beta(\rho^{-n}t) + L/(1 + |\text{Log} \; t|)^{\xi}\} \; .$$

We want to apply recursively this identity to

$$\beta_0(t) = L/(1 + |\text{Log} \; t|)^{\xi} \; .$$

Define B to be the smallest integer for which

$$B > [(1 + A)^2/A \;]^{1/\xi}/\text{Log} \; \rho \; , \text{ and}$$

set $n_\ell = B^\ell$. We define

$$\beta_\ell(t) = (A_{n_\ell} \beta_{\ell-1}) \; (t) \; , \quad \ell = 1,2,\ldots \; .$$

We claim that

$$\beta_\ell(\rho^{-n_{\ell+1}}) \leq L, \quad \text{for} \quad \ell = 0,1,2,\ldots .$$

This is obviously true for β_0. We observe that

$$\beta_\ell(t) = (1 + A) \; L/(1 + |\text{Log} \; t|)^{\xi}$$
$$+ L \sum_{j=1}^{\ell} (1 + A)^j/[1 + \text{Log} \; \rho \sum_{s=0}^{j} n_{\ell-s} + |\text{Log}(t)| \;]^{\xi} .$$

Therefore

$$\beta_\ell(\rho^{-n_{\ell+1}}) = (1 + A)\, L/(1 + n_{\ell+1}\, \mathrm{Log}\rho)^\xi$$

$$+ L \sum_{j=1}^{\ell} (1 + A)^j \,[1 + n_{\ell+1}\, \mathrm{Log}\rho + \mathrm{Log}\, \rho \sum_{s=0}^{j} n_{\ell-s}]^{-\xi}$$

$$\leqslant A^{-1}\, L(1 + A)^{\ell+2}\, [1 + n_{\ell+1}\, \mathrm{Log}\, \rho]^{-\xi}$$

$$\leqslant L\, (1 + A)/A \,.\, ((1 + A)/[B\, \mathrm{Log}\, \rho]^\xi)^{\ell+1}$$

$$\leqslant L$$

from our choice of B.
We now define a sequence g_ℓ by $g_0 \in \mathcal{R}_\xi$, $\mathrm{Dis}(g_0) \leqslant L$, $\int_0^1 g_0\, d\lambda = 1$

$$g_{\ell+1} = P^{n_\ell} g_\ell - \eta h \,\| g_\ell \|_{L^1}$$

(here also η depends on ℓ , however, $\eta \geqslant \exp - [\,3L + 1]$).
We observe that from our choice of n_ℓ , we have $g_\ell \geqslant 0$, and from our previous computation

$$g_\ell(x)/g_\ell(y) \leqslant \exp \beta_\ell(|x-y|) \,.$$

Let $\eta_0 = \exp - [\,3L + 1]$, we have

$$\| g_{\ell+1} \|_{L^1} = \| g_\ell \|_{L^1} (1 - \eta) \leqslant (1 - \eta_0)\, \| g_\ell \|_{L^1} \,,$$

therefore

$$\| g_\ell \|_{L^1} \leqslant (1 - \eta_0)^\ell\, \| g_0 \|_{L^1} \,.$$

Let $N_\ell = n_1 + n_2 + \ldots + n_\ell$, it is easy to verify that

$$\| P^{N_\ell} g_0 - h \|_{L^1} \leqslant (1 - \eta_0)^\ell.$$

From the contractivity of P in L^1, we have for $0 \leqslant j \leqslant n_{\ell+1}$

$$\| P^{N_\ell + j} g_0 - h \|_{L^1} \leqslant (1 - \eta_0)^\ell \leqslant B^{-\ell\, [|\mathrm{Log}(1-\eta_0)|/\mathrm{Log}\, B]}$$

$$\leqslant B^\gamma\, n^{-\gamma} \quad \text{where} \quad n = N_\ell + j,$$

and

$$\gamma = |\mathrm{Log}(1-\eta_o)| / \mathrm{Log}\, B \simeq \xi \exp - [3L + 1] / \mathrm{Log}((1+A)^2/A)$$

for ξ large .

QED

One can easily prove a similar result without the hypothesis $\mathrm{Dis}(g) \leq L$. However, the constant R will now be a function of $\mathrm{Dis}(g)$.

If θ is a positive real number, we shall denote by Y_θ the set of continuous functions g which, for some positive number M satisfy

$$|g(x) - g(y)| \leq M (1 + |\mathrm{Log}\, |x - y||)^{-\theta}$$

for all x and y in I. We shall refer to M as the Y_θ norm of g, although strictly speaking this is not a norm.

We now prove the decay of correlations.

Theorem 7 : Let γ be the number given in Lemma 6, for any $g_1 \in Y_\theta$ there is a positive number S, which for ρ, ξ and σ fixed depends only on $\|g_1\|_{L^\infty}$ and on the Y_θ norm of g_1 such that for any $g_2 \in L^\infty$,

$$|\langle g_2 \circ f^n g_1 \rangle - \langle g_1 \rangle \langle g_2 \rangle| \leq n^{-\gamma} S \|g_2\|_{L^\infty}$$

Proof : We shall assume first $g_1 \in \mathcal{R}_\xi$. We have

$$\langle g_2 \circ f^n g_1 \rangle = \int_0^1 g_2(f^n(x))\, g_1(x)\, h(x)\, d\lambda(x)$$

$$= \int_0^1 g_2(x)\, (P^n(g_1 h))(x)\, d\lambda(x).$$

Therefore, from Lemma 6 we have

$$| \langle g_2 \circ f^n g_1 \rangle - \langle g_1 \rangle \langle g_2 \rangle | \leq Rn^{-\gamma} \|g_1\|_{L^\infty} \|g_2\|_{L^\infty} .$$

Assume now $g_1 \in Y_\xi$. Let G_1 be defined by

$$G_1 = g_1 + 2 \|g_1\|_{L^\infty} .$$

It is easy to verify that $G_1 \in \mathcal{R}_\xi$, and to give an explicit bound on $\mathrm{Dis}(G_1)$. We now observe that

$$< g_2 \circ f^n \; G_1 > - < G_1 > \; < g_2 >$$

$$= < g_2 \circ f^n \; g_1 > - < g_1 > \; < g_2 > + 2 \| g_1 \|_{L^\infty} (< g_2 \circ f^n > - < g_2 >)$$

$$= < g_2 \circ f^n g_1 > - < g_1 > \; < g_2 > .$$

The result now follows from $\| G_1 \|_{L^\infty} \leqslant 3 \| g_1 \|_{L^\infty}$.

QED

It is easy to verify that if f' is Hölder, one gets by this method an exponential decay of correlations.
We now have enough information to answer the following question. From the ergodicity of f , and Birkhoff's ergodic theorem, we have for any g in L^1 and a.e. x in $[0,1]$

$$\lim_{n \to \infty} n^{-1} \sum_{j=0}^{n-1} g(f^j(x)) = \int_0^1 g(y) \, d\lambda(y).$$

But what is the rate of convergence ? The following result, known as the central limit theorem, gives an answer to this question for functions g in Y_ξ . We assume ξ is sufficiently large, so that $\gamma > 1$.

Theorem 8 : Let g be a function in Y_ξ , and assume for simplicity $< g > = 0$. If we define $\sigma^2 = \sum_0^\infty < g \; g \circ f^n >$, then

$$\lim_{n \to \infty} \mu \{ x \mid \sum_{j=0}^{n} g(f^j(x))/\sigma\sqrt{n} < t \} = \int_{-\infty}^{t} e^{-u^2} \, du/\sqrt{\pi} .$$

Proof : We shall denote by $\Sigma_n(x)$ the quantity $\sum_{j=0}^{n-1} g(f^j(x))$.

Let

$$G_n(\lambda) = < \exp(i\lambda \; \Sigma_n) > \quad ,$$

it is well known [Fe] that the central limit theorem follows if one can show that uniformly on compact sets

$$G_n(\lambda/\sqrt{n}\sigma) \to \exp(-\lambda^2).$$

Let m be the unique integer for which $\sqrt{n}-1 < m \leqslant \sqrt{n}$.
We have $\Sigma_p(x) = \Sigma_m(x) + \Sigma_{p-m}(f^m(x))$ if $p > m$ and we shall compare $G_p(\lambda)$ to $G_{p-m}(\lambda) \cdot G_m(\lambda)$.
We can apply Theorem 7 with $g_2 = \exp(i\lambda\, \Sigma_{p-m}/\sigma\sqrt{n})$, and $g_1 = \exp(i\,\lambda\, \Sigma_m/\sigma\sqrt{n})$. Notice that $\|g_1\|_{L^\infty}$ and the Hölder constant of g_1 are bounded uniformly in m and n.
We deduce that for some number γ , $0 < \gamma$, and some constant T,

$$|G_p(\lambda/\sigma\sqrt{n}) - G_{p-m}(\lambda/\sigma\sqrt{n})\, G_m(\lambda/\sigma\sqrt{n})| \leqslant T\, m^{-\gamma}$$

uniformly for λ varying in a compact set (we have used $\|g_1\|_{L^\infty} \leqslant 1$ and $\|g_2\|_{L^\infty} \leqslant 1$). Iterating the above argument, we obtain (using $|G_m(\lambda)| \leqslant 1$),

$$|G_n(\lambda/\sigma\sqrt{n}) - G_m(\lambda/\sigma\sqrt{n})^q\, G_r(\lambda/\sigma\sqrt{n})| \leqslant T\, q\, m^{-\gamma}$$

where $n = qm + r$, $0 \leqslant r \leqslant m-1$. Notice that $\sqrt{n}-1 < q < \sqrt{n}+2$.

We shall now show that for $n \to +\infty$, $G_r(\lambda/\sigma\sqrt{n}) \to 1$ and $G_m(\lambda/\sigma\sqrt{n})^q \to e^{-\lambda^2}$ uniformly for λ in a compact set. The result will follow from $T\,q\,m^{-\gamma} \to 0$ since $\gamma > 1$. To verify these assertions, we have to make use of some sort of Lindeberg's condition. It is easy to verify that

$$\begin{aligned} G_m(\lambda/\sigma\sqrt{n}) &= \langle \exp(i\lambda\Sigma_m/\sigma\sqrt{n}) \rangle \\ &= 1 - \lambda^2 \langle \Sigma_m^2 \rangle /2\sigma^2 n - i\,\lambda^3 \langle \Sigma_m^3 \rangle / \sigma^3 n^{3/2} + \varepsilon \end{aligned}$$

where $|\varepsilon| \leqslant C(\lambda) \langle \Sigma_m^4 \rangle / n^2$ and $C(\lambda)$ is a uniformly bounded constant if λ varies in a bounded set. The result will follow from $\langle \Sigma_m^2 \rangle /m \to \sigma^2$, $\langle \Sigma_m^3 \rangle /m^2 \to 0$, and $\langle \Sigma_m^4 \rangle /m^3 \to 0$ if $m \to +\infty$.

We shall only indicate the proof of $\langle \Sigma_m^3 \rangle /m^2 \to 0$, the other results being proven similarly. We have :

$$\langle \Sigma_m^3 \rangle = 8 \sum_{0 \leqslant j_1 < j_2 < j_3 \leqslant m} \langle g \circ f^{j_1}\; g \circ f^{j_2}\; g \circ f^{j_3} \rangle$$

+ corresponding terms where some of the j_e's are equal.

We shall only treat the first sum, the others are dealt with in a similar fashion. From the invariance of the expectation, we deduce that the first sum is equal to

$$\sum_{\substack{0 \leqslant j_1 < m \\ 0 < j_2 < j_3 < m - j_1}} \langle g \; g \circ f^{j_2} \; g \circ f^{j_3} \rangle \; .$$

We now give two estimates for the quantity $|\langle g \; g \circ f^{j_2} \; g \circ f^{j_3} \rangle|$.

We first apply Theorem 7 to the functions $g_2 = g \; g \circ f^{j_3 - j_2}$, $g_1 = g$,

and $n = j_2$. We obtain using $\langle g \rangle = 0$

$$|\langle g \; g \circ f^{j_2} \; g \circ f^{j_3} \rangle| \leqslant T/j_2^{\gamma}$$

where T does not depend on j_2 or j_3.

Applying Theorem 7 to $g_2 = g$, $g_1 = P^{j_2}(g \; g \circ f^{j_2} \; h)/h$ and $n = j_3 - j_2$, we obtain :

$$|\langle g \; g \circ f^{j_2} \; g \circ f^{j_3} \rangle| \leqslant T/(j_3 - j_2)^{\gamma} \; .$$

We now distinguish three cases. Let $\delta = 2/(\gamma + 3)$

Case 1 : $j_3 \leqslant 2m^{\delta}$. Then we have

$$\sum_{\substack{0 \leqslant j_1 < m \\ 0 < j_2 < j_3 < m^{\delta}}} |\langle g \; g \circ f^{j_2} \; g \circ f^{j_3} \rangle| \leqslant 4 \; m^{1+2\delta} \, \|g\|^3_{L^{\infty}}$$

Case 2 : $j_3 > 2m^{\delta}$, $j_2 > m^{\delta}$.

Combining the two estimates, we obtain

$$|\langle g \; g \circ f^{j_2} \; g \circ f^{j_3} \rangle| \leqslant T \, j_2^{-(\gamma-1)/2} (j_3 - j_2)^{-(\gamma+1)/2} \; .$$

This implies

$$\sum_{\substack{0 \leq j_1 < m \\ m^\delta < j_2 < j_3}} |\langle g\; g\circ f^{j_2}\; g\circ f^{j_3}\rangle| \leq T'\, m^{2-\delta(\gamma-1)/2}$$

where T' does not depend on m.

Case 3 : $j_3 > 2m^\delta$, $(j_3 - j_2) > m^\delta$, $j_2 \leq m^\delta$.

Using the second estimate we obtain

$$\sum_{\substack{0 < j_1 < m \\ 0 < j_2 < m^\delta < j_3}} |\langle g\; g\circ f^{j_2}\; g\circ f^{j_3}\rangle| \leq T''\, m^{1+\delta},$$

where T'' does not depend on m.

Combining these three cases, we obtain $\langle \Sigma_m^3 \rangle / m^2 \to 0$ if $m \to +\infty$.

Using the above results we have

$$G_m(\lambda/\sigma\sqrt{n})^q = \exp\, [-\lambda^2 \langle \Sigma_m^2 \rangle\; q/2\sigma^2 n + T_m(\lambda)]$$

where $T_m(\lambda)$ converges to zero uniformly on compact sets if $n \to +\infty$.
Similar estimates imply $G_r(\lambda/\sigma\sqrt{n}) \to 1$ if $n \to +\infty$ uniformly on compact sets.

QED

We now give a formula for the entropy of the map f.

Theorem 9 : Under the above hypothesis, we have

$$h(\mu) = \int_0^1 \mathrm{Log}\; |f'(x)|\; d\mu(x).$$

Proof : Let V be the partition of [0,1] generated by the points a_0, $a_1, \ldots, a_p$. The Kolmogoroff Sinai theorem [W2] tells us that

$$h(\mu) = \lim_{n \to +\infty} n^{-1}\, H(V_n)$$

where $V_n = \bigvee_0^n T^{-j} V$, and

$$H(V_n) = -\sum_{\zeta \in V_n} \mu(\zeta) \operatorname{Log}\mu(\zeta) \quad \text{(see [W2] for example).}$$

As we have seen during the proof of Proposition 3, if y and x belong to the same atom of V_n we have

$$|f^{n'}(x)|/|f^{n'}(y)| \leqslant e^L.$$

From $\zeta \in V_n$ and the Markoff property, we have $f^n(\zeta) = [0,1]$, and $f^n|_\zeta$ is a bijection. Let h be the density of μ, we have

$$e^{-L} \leqslant h(x) \leqslant e^{-L} \quad \text{for any } x \text{ in } [0,1].$$

Therefore, for any x in ζ we get

$$e^{-2L} \leqslant \mu(\zeta) \cdot |f^{n'}(x)| \leqslant e^{2L}$$

wich implies

$$-2L/n + n^{-1} \operatorname{Log} |f^{n'}(x)| \leqslant -n^{-1} \operatorname{Log} \mu(\zeta) \leqslant 2L/n + n^{-1} \operatorname{Log} |f^{n'}(x)|.$$

Integrating over ζ with respect to μ we get

$$-2L\mu(\zeta)/n + \int_\zeta n^{-1} \operatorname{Log} |f^{n'}(x)| \, d\mu(x) \leqslant -n^{-1} \mu(\zeta) \operatorname{Log}\mu(\zeta)$$

$$\leqslant 2L \mu(\zeta)/n + \int_\zeta n^{-1} \operatorname{Log} |f^{n'}(x)| \, d\mu(x).$$

If we sum over $\zeta \in V_n$, we get

$$-2L/n + \int_0^1 \operatorname{Log}|f'(x)| d\mu(x) \leqslant n^{-1} H(V_n)$$

$$\leqslant 2L/n + \int_0^1 \operatorname{Log}|f'(x)| d\mu(x)$$

and we obtain the result if $n \to +\infty$.

<u>QED</u>

III. EXPANDING MAPPINGS, THE GENERAL CASE

In this paragraph, we shall briefly review the general results about expanding mappings. One of the main theorems is due to Lasota and Yorke [La. Y1]. They assumed the map is piecewise expanding and piecewise C^2. Their result was then extended by S. Wong [Wo] and Rychlik [Ry] and the following theorem summarizes various results which also apply to the case treated by Wong and Rychlik. We shall only give the version of Lasota and Yorke, and refer the reader to the original paper of Wong for the more general and more technical condition.

Theorem 10 : Assume the map f is differentiable except at a finite number of points, is piecewise expanding, and $1/f'$ is of bounded variation. Then

i) f has an absolutely continuous invariant measure ([Wo 1], [Ry]).
ii) f has only finitely many ergodic a.c.i.m, [Li,Y] .
iii) Each of the ergodic dynamical systems in ii) is weak Bernoulli ([B.3] , [H K]).
iv) Each of these systems has exponential decay of correlations ([H K]).
v) For each system the central limit theorem holds for sufficiently regular functions ([Wo2] , [H K] , [R]).

In this theorem, one need not assume that f' is bounded. The proof is based on the following estimate for the Perron-Frobenius operator. If $|||\ |||$ denotes the variation norm, there is a number $\alpha \in [0,1[$ and a positive constant C such that if g is of bounded variation we have

$$|||\, Pg\, ||| \leqslant \alpha\ |||\, g\, ||| + C\, \|g\|_{L^1} .$$

It is now easy to verify that

$$|||Q_n\, g\, ||| \leqslant (1 + C/(1-\alpha))\ |||\, g\, ||| .$$

Since any bounded set in the $|||\ |||$ norm is precompact in L^1, the sequence $(Q_n g)_{n\in\mathbb{N}}$ is precompact and i) follows. We refer the reader to the original papers for the rest of the proof of Theorem 10.

At least in the case where f is piecewise expanding and $|f'|$ is bounded, one can prove the existence of an absolutely continuous invariant measure under the weaker hypothesis H (see § II). We do not know how to prove the analog of the preceding estimate for the Perron-Frobenius operator, and we use a method which is quite different. One first shows that there are two strictly positive numbers ε_o and γ such that if J is any interval for which $\lambda(J) < \varepsilon_o$, then

$$\int_J (P^n 1)(x)\, d\lambda(x) \leqslant \lambda(J)^\gamma .$$

One can now adapt the techniques used in § II to show that for some constant C, for any $\varepsilon \in [0, \varepsilon_o[$, and for any integer n,

$$\int_0^1 |P^n 1\ (x) - P^n 1(x+\varepsilon)|\, d\lambda(x) \leqslant C(1 + |\mathrm{Log}\varepsilon|)^{-\xi} .$$

These two results will be proven in the appendix .
The sequence $(Q_n 1)_{n \in \mathbb{N}}$ satisfies the same estimate and Kolmogoroff's theorem (see [Y] for example) implies that the sequence $(Q_n 1)_{n \in \mathbb{N}}$ is precompact in L^1.

Lasota and York [La.Y2] have proven the existence of a.c.i.m. under an even weaker assumption for some particular maps. One may wonder whether piecewise C^1 and expansiveness are enough to have an a.c.i.m., however, analogies with statistical mechanics suggest that one may have counter examples if condition H is violated.

Ledrappier [Le] has obtained a criterion to check whether an ergodic invariant measure is an a.c.i.m. We shall come back to this result in the next chapter.

One can ask how crucial the expansivity condition is. It is obvious that one has similar results if some iterate of the map satisfies one of the above hypothesis. One can obtain some interesting generalizations by constructing the so-called return maps and then use results of Walters([W1]). If there is a fixed point (or a periodic point) which is not expanding the situation is more involved. Lasota and York [La.Y1] gave an example where there is no finite a.c.i.m. One can sometimes construct σ finite a.c.i.m. [B] , [S] , [T] , [C.F.] , [M], [A].

Other results are concerned with parameter dependence of the density : [Mc] , [K] , [C.T.] , or with stochastic perturbations [Go] .

IV. UNIMODAL MAPPINGS

For this class of mappings, there is a good topological theory (see [M.T], [C.E.1], [N], [P], for example). Much less is known about ergodic theory. As we have seen in the preceding paragraphs, the expansive character of a map seems to play an important role for the existence of an a.c.i.m. If the map has a critical point, expansivity is violated, however some sort of non-uniform expansivity may hold. We shall first give two general results which are based on such a notion. See [Mi], [SZ], [Pi], [R2].

Theorem 11 : Assume there is a neighborhood U of the critical point such that the orbit of the image of the critical point does avoid U. Then f has a unique a.c.i.m. (which is therefore ergodic).

This theorem provides a nice and efficient criterion for proving that a map has an a.c.i.m. Unfortunately, if one looks at a one parameter family of unimodal maps $a \to f_a$, the condition is probably fulfilled only on an exceptionel set of values of a (which is however uncountable). The following theorem extends the preceding one and in some one parameter families it can be used to prove existence of an a.c.i.m. for a set of values of the parameter of positive Lebesgue measure.

Theorem 12 [C.E2] : Assume one can find two strictly positive constants C and χ such that

i) for any integer n, $|f^{n'}(1)| > Ce^{n\chi}$

ii) If z satisfies $f^p(z) = \frac{1}{2}$, then $|f^{p'}(z)| \geq C\, e^{p\chi}$.

Then f has an a.c.i.m..

We shall also mention the following interesting result which is due to Ledrappier, and is also valid for piecewise expanding mappings.

Theorem 13 : [Le] Let μ be an invariant measure, ergodic, and assume the entropy of μ $(h(\mu))$ is strictly positive, and assume also that $\int \mathrm{Log}|f'(x)| \quad \mu(x) > -\infty$. Then μ is an a.c.i.m. if and only if the Rohlin formula holds, i.e.

$$h(\mu) = \int_I \mathrm{Log}|f'(x)|\, d\mu(x).$$

If one considers particular one parameter families, for example $x \to ax(1-x)$, $a \in [0,4]$, some results are known about the set of values of a for which there is an a.c.i.m. We have the following theorem.

Theorem 14 : ([J]), [G.], [B.C.]) For the one parameter family $x \to ax(1-x)$, $a \in [0,4]$, the set of values of a for which there is

an a.c.i.m. is of positive Lebesgue measure.
For a different one parameter family, see [C.E.3] . The (very rough) idea of the proof is to establish an estimate of the form of Theorem 12. One looks at the orbit of 1 and selects a set of values of a such that this orbit does not come back too fast too close the critical point. We refer the reader to the original papers for more details.

APPENDIX

In this appendix we shall prove the following theorem.

Theorem A : Let f be piecewise expanding, and assume condition H holds, namely, there are two positive constants M and ξ such that if x and y belong to some interval $[a_k, a_{k+1}]$ of continuity of f', then

$$|f'(x) - f'(y)| \leqslant M/(1+|\text{Log}\,|x-y||)^{1+\xi} .$$

Then f has an a.c.i.m.
We shall only give the proof in the case where none of the a_k's is periodic. The general case is similar. We shall also assume $\rho > 2$ which is always possible by considering some iterate of f. The proof will rely on the following estimate which was already used in the proof of Proposition 3.

Let y and z be such that for any ℓ, $0 \leqslant \ell \leqslant n-1$, $f^{\ell}(y)$ and $f^{\ell}(z)$ belong to the same interval of uniform continuity of f'. Then

$$|f^{n'}(y)/f^{n'}(z)| \leqslant 1 + K/(1 + |\text{Log}\,|f^n(y) - f^n(z)||)^{\xi}$$

and K is independent of y, z and n. We shall refer to this estimate as (E).
The proof of Theorem A will rely on the following lemma.

Lemma B : There are two strictly positive numbers ε_o and γ such that for $\varepsilon < \varepsilon_o$,

$$\forall\, x \in [0,1], \quad \int_{[0,1]\cap[x-\varepsilon,\, x+\varepsilon]} (P^n 1)(x)\, dx \leqslant \varepsilon^{\gamma} .$$

This lemma is essentially a word for word adaptation of a result of [C.L]. We now give the proof for the convenience of the reader. We shall denote by Σ the set $\{a_o, a_1, \ldots a_p\}$ defined in the first chapter.

Proof of Lemma B

The proof is recursive . Using the notations of chapter 1 we have

$$(P1)(x) = \sum_{y,f(y)=x} |f'(y)|^{-1} \leqslant \rho^{-1} p$$

and the result follows for $n = 1$ if ε_0 is small enough. We shall now assume that the result has been proved for P^S, $S = 1,2,\dots, n-1$, and we shall prove it for $S = n$.

Let J_n be the set of connected components of $I \setminus \bigcup_0^{n-1} f^{-\ell}(\Sigma)$.

For a given x, we shall denote by A_ε the interval $[x-\varepsilon, x+\varepsilon] \cap I$. We have

$$\lambda(f^{-n}(A_\varepsilon)) = \sum_{M \in J_n} \lambda(f^{-n}(A_\varepsilon) \cap M) .$$

Let $0 < \delta < 1$ and $C > 0$ be such that

$$3/\mathrm{Log}\ 2 > C(1-\delta) > (\mathrm{Log}\ \rho)^{-1} .$$

We have $J_n = E_1 \cup E_2 \cup E_3$, where

$$E_1 = \{M \in J_n \mid \lambda(f^n(M)) \leqslant \varepsilon^\delta\} .$$

$E_2 \subset J_n \setminus E_1$ will be defined later on, and $E_3 = J_n \setminus (E_1 \cup E_2)$.

Consequently we have

$$\lambda(f^{-n}(A_\varepsilon)) = \sum_{M \in E_1} \lambda(f^{-n}(A_\varepsilon) \cap M) + \sum_{M \in E_2} \lambda(f^{-n}(A_\varepsilon) \cap M)$$

$$+ \sum_{M \in E_3} \lambda(f^{-n}(A_\varepsilon) \cap M) .$$

We shall estimate the three sums separately, the first one being a warming up for the estimation of the second one. We now make the very important remark that $f^n_{|M}$ is a bijection from M to $f^n(M)$.

Using this observation, if $M \in E_1$ we have

$$\lambda(f^{-n}(A_\varepsilon) \cap M)/\lambda(M) \leqslant \sup_{x \in M,\ y \in M} f^{n'}(x) \ / \ f^{n'}(y)$$

$$\cdot\ \lambda(A_\varepsilon \cap f^n(M))/\lambda(f^n(M))$$

$$\leqslant C_1 \ \varepsilon^{1-\delta} ,$$

where $C_1 = K+1$ by the estimate (E). Therefore

$$\sum_{M \in E_1} \lambda(f^{-n}(A_\varepsilon) \cap M) \leq C_1\ \varepsilon^{1-\delta} \sum_{M \in E_1} \lambda(M) \leq C_1\ \varepsilon^{1-\delta}.$$

We now define E_2.

If $M \in J_n$ we shall denote by $q(M)$ and $r(M)$ (q and r for short) the two integers smaller than or equal to n such that

$$\overline{f^{n-q}(M)} \cap \Sigma \neq \phi \quad , \quad \text{and} \quad \overline{f^{n-r}(M)} \cap \Sigma \neq \phi$$

(q has to do with, say, the left endpoint of M, and r with the right endpoint).

Let $k(M) = \inf(q,r)$, and let $R_\ell(M)$ be the element of $J_{n-\ell}$ containing $M (1 \leq \ell \leq n)$. We define recursively a finite sequence of integers $k_i(M)$ by

$$k_o(M) = 0,\ k_1(M) = k(M),$$

$$k_{i+1}(M) = k_i(M) + k(R_{k_i(M)}(M))$$

as long as $k_{i+1}(M) \leq n$.

We now define $E_2 \subset J_n \setminus E_1$ by

$$E_2 = \{M \in J_n \ |\lambda(f^n(M)) < \varepsilon^\delta,\ \exists i \in \mathbb{N},\ k_i(M) < C\ |\mathrm{Log}\ \varepsilon|,$$
$$\lambda(f^{n-k_i(M)}(R_{k_i(M)}(M))) > \varepsilon^\delta\ \rho^{-k_i(M)}\}.$$

The estimation of the sum for $M \in E_2$ will be a refinement of the argument used for E_1.

For $\tilde{M} \in J_{n-q}$ with $\overline{f^{n-q}(\tilde{M})} \cap \Sigma \neq \phi$, let $N(\tilde{M},q)$ be the number of elements $M \in E_2$ such that $M \subset \tilde{M}$ and $k_{\sigma(M)} \geq q$ where $\sigma(M)$ is the smallest integer for which

$$\lambda(f^{n-k_i(M)}(R_{k_i(M)}(M))) > \varepsilon^\delta\ \rho^{-k_i(M)}.$$

Let $L_q = \sup_{0 \leq k \leq q} \sup_{\substack{\tilde{M} \in J_n \\ \overline{f^{n-k}(\tilde{M})} \cap \Sigma \neq \phi}} N(\tilde{M},k)$,

we shall see later on that $L_q \leq C_2\, 2^{q/3}$ where C_2 does not depend on q.

Let $M \in E_2$ and $\tilde{M} \in J_{n-q}$ such that $M \subset \tilde{M}$ and $k_{\sigma(M)} = q$, and $\overline{f^{n-q}(\tilde{M})} \cap \Sigma \neq \phi$. We have, using that $f^{n-q}_{|\tilde{M}}$ is a diffeomorphism,

$$\lambda(f^{-n}(A_\varepsilon) \cap M)/\lambda(\tilde{M}) \leq C_1 \lambda(f^{-q}(A_\varepsilon) \cap f^{n-q}(M))/\lambda(f^{n-q}(\tilde{M}))$$

$$\leq C_1 \;\varepsilon\rho^{-q} / \varepsilon^{\delta}\rho^{-q} \leq C_1 \;\varepsilon^{1-\delta},$$

i.e.

$$\lambda(f^{-n} (A_\varepsilon) \cap M) \leq C_1\, \varepsilon^{1-\delta}\, \lambda(\tilde{M}).$$

Therefore

$$\sum_{M \in E_2} \lambda(f^{-n}(A_\varepsilon) \cap M) =$$

$$= \sum_{q=0}^{[C|\mathrm{Log}\varepsilon|]+1} \; \sum_{\substack{\tilde{M} \in J_{n-q} \\ \overline{f^{n-q}(\tilde{M})} \cap \Sigma \neq \phi}} \; \sum_{\substack{M \in E_2 \\ M \subset \tilde{M} \\ k_{\sigma(M)}(M) = q}} \lambda(f^{-n}(A_\varepsilon) \cap M)$$

$$\leq \sum_{q=0}^{[C|\mathrm{Log}\varepsilon|]+1} \; \sum_{\substack{\tilde{M} \in J_{n-q} \\ \overline{f^{n-q}(\tilde{M})} \cap \Sigma \neq \phi}} C_1\, \varepsilon^{1-\delta}\, \lambda(\tilde{M})\; L_q$$

$$\leq C_3\, \varepsilon^{1-\delta-(C\,\mathrm{Log}\,2)/3} .$$

We now prove the bound on L_q. This bound will follow from the simple observation that if Σ does not contain any periodic point, then one can find a positive number θ such that for every interval L with $\lambda(L) < \theta$ and $\bar{L} \cap \Sigma \neq \phi$, one has $\overline{f^{-j}(L)} \cap \Sigma = \phi$ for $1 \leq j \leq 4$ card Σ.

Let $\tilde{M} \in J_{n-q}$ and assume that $\overline{f^{n-q}(\tilde{M})} \cap \Sigma \neq \phi$. Let $\{x\} = \overline{f^{n-q}(\tilde{M})} \cap \Sigma$ (one of two points if necessary) and define $\tilde{M}_L$ by $f^{n-q}(\tilde{M}_L) = f^{n-q}(\tilde{M}) \cap [0,x]$ and $\tilde{M}_R = \tilde{M} \setminus \tilde{M}_L$ (recall that f^{n-q} is a bijection on $\tilde{M}$).

Let $E_L(q) = \{M \in E_2 \mid M \subset \tilde{M}_L \, , \, k_{\sigma(M)}(M) \geq q\}$, and define similarly $E_R(q)$ (Notice that they both depend on $\tilde{M}$). We shall assume $E_L(q) \neq \phi$, otherwise one should consider directly $E_R(q)$. (In general both $E_L(q)$ and $E_R(q)$ are not empty). If both $E_L(q)$ and $E_R(q)$ are empty, we have $N(\tilde{M},q) = 0$ and there is no estimation to be done. Let r be the smallest integer such that $\overline{f^{n-q+r}(\tilde{M}_L)} \cap \Sigma \neq \phi$. There are two cases :

Case 1 $r<q$. This implies $\tilde{M}_L \in J_{n-q+r}$, and for any $M \in E_L(q)$ we have $R_{k_i}(M) = \tilde{M}_L$ for some integer i. Moreover,

$$k_i(M) = q-r < q \leq k_{\sigma(M)}(M).$$

This implies $i < \sigma(M)$, and

$$\lambda(f^{n-q+r}(\tilde{M}_L)) \leq \varepsilon^{\delta} \rho^{-(q-r)}.$$

If ε_0 is small enough, we have $\varepsilon^{\delta} \rho^{-(q-r)} \leq \varepsilon^{\delta} \leq \varepsilon_0^{\delta} \leq \theta$. Therefore $r > 4$ Card Σ. Consequently

$$\begin{aligned} \text{Card } E_L(q) &= \text{Card } \{M \in E_2 \mid M \subset \tilde{M}_L, \, k_{\sigma(M)}(M) \geq q\} \\ &\leq \text{Card } \{M \in E_2 \mid M \subset \tilde{M}_L, \, k_{\sigma(M)}(M) \geq q-r\} \\ &\leq \sup_{0 \leq k \leq q-4 \,\text{Card}\Sigma} \; \sup_{\substack{M' \in J_{n-k} \\ \overline{f^{n-k}(M')} \cap \Sigma \neq \phi}} N(M',k) = L_{q-4 \,\text{Card}\, \Sigma} \end{aligned}$$

(where we have assumed $q > 4$ Card Σ).

Case 2 $r \geq q$. In this case we have $\tilde{M}_L \in J_n$, and Card $E_L(q) = 1$.

We have similar estimates for $E_R(q)$, and we have finally

$$N(\widetilde{M},q) = \text{Card } E_L(q) + \text{Card } E_R(q) \leqslant \text{Sup }(2, 2L_{p-4} \text{ Card } \Sigma).$$

This implies

$$L_q \leqslant R \text{ Sup }(2, 2L_{q-4R}) \qquad \text{where } R = \text{Card } \Sigma,$$

which implies

$$L_q \leqslant C_2 \; 2^{q/3}$$

where C_2 does not depend on q.

We now come to the estimation of the sum corresponding to E_3. We first observe that $E_3 \neq \phi$ implies $n \geqslant C\,|\text{Log}\,\varepsilon|$, since for ε_0 small enough, each element M of J_0 satisfies $\lambda(M) > \varepsilon_0^\delta$. For $M \in E_3$, let i be the unique integer such that

$$k_i(M) < C|\text{Log}\,\varepsilon| \leqslant k_{i+1}(M).$$

From the preceding argument one also gets that there is always such an i, which may eventually be equal to zero. We also observe that

$$\overline{f^{n-k_{i+1}(M)}(R_{k_i}(M))} \cap \Sigma \neq \phi,$$

and notice that $f^{k_{i+1}-k_i-1}$ is a bijection on $f^{n-k_i+1}(R_{k_i}(M))$.

Therefore

$$\begin{aligned}\lambda(f^{n-k_{i+1}+1}(R_{k_i}(M))) &\leqslant \rho^{-(k_{i+1}-k_i-1)}\,\lambda(f^{n-k_i}(R_{k_i}(M)))\\ &\leqslant \rho^{-(k_{i+1}-k_i-1)}\,\varepsilon^\delta \rho^{-k_i}\\ &\leqslant \rho\,\varepsilon^\delta\,\rho^{-k_{i+1}}.\end{aligned}$$

Let $B_\eta = \{x \in I \mid d(x,\Sigma) \leqslant \eta\}$ where d is the usual distance and η a positive number. We have

$$f^{n-k_{i+1}}(R_{k_i}(M)) \subset B_{\varepsilon^{\delta}\rho^{-k_{i+1}}} .$$

We now use the recursion hypothesis. We have

$$\sum_{\substack{M \in E_3,\ k_{i+1}(M) = q \\ k_i(M) < C|\mathrm{Log}\,\varepsilon| \leqslant k_{i+1}(M)}} \lambda(f^{-n}(A_\varepsilon) \cap M) \leqslant$$

$$\leqslant \sum_{\substack{M \in E_3,\ k_{i+1}(M)=q \\ k_i(M) < C|\mathrm{Log}\,\varepsilon| \leqslant k_{i+1}(M)}} \lambda(f^{-(n-q)}(f^{-q}(A_\varepsilon) \cap f^{n-q}(M)))$$

$$\leqslant \lambda(f^{-(n-q)}(B_{\varepsilon^{\delta}\rho^{-q}}))$$

$$\leqslant 2 \text{ Card } \Sigma\ \varepsilon^{\gamma\delta} \rho^{-q\gamma} .$$

Therefore

$$\sum_{M \in E_3} \lambda(f^{-n}(A_\varepsilon) \cap M) \leqslant \sum_{q=C|\mathrm{Log}\,\varepsilon|}^{n} 2 \text{ Card } \Sigma\ \varepsilon^{\gamma\delta} \rho^{-\gamma q}$$

$$\leqslant \varepsilon^{\gamma}/3 ,$$

if ε_0 is small enough since $C \log \rho + \delta > 1$.
Let γ be such that $0 < \gamma < 1 - \delta - \frac{C}{3} \mathrm{Log}\, 2$, we have for ε_0 small enough

$$\sum_{M \in J_n} \lambda(f^{-n}(A_\varepsilon) \cap M) \leqslant \varepsilon^{\gamma} .$$

QED

The following lemma will provide the compactness argument needed to prove theorem A.

Lemma C : There are two strictly positive numbers C_4 and ε_1 such that if $\varepsilon \in [0, \varepsilon_1[$, and $n \in \mathbb{N}$, we have

$$\int_0^1 |P^n 1(x) - P^n 1(x+\varepsilon)| \, d\lambda(x) \leq C_4(1 + |\mathrm{Log}\,\varepsilon|)^{-\xi}.$$

Proof . Let $h_{n,q}(x)$ be defined for $0 \leq q \leq n$ by

$$h_{n,q}(x) = \sum_{\substack{y, f^n(y) = x \\ d(f^{n-q}(y), \Sigma) \leq 3\varepsilon\rho^{-q}}} |f^{n'}(y)|^{-1}.$$

We have with J_n the set of connected components of $I \setminus \bigcup_0^{n-1} f^{-\ell}(\Sigma)$,

$$|P^n 1(x) - P^n 1(x+\varepsilon)| \leq \sum_{\substack{L \in J_n \\ \exists y \in L, \exists y' \in L \\ f^n(y) = x, \; f^n(y') = x+\varepsilon}} \left| |f^{n'}(y)|^{-1} - |f^{n'}(y')|^{-1} \right|$$

$$+ \sum_{q=0}^{n} [h_{n,q}(x) + h_{n,q}(x+\varepsilon)].$$

Using estimate E, we have for y and y' appearing in the first sum

$$|f^{n'}(y)|/|f^{n'}(y')| \leq 1 + M/(1 + |\mathrm{Log}\,\varepsilon|)^{\xi}.$$

Therefore the first sum is bounded by

$$M P^n 1(x)/(1 + |\mathrm{Log}\,\varepsilon|)^{\xi}.$$

We now estimate $\int_0^1 h_{n,q}(x) d\lambda(x)$. We have

$$h_{n,q}(x) = \sum_{z, f^q(z) = x} \chi_{\{z | d(z, \Sigma) \leq 3\varepsilon\rho^{-q}\}}(z) \, |f^{q'}(z)|^{-1}.$$

$$\cdot \sum_{y, f^{n-q}(y) = z} |f^{n-q'}(y)|^{-1}$$

where $\chi_A(.)$ denotes the characteristic function of the set A.

Using theorem 1, we obtain

$$\int_0^1 h_{n,q}(x)\, d\lambda(x) = \int_{d(x,\Sigma) \leqslant 3\varepsilon\rho^{-q}} P^{n-q}\, 1\, (x)\, dx \leqslant (3\varepsilon\rho^{-q})^{\gamma}$$

B.C. M. Benedicks, L. Carleson, On iterations of $1 - ax^2$ on $(-1,1)$. Preprint, Institut Mittag Leffler 1983.

CE1 P. Collet, J-P. Eckmann, "Iterated maps on the interval as dynamical systems", Birkhaüser, Basel, Boston, 1980.

CE2 P. Collet, J-P. Eckmann, Positive Lyapunov exponents and absolute continuity for maps of the interval. Ergodic theory and dynamical systems, to appear.

CE3 P. Collet, J-P. Eckmann, On the abundance of aperiodic behaviour for maps on the interval. Commun. Math . Phys., 73, (1980) 115-160.

CEL P. Collet, J-P. Eckmann, O.E. Lanford III : Universal properties of maps on an interval. Commun. Math. Phys., 76, (1980) 211-254.

C.F. P. Collet, P. Ferrero, In preparation.

C.L. P. Collet, Y. Lévy, Ergodic properties of the Lozi Mappings. Preprint Ecole Polytechnique, Paris 1983.

C.T. P. Collet, C. Tresser, In preparation.

Fe. W.F. Feller, "An introduction to probability theory and applications", John Wiley and Sons 1971.

G. J. Guckenheimer, Renormalization of one dimensional mappings and strange attractors. Unpublished.

H.K. F. Hofbauer, G. Keller, Ergodic properties of invariant measures for piecewise monotonic transformations. Math. Zeit. 180, (1982) 119-140.

J. M. Jakobson, Absolutely continuous invariant measures for one parameter families of one dimensional maps. Commun. Math. Phys. 81, (1981) 39-88.

K. G. Keller, Stochastic stability in some chaotic dynamical systems. Preprint, Univ. Heidelberg 1981.

L. O.E. Lanford III, CIME Lecture 1978.

La.Y1 A. Lasota, J. Yorke, On the existence of invariant measures for piecewise monotonic transformations. Trans. Amer. Math. Soc. 186 (1973) 481-488.

La.Y2 A. Lasota, J. Yorke, Exact dynamical systems and the Frobenius-Perron operator. Trans. Amer. Math. Soc., 273 (1982) 375-384.

Le. F. Ledrappier, Some properties of absolutely continuous invariant measures on an interval. Ergod. Th. and Dynam. Sys. 1, (1981) 77-93.

Li.Y.T.Y. Li, J. Yorke, Ergodic transformations from an interval into itself. Trans. Amer. Math. Soc. 235 (1978) 183-193.

M. P. Manneville, Intermittency, self-similarity and 1/f spectrum in dissipative dynamical systems. Journ. de Phys. Paris 41, (1980) 1235-1243.

Me. C. Meunier, Continuity of type I Intermittency from a measure theoretical point of view. Preprint, Ecole Polytechnique, Palaiseau 1982.

Mi. M. Misiurewicz, Absolutely continuous measure for certain maps of an interval. Publ. Sci. IHES. 53 (1981) 17-52.

M.T. J. Milnor , P. Thurston, On iterated maps of the interval, I, II. Preprint, Princeton University 1977.

N. Z. Nitecki, Topological Dynamics on the interval,"Ergodic theory and dynamical systems", Vol. II, A. Katok ed. Birkhäuser, Boston, Basel 1981.

P. C. Preston, Iterates of maps on an interval, Lecture Notes in Mathematics 999. Springer-Verlag, Berlin, Heidelberg, New York 1983.

Pi. G. Pianigiani, First return map and invariant measures, Israël J. Math., 35 (1980) 32-48.

Re. A. Renyi, Representations of real numbers and their ergodic properties. Acta. Math. Akad. Sci. Hungar., 8 (1957) 477-493.

R1. D. Ruelle,"Thermodynamic formalism", Addison-Wesley, London, Amsterdam 1978.

R2. D. Ruelle, Applications conservant une mesure absolument continue par rapport à dx sur [0,1] Commun. Math. Phys. 55 (1977) 47-51.

R. J. Rousseau-Egele, Un théorème de la limite locale pour une classe de transformations dilatantes et monotones par morceaux. The Ann. of Prob. 11,(1983) 772-788.

Ry. M. Rychlik, Invariant measures for piecewise monotonic, piecewise C^{1+} transformations. To appear.

S. F. Schweiger, Number theoretical endomorphisms with σ finite invariant measure. Israël J. Math. 21 (1975) 308-318.

Sz. W. Szlenk, Some dynamical properties of certain differentiable mappings of an interval I.II Preprint IHES 1980.

T. M. Thaler, Tranformation on [0,1] with infinite invariant measures. Preprint, University of Salzburg 1983. Estimates of the invariant densities of endomorphisms with indifferent fixed points. Isräel J. Math., 37 (1980) 303-314.

U. S. Ulam, "A collection of mathematical problems", Interscience New York 1960.

Wo1. S. Wong, Hölder continuous derivatives and ergodic theory. Proc. London Math. Soc. 22 (1980) 506-520.

Wo2. S. Wong, A central limit theorem for piecewise monotonic mappings of the unit interval. Ann. of Prob. 7 (1979) 500-514.

W1. P. Walters, Invariant measures and equilibrium states for some mappings which expand distances. Trans. Amer. Math Soc. 236 (1978) 121-153.

W2. P. Walters, Ergodic theory introductory lectures. Lecture Notes in Mathematics 458, Springer-Verlag, Berlin, Heidelberg, New York 1975.

Y. K. Yosida,"Functional Analysis", Springer-Verlag Berlin, Heidelberg, New York 1968.

INTEGRABLE DYNAMICAL SYSTEMS

Eugene Trubowitz (Lecture given by)

(Notes by A.S. Wightman)
Forschungs Institut fur Mathematik
ETH - Zentrum CH 8092
Zurich, Switzerland

Lecture I Solvable Systems

What does it mean to "solve" the equations of a dynamical system? We answer the question by giving some examples and some general theorems which cover the examples.

Example 1

$$\dot{x} = \sqrt{1 - x^2}$$

We guess

$$x = \sin \theta$$

Then

$$\dot{\theta} \cos \theta = \sqrt{1 - \sin^2\theta} = \cos \theta$$

so the equation redues to

$$\dot{\theta} = 1$$

so $\theta = t +$ constant. The trick was to know the trigonometric functions.

Example 2

$$\dot{x} = [x(x-1)x-\lambda)]^{\frac{1}{2}} \qquad \lambda \text{ real}, \neq 0,1$$

Here the appropriate guess is not a trigonometric function but a Weierstrass function, . It may be defined by the formula

$$z = \int_{\wp(z)}^{\infty} [4t^3-g_2t-g_3]^{-\frac{1}{2}}\, dt$$

i.e. it is the inverse of the function $z(\zeta)$ defined by the indefinite integral

$$z = \int_{\zeta}^{\infty} [4t^3 - g_2 t - g_3]^{-\frac{1}{2}}\, dt$$

$[x(x-1)(x-\lambda)]^{\frac{1}{2}}$ can be brought to the form appearing in these definitions by a shift and a rescaling. The essential point is that the Riemann surface of the irrational function $[x(x-1)(x-\lambda)]^{\frac{1}{2}}$ is topologically a torus

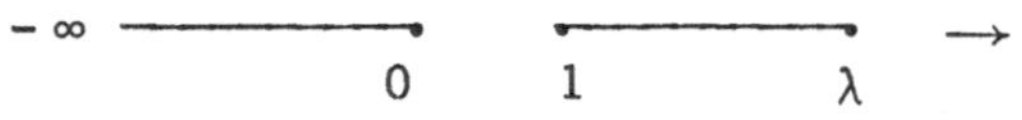

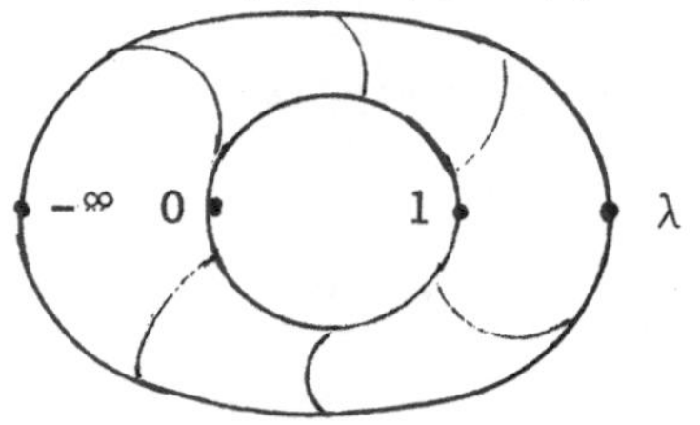

Once this has been recognized, one can treat more complicated cases, for example:

<u>Example 3</u>

$$P(x) = (x-\lambda_1)(x-\lambda_2)\ldots(x-\lambda_4)$$

and ask for the solution of

$$\dot{x}_1 = \frac{\sqrt{P(x_1)}}{x_2 - x_1} \qquad \dot{x}_2 = \frac{\sqrt{P(x_2)}}{x_1 - x_2}$$

This can again be treated using elliptic functions.

<u>Example 4</u>

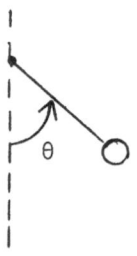

The equation of a pendulum

$$\ddot{\theta} = -\sin\theta$$

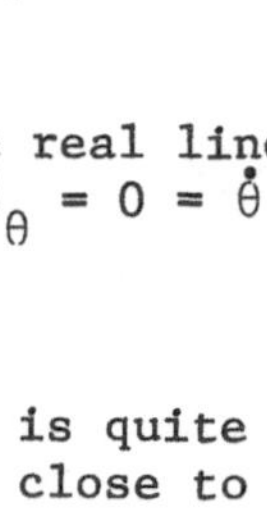

can be written as a Hamiltonian system

$$\dot{p}_\theta = -\frac{\partial H}{\partial \theta} \qquad \dot{\theta} = \frac{\partial H}{\partial p_\theta}$$

if

$$H = \frac{p_\theta^2}{2} - \cos\theta$$

θ varies on the unit circle and p_θ over the whole real line. This Hamiltonian has two equilibrium points where $\dot{p}_\theta = 0 = \dot{\theta}$,

$$\theta = 0 \quad \text{and} \quad \theta = \pi$$

The qualitative behavior of the flow in phase space is quite different near the two equilibrium points. The orbits close to $\theta = 0,\ p_\theta = 0$ circle it while those near $\theta = \pi,\ p_\theta = 0$ run away. If we unwrap θ on the line we have

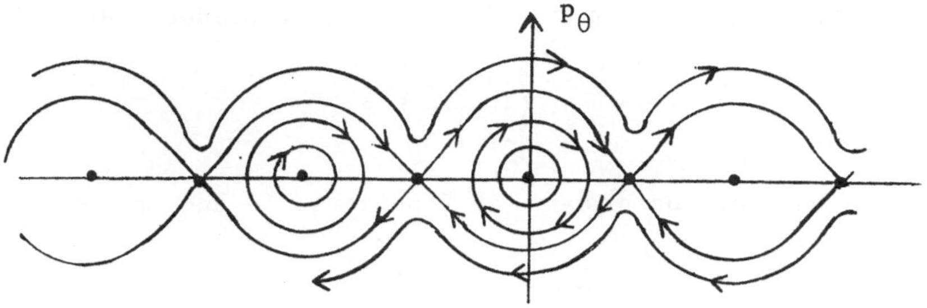

although the actual phase space is a cylinder.

More generally, if the phase space portrait of a system of differential equations can be displayed, one should in some sense regard the system as one which can be "solved". Then one can often find a system of coordinates in which the solution is explicit. What is the method for finding this system? Guess!

Some of this general talk can be made more concrete as follows. For simplicity, consider the case of Hamiltonian systems in $\mathbb{R}^{2n}$. We write

$$\{x,y\} \quad \mathbb{R}^{2n} , \quad \dot{x}_i = \frac{\partial H}{\partial y_i} , \quad \dot{y}_i = -\frac{\partial H}{\partial x_i} \qquad i = 1,\ldots n$$

where H is the Hamiltonian expressed as a function of x and y. Then we have the

Rectification Theorem for Hamiltonian Systems

In a sufficiently small neighborhood of any non-equilibrium point there is a canonical change of variables so that the new Hamiltonian is y_1.

(Here as usual an equilibrium point is defined as one where the gradient of H vanishes $\nabla H = \{\partial H/\partial x_i , \partial H/\partial y_i\} = 0$ a canonical change of coordinates is one which preserves the Hamiltonian form of the equations of motion.)

In the new coordinates the time evolution of the system is locally a translation along the y_1 axis and any function depending on $x_2 \ldots x_n$ and $y_2 \ldots y_n$ is a local integral of motion. The usefulness of these integrals of motion to describe the phase space portrait of the flow is determined by the global nature of the flow. When an orbit returns to a neighborhood of its starting point do the values of the integral of motion serve to separate orbits in some smooth way?

To pursue this idea quantitatively, we introduce the $2n \times 2n$ matrix

$$J = \left\{\begin{array}{c|c} 0 & I \\ \hline -I & 0 \end{array}\right\}$$

and define the Poisson bracket of functions f and g on R^{2n}.

$$\{f,g\} = \langle \nabla f, J\nabla g \rangle$$

where $\langle\ ,\ \rangle$ is the ordinary Euclidean scalar product. Then f is an integral of motion if $\{f,H\} = 0$. We propose a definition of integrability and then justify it.

Definition

Let Ω be an open set in $\mathbb{R}^{2n}$ and H be a function in $C^2(\mathbb{R}^{2n})$. Then the Hamiltonian vector field

$$\left(\frac{\partial H}{\partial y}, -\frac{\partial H}{\partial x}\right)$$

is integrable in Ω if there exist n functions $F_1 \ldots F_n \in C^2(\Omega)$ such that

i) $\nabla F \ldots \nabla F_n$ are linearly independent at each point of Ω

ii) $\{H,F_j\} = 0 \quad j = 1,\ldots n$

iii) $\{F_j,F_k\} = 0 \quad j,k = 1,\ldots n$

Then $F_1 \ldots F_n$ are said to be in involution.

To understand this definition pick a vector $c \in \mathbb{R}^n$ and define

$$M_c = \{\{x,y\} | \in \Omega\ ;\ F_j(x,y) = c_j,\ j = 1,\ldots n\}$$

Then i) guarantees that M_c is a differentiable submanifold of dimension n, ii) assures that M_c is invariant under the flow and iii) says that the vectors $J\nabla F_k$ lie in the tangent space of M_c. They must be linearly independent because the ∇F_k are. Thus, they can be used to define flows on the manifold M_c. These flows commute by condition iii) and so acting jointly on a given point of M_c, they cover a neighborhood of the point.

To get further we have to make more assumptions. Suppose M_c lies entirely in Ω and is compact. Then we can prove that the flow just defined has a discrete subgroup, Γ, a lattice with n independent basis vectors, which leaves a point fixed and the points of M_c can be labeled by the quotient $\mathbb{R}^n/\Gamma$. In other words, M_c is an n-torus. The detailed display of this geometrical fact using angle and action variables is to be found in V.I. Arnold <u>Mathematical Methods of Classical Mechanics</u> pp. 271-285.

We will apply the ideas just described to the system of equations

$$\dot{x}_k = -\tfrac{1}{2}\left(e^{x_k - x_{k+1}} + e^{x_{k-1} - x_k}\right)$$

$k = 0,1,\ldots 2n+1$ but $x_0 = -\infty$, $x_{2n+1} = +\infty$

<u>Theorem</u>

$$x_k(t) \sim \alpha_k^+ t + \beta_k^+ \quad \text{as} \quad t \to +\infty$$

$$x_k(t) \sim \alpha_k^- t + \beta_k^- \quad \text{as} \quad t \to -\infty$$

and

$$\alpha_1^+ = \alpha_2^+ , \qquad \alpha_3^+ = \alpha_4^+ , \ldots$$

$$\alpha_{2n-1}^- = \alpha_{2n}^- , \ \alpha_{2n-3}^- = \alpha_{2n-2}^- , \ldots$$

<u>Proof</u>

Change variables

$$4a_k^2 = e^{x_k - x_{k+1}} \qquad a_0 = 0 = a_{2n+1}$$

Then

$$\dot{a}_k = a_k \left(a_{k+1}^2 - a_{k-1}^2\right)$$

Now we look for integrals of motion. For example,

$$\frac{d}{dt}\sum_{k=1}^{2n-1} a_k = 2\sum_{k=1}^{2n-1} a_k\dot{a}_k = 2\lfloor a_1 a_1 (a_2^2 - a_0^2)$$
$$+ a_2 a_2 (a_3^2 - a_1^2)$$
$$+ \ldots\ ] = 0$$

a telescoping sum. Thus $\sum_{k=1}^{2n-1} a_k^2$ is a constant of the motion.

That implies a_k is bounded which in turn implies $\dot{a}_k$ is bounded.

Next look at $\prod_{j=1}^{n} a_{2j-1}^2$. We claim it is also an integral of motion and so is bounded away from zero. Consider first

$$\frac{d}{dt} \log a_1 = a_2^2$$

It implies

$$\lim_{t\to\infty} \int_0^t a_2^2\, d\tau = -\log a_1(0) + \lim_{t\to\infty} \log a_1(t)$$

which gives, since the integral is increasing in t and therefore has either $+\infty$ or a finite number as limit,

$$\int_0^\infty a_2^2\, d\tau < \infty \quad ,$$

and therefore

$$a_2(t) \to 0 \quad \text{as} \quad t \to \infty \; .$$

Next note

$$\frac{d}{dt} (\log a_1 a_3) = a_4^2$$

$$\frac{d}{dt} (\log a_1 a_3 a_5) = a_6^2$$

$$\vdots$$

a sequence of equations obtained by addition:

$$\frac{\dot{a}_1}{a_1} + \frac{\dot{a}_3}{a_3} = a_2^2 + (a_4^2 - a_2^2), \; \frac{\dot{a}_1}{a_1} + \frac{\dot{a}_3}{a_3} + \frac{\dot{a}_5}{a_5} = a_4^2 + (a_6^2 - a_4^2), \ldots \; .$$

From these equations there follows by the above argument

$$\lim_{t\to\infty} a_{2j} = 0 \quad \text{and} \quad \lim_{t\to\infty} a_{2j-1} \text{ exists and is } \neq 0$$

$$\lim_{t\to\infty} \dot{a}_{2j-1}(t) = 0 \quad .$$

Now

$$8\, a_k\, \dot{a}_k = (4\, \dot{a}_k^2) = (\dot{x}_k - \dot{x}_{k+1}) 4\, a_k^2$$

which implies

$$\lim_{t\to\infty} [\dot{x}_k - \dot{x}_{k+1}] = 0$$

for k odd. If we go back to the equations for the x_t we see that $\dot{x}_k$ approaches a limit as $t \to \infty$ because the right-hand side of the differential equations for it is expressible in terms of the a_k's and they approach a limit. We define $\lim_{t\to\infty} \dot{x}_k = \alpha_k^+$. The statement $\lim_{t\to\infty} [\dot{x}_k - \dot{x}_{k+1}] = 0$ for k odd then gives half the theorem

$$\alpha_1^+ = \alpha_2^+ \quad , \quad \alpha_3^+ = \alpha_4^+ \; , \; \ldots$$

The other half is proved in a similar way.

The quantities α_{2j}^+ $j = 1,\ldots n$ are the eigenvalues of the matrix

$$L = \begin{pmatrix} 0 & a_1 & 0 & 0 & \\ a_1 & 0 & a_2 & 0 & \\ 0 & a_2 & 0 & a_3 & \cdots \\ 0 & 0 & a_3 & 0 & \\ & & \vdots & & \end{pmatrix}$$

More precisely,

Theorem

Let $\lambda_1\ldots\lambda_n$ be the eigenvalues of L at time $t = 0$ written in decreasing order. Then

$$\lambda_1 > \lambda_2 >\ldots> \lambda_n \; ,$$

$$\alpha_{2j}^+ = -2\lambda_j^2$$

and the λ_j are the same if L is evaluated at any other time.

We look for a unitary one-parameter family of operators satisfying

$$L(0) = U(t)\; L(t)\; U(t)^{-1}$$

Notice that, because $L(t)$ is self-adjoint, unitarity of $U(t)$ is consistent with the self-adjointness of the L's. Differentiating with respect to t, we have

$$0 = \dot{U}LU^{-1} + U\dot{L}U^{-1} + UL\dot{U}^{-1}$$

i.e. if we write

$$\dot{U}(t) = U(t)B(t)$$

and note $B(t)^* = - B(t)$

$$0 = UBLU^{-1} + U\dot{L}U^{-1} - ULU^{-1}\, UBU^{-1}$$

i.e.

$$\dot{L} = [L,B] \quad .$$

Here we have used

$$\dot{U}^{-1} = - \dot{U}^{-1}\, U\, U^{-1}$$

which is obtained by differentiating $UU^{-1} = 1$. In our present case B turns out to be

$$B = \begin{pmatrix} 0 & 0 & a_1a_2 & 0 & 0 & \cdots \\ 0 & 0 & 0 & a_2a_3 & 0 & \\ -a_1a_2 & 0 & 0 & 0 & & \\ 0 & -a_2a_3 & 0 & 0 & & \\ & \vdots & & & & \end{pmatrix}$$

We have arrived at the ideas of Lax who relates the existence of integrals of motion to isospectral transformations.

Lecture II <u>The Korteweg deVries Equation</u>

The KdV equation is a partial differential equation for a function of two real variables $q(x,t)$

$$q_t = 3q\, q_x - \frac{1}{2}\, q_{xxx} \quad .$$

It was studied by Kruskal, Zabusky et al. Many integrals were found e.g.

$$H_1 = \int_0^1 q\, dx, \quad H_2 = \int_0^1 q^2 dx, \quad H_3 = \int_0^1 \left(\frac{(q_t)^2}{4} + \frac{q^3}{2}\right) dx \quad .$$

To understand this, we study how

$$- \frac{d^2}{dx^2} + q(x)$$

acts on $L^2(\mathbb{R}^1)$ when $q(x) = q(x+1)$. Note that it has a real

spectrum which occurs in bands

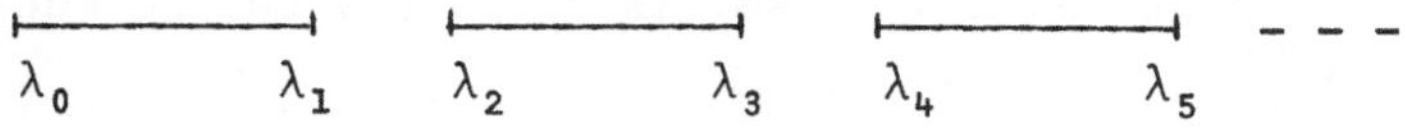

where the corresponding eigenfunctions $\psi_0, \psi_2, \psi_4, \psi_6, \psi_8, \ldots$ are periodic and $\psi_1, \psi_3, \psi_5, \psi_7, \ldots$ are antiperiodic. We thus look at all solutions of

$$-\frac{d^2y}{dx^2} + q(x)y = \lambda y \qquad y(x+1) = \pm\, y(x) \quad .$$

We have

Theorem

Let $q(x,t)$ be a spatially periodic solution of

$$q_t = 3\,q\,q_x - \frac{1}{2}\,q_{xxx}$$

Then for all $n \geq 0$

$$\frac{d}{dt}\,[\lambda_n\,(q(t,\cdot))] = 0$$

Proof

In order to evaluate the time derivative appearing in the theorem, we need a formula for the derivative of a functional of a function. We define it as a linear functional by

$$\frac{\partial}{\partial\varepsilon}\;\lambda_n(q+\varepsilon v)\Big|_{\varepsilon=0} = \left< \frac{\partial\lambda_n}{\partial q}\,,\,v \right>$$

Then we have the formula

$$\frac{d}{dt}\,\lambda_n(q(t,\cdot)) = \int_0^1 \frac{\partial\lambda_n(q(t,\cdot))}{\partial q(t,x)}\;\frac{\partial q}{\partial t}\,(t,x)\;dx$$

Now we assert that

$$\frac{\partial\lambda_n}{\partial q(t,x)} = \psi_n^{\,2}(x,q(t,x))$$

where ψ_n is the eigenfunction of $-\frac{d^2}{dx^2} + q(t,x)$ belonging to the eigenvalue $\lambda_n(q(t,x))$. In order to prove this, differentiate the equation

$$- \psi_n''(x,q+\varepsilon v) + (q+\varepsilon v)\psi_n(x,q+\varepsilon v) = \lambda_n(q+\varepsilon v)\psi_n(x,q+\varepsilon v)$$

with respect to ε . Then, if we introduce the notation $\bullet$ for $\frac{\partial}{\partial\varepsilon}(\ )\Big|_{\varepsilon=0}$, we get

$$-\dot{\psi}_n''(x,q) + v\psi_n(x,q) + q\dot{\psi}_n(x,q) = \dot{\lambda}_n(q(t,\cdot))\psi_n(x,q) + $$

$$+ \lambda_n(q(t,\cdot))\dot{\psi}_n(x,q) \quad .$$

When this is multiplied by $\psi_n(x,q)$ and integrated over the unit interval, it yields

$$<\psi_n,\dot{\psi}_n''> + <\psi_n,v\psi_n> + <\psi_n,q\dot{\psi}_n> =$$

$$= \dot{\lambda}_n <\psi_n,\psi_n> + \lambda_n <\dot{\psi}_n,\psi_n>$$

Now we integrate by parts in the first term and use the differential equation for ψ_n . Four terms then cancel leaving

$$\dot{\lambda}_n <\psi_n,\psi_n> = <\psi_n,v\psi_n>$$

which is just

$$\dot{\lambda}_n(v) = <\psi_n^2,v>$$

if we assume, as we do, that ψ_n is so normalized that $<\psi_n,\psi_n> = 1$. This justifies

$$\frac{d}{dt}\ \lambda_n(q(t,\cdot)) = \int_0^1 \psi_n^2(x,q(t,x))\ \frac{\partial q(t,x)}{\partial t}\ dx \quad .$$

Now we use the differential equation for q to write the right-hand side as

$$\int_0^1 \psi_n^2(x,q(t,x))(3\,q\,q_x - \frac{1}{2}\,q_{xxx})\ dx = <\psi_n^2,L_q q>$$

where by definition

$$L_q q = [q\,\frac{d}{dx} + \frac{d}{dx}\,q - \frac{1}{2}\,\frac{d^3}{dx^3}]q = 3qq_x - \frac{1}{2}\,q_{xxx}$$

Notice that L_q is skew symmetric with respect to the scalar product $<\cdot,\cdot>$ and

$$L_q \psi_n^2 = 2\lambda_n \frac{d}{dx} \psi_n^2 \quad .$$

Therefore,

$$<\psi_n^2, L_q q> = - <L_q \psi_n^2, q> = -2\lambda_n <\frac{d}{dx} (\psi_n^2), q>$$

$$= - 4 \lambda_n \int_0^1 \psi_n \frac{d\psi_n}{dx} q \, dx \quad .$$

But

$$q \psi_n = \frac{d^2}{dx^2} \psi_n + \lambda_n \psi_n$$

so

$$<\psi_n^2, L_q q> = - 4\lambda_n \int_0^1 \frac{d}{dx} \left[\frac{1}{2} \left(\frac{d\psi_n}{dx}\right)^2 + \lambda_n \psi_n^2 \right] dx$$

$$= - 4\lambda_n \left[\frac{1}{2} \left(\frac{d\psi_n}{dx}\right)^2 + \lambda_n \psi_n^2 \right] \Big|_0^1 = 0$$

In the last step we have used the fact that ψ_n is either periodic or antiperiodic.

This theorem establishes that a time-dependent potential $q(t,x)$ when inserted in the Schrödinger operator $-d^2/dx^2 + q(t,x)$ yields a spectrum independent of t, if $q(t,x)$ satisfies the KdV equation as a function of t and x. Next we ask: which potentials yield the same spectrum? The answer is provided by the following

Theorem

Let p be a potential in $L^2_{\mathbb{R}}(0,1)$.

The set

$$L(p) = \{q \mid q \in L^2_{\mathbb{R}}(0,1), \ \lambda_i(q) = \lambda_i(p) \quad i = 0,1,\ldots\}$$

is homeomorphic to a product of circles, one circle for each strict inequality, $\lambda_{2j-1} < \lambda_{2j}$.

For a proof of this and the following assertions see

H.P. McKean and E. Trubowitz Hill's Operator and Hyperelliptic Function Theory Commun. Pure and Appl. Math. 29 (1976) 143-226

Thus to parametrize the potentials in $L^2\mathbb{R}(0,1)$, we use the λ_i $i = 0,1,\ldots$ which define the band spectrum and the $L(p)$ which give the fiber above a given band spectrum. We have yet to describe exactly which band spectra actually occur. One thing which is known is

$$\lambda_{2n}, \lambda_{2n-1} \sim n^2\pi^2 + \int_0^1 q\,dx + \ell^2(n)$$

where $\ell^2(n)$ stands for any real sequence a_n such that

$$\sum_{n=0}^{\infty} a_n^2 < \infty .$$

To make a more precise statement it is convenient to introduce a description of the band widths and gaps

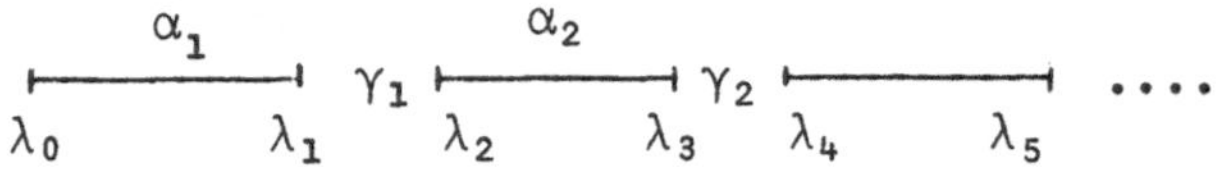

We have three theorems

<u>Theorem</u>

$$\alpha_n(z) \leq (2n-1)\pi^2 \qquad n \geq 1$$

and if, for a single n, $\alpha_n = (2n-1)\pi^2$ then $q = 0$.

<u>Theorem</u>

A necessary and sufficient condition that the sequence $\{\gamma_1,\gamma_2,\ldots\}$ be the gaps for some $q \in L^2(0,1)$ is

$$\sum_n \gamma_n^2 < \infty \qquad \gamma_n \geq 0$$

For any such sequence there is a unique way of choosing the band widths $\{\alpha_1,\alpha_2,\ldots\}$ so that the resulting band spectrum belongs to a potential $q \in L^2(0,1)$

<u>Theorem</u>

The sequences of gaps $\{\gamma_1,\gamma_2,\ldots\}$ such that $\gamma_j = 0$ for all sufficiently large j yield a dense set of q's.

For further information, see E. Trubowitz <u>The Inverse Problem for Periodic Potentials</u> Commun. in Pure and Applied Math. XXX (1977) 321-337

Lecture III Sonja Kovalevskaya and Rigid Body Motion as an Integrable System

In 1889, Sonja Kovalevskaya published her treatment of integrable cases of rigid body motion. The paper had won the Bordin Prize of the French Academy of Sciences. The reference is S. Kovalevski Sur le problème de la rotation d'un corps solide autour d'un point fixe Acta Math. 12 (1889) 177-232. We will explain her result here in the light of modern knowledge of algebraic geometry.

We begin by discussing the effects of constraints on a set of simple harmonic oscilaltors. Suppose the motion of the oscillators is given by the differential equations

$$\ddot{x}_i + \sigma_i x_i = 0 \qquad i = 1 \ldots n$$

but they are constrained to move on the sphere

$$\sum_{i=1}^{n} x_i^2 = 1 \ .$$

The effect of the constraint forces is to change the equations of motion to

$$\ddot{x}_i + \sigma_i x_i = F(x,x) x_i \quad .$$

To determine F differentiate the constraint to obtain

$$\sum_{i=1}^{n} x_i \dot{x}_i = 0 \qquad \text{and} \qquad \sum_{i=1}^{n} (x_i \ddot{x}_i + (\dot{x}_i)^2) = 0$$

and multiply the equation of motion by x_i and sum:

$$\sum_{i=1}^{n} x_i \ddot{x}_i + \sum_{i=1}^{n} \sigma_i (x_i)^2 = F(x,\dot{x}) \sum_{i=1}^{n} x_i^2 \ .$$

Using the constraint and the second identity derived from it we get

$$F(x,\dot{x}) = \sum_{i=1}^{n} [-(\dot{x}_i)^2 + \sigma_i (x_i)^2] \quad .$$

Now write the equations as a first order system in the phase space $\mathbb{R}^{2n}$, with coordinates $x_1 \ldots x_n \; y_1 \ldots y_n$ and $\dot{x}_j = y_j$, $j = 1,2,\ldots n$

$$\dot{y}_i = -\sigma_i x_i + \sum_{i=1}^{n} [-(y_i)^2 + \sigma_i (x_i)^2] x_i$$

and the constraints

$$\sum_{i=1}^{n} x_i^2 = 1 \quad \text{and} \quad \sum_{i=1}^{n} x_i y_i = 0 \ .$$

Remarkably, it turns out that this system is integrable. More precisely,

<u>Theorem</u>

For $k = 1,\dots n$ set

$$F_k = x_k^2 + \sum_{\ell \neq k} \frac{(x_k y_\ell - x_\ell y_k)^2}{(\sigma_k - \sigma_\ell)}$$

Then the Poisson brackets

$$\{F_j, F_k\} = 0$$

and on the manifold defined by the constraints

$$\sum_{j=1}^{n} F_j = 1$$

$$\sum_{j=1}^{n} \sigma_j F_j = 2H$$

where H is the Hamiltonian

$$H_0 = \sum_{j=1}^{n} \frac{1}{2} (y_j^2 + \sigma_j x_j^2) \ .$$

It can be shown that this integrable dynamical system gives the KdV equation when $n \to \infty$. See

P. Deift, F. Lund, E. Trubowitz Non linear wave equations and constrained harmonic motion Commun. Math. Phys. <u>74</u> (1980) 141-188.

<u>Digression on the zero sets of polynomials and algebraic geometry</u>

An <u>affine algebraic variety</u> is the set of common roots of a family of polynomials $f_i(z_1,\dots z_n)$, $i = 1,\dots N$.

Now form the k dimensional complex torus

$$T = \mathbb{C}^k/\Lambda \quad \text{where} \quad \Lambda = \{ \sum_{j=1}^{2k} n_i \lambda_i \,|\, (n_1 \dots n_k) \in \mathbb{Z}^{2k} \}$$

where $\lambda_1 \ldots \lambda_{2k}$ are vectors in C^k linearly dependent over $\mathbb{R}$. You cannot imbed T in $\mathbb{C}^n$. (To see this apply the maximum principle to conclude that the coordinates are constant on T.) For example, non-trivial analytic functions on $E = \mathbb{C}/_{\{n\omega_1 + n\omega_2\}}$ have poles, and are elliptic functions like the Weierstrass function

$$p(z) = \frac{1}{z^2} + \sum_{\substack{m_1, m_2 \in \mathbb{Z} \\ \text{not both } 0}} \left[\frac{1}{(z - m_1\omega_1 - m_2\omega_2)^2} - \frac{1}{(m_1\omega_2 + m_2\omega_2)^2} \right] .$$

Next we come to the definition of complex projective n dimensional space $\mathbb{C}P^n$. It is defined as the quotient space $\mathbb{C}^{n+1} \setminus \{0\}/\sim$ of $\mathbb{C}^{n+1}$ minus the origin by the equivalence relation

$$w \sim z \quad \text{iff} \quad w = \gamma z \quad \text{for some} \quad \gamma \in \mathbb{C} .$$

It is a complex manifold, defined by the charts $U_k \subset \mathbb{C}P^n$ where U_k consists of the equivalence classes of points $\{z_0, \ldots z_n\} \in \mathbb{C}^{n+1}$ with $z_k \neq 0$ (all such belong to $\mathbb{C}^{n+1} \setminus \{0\}$). The mapping $U_k \to \mathbb{C}^n$ which defines U_k as a chart is for the representative point $\{z_0, \ldots z_n\}$

$$\{z_0, z_1, \ldots z_n\} \mapsto \left\{\frac{z_0}{z_k}, \frac{z_1}{z_k}, \ldots \frac{z_n}{z_k}\right\} .$$

(Ignore the 1 which occurs in the kth place; then the right-hand side defines a point of $\mathbb{C}^n$.) With this definition

$$\mathbb{C}P^n = \bigcup_{k=1}^{n} U_k$$

and on the overlap $U_k \cap U_\ell$ the mappings associated with U_k and U_ℓ differ by the multiplication by z_ℓ / z_k

$$\begin{array}{ccc} & U_k \cap U_\ell & \\ \swarrow & & \searrow \\ \mathbb{C}^k & \xrightarrow{\hspace{2em}} & C^n \\ & z_k/z_\ell & \end{array}$$

Notice that the representative points, $\{z_0, z_1, \ldots z_n\}$, of the equivalence classes in U_0 satisfy $z_0 \neq 0$ but have $z_1, \ldots z_n$ running over $\mathbb{C}^n$. Thus, $\mathbb{C}P^n \setminus U_0$ has representative points $\{0, z_1, \ldots z_n\}$ with at least one $z_j \neq 0$, $j = 1, \ldots n$, and so is just $\mathbb{C}P^{n-1}$. Thus $\mathbb{C}P^n$ is obtained from $\mathbb{C}P^{n-1}$ by gluing on $\mathbb{C}^n$ at infinity.

Definition

A projective algebraic variety is the set of common zeros in $\mathbb{C}P^n$ of a family $f_i(z_0,\ldots z_n)$ $i = 1,2,\ldots$ of homogeneous polynomials.

Notice that $f_i(z_0,z_1,\ldots z_n) = 0$ implies $f_i(\lambda z_0,\lambda z_1\ldots\lambda z_n)=0$ so the vanishing of a homogeneous polynomial in $z_0,\ldots z_n$ determines a set of points in $\mathbb{C}P^n$.

Now we have an extraordinary general fact about $\mathbb{C}P^n$.

Chow's Theorem

Let X be any complex analytic submanifold of $\mathbb{C}P^n$. Then X is a projective algebraic variety.

Thus, if we can embed the complex torus T in $\mathbb{C}P^n$ for some n then it must be given by the roots of a homogeneous polynomial. When the dimension k of T, is one it always can be embedded in $\mathbb{C}P^2$. In fact, the embedding can be given explicitly in terms of the Weierstrass $\wp$ function

$$E = \mathbb{C}/\{m_1\omega_1 + m_2\omega_2\} \hookrightarrow (1, \wp(z), \wp'(z)) \quad .$$

On the other hand, when $k \geq 2$ not every torus can be embedded in some $\mathbb{C}P^n$. Conditions for this to hold are given by the following theorem

Theorem

$A = \mathbb{C}^n/\Lambda$ is embeddable in some $\mathbb{C}P^n$ if and only if there exists a positive hermitean form $H(z,\omega)$ on $\mathbb{C}^n$ such that

$$\operatorname{im}\ H(\lambda,\lambda') \in \mathbb{Z}$$

for all $\lambda,\lambda' \in \Lambda$. A is then called an abelian variety.

Example

Assume the $2n$ vectors $\{I,R,\ldots R_n\}$ of $\mathbb{C}^n$ are linearly independent over $\mathbb{R}$, and that the $n \times n$ matrix

$$R = \{R_1,\ldots R_n\}$$

is symmetric with imaginary part, im R, positive definite. Then the series

$$\Theta(\underline{z}) = \sum_{\underline{m} \in \mathbb{Z}^n} \exp 2\pi i\underline{m}\cdot\underline{z} \exp \pi i \langle\underline{m},R\underline{m}\rangle$$

converges and defines a function of n variables $\{z_1\ldots z_n\} = \underline{z}$ of period 1 in each of its arguments. We have also

$$\Theta(\underline{z} + R_j) = \exp 2\pi i(z_j + R_{jj}/2)\,\Theta(z)$$

These functions can be used, together with some others to define the embedding of A in some $\mathbb{C}P^n$.

Now let us return to Sonja Kovalevaskaya's problem. She was looking for special cases of the motion of rigid body, supported at a point in a uniform gravitational field. The special feature required was that the equations be algebraically integrable. That means that there are integrals of motion which are polynomials that can be complexified to yield projective varieties, and that the flow defined by the Hamiltonian becomes a straight line motion. Formally, we expect $H(x,y)$, $x, y \in \mathbb{C}^n$, and $\dot{x}_i = \partial H/\partial y_i$, $\dot{y}_i = -\partial H/\partial x_i$, $i = 1,\ldots n$. As functions of t, $x_i(t)$ and $y_i(t)$ might have poles. That means one is approaching a point at infinity. The Laurent expansion near such a pole may be taken as

$$x_i(t) = t^{-n_i} \sum_{j=0}^{\infty} \hat{x}_{ij}\, t^j, \quad y_i(t) = t^{-m_i} \sum_{j=0}^{\infty} \hat{y}_{ij}\, t^j$$

$\hat{x}_{ij}$ and $\hat{y}_{ij}$ should depend on $2n-1$ parameters.

Euler's equations for a rigid body supported at the origin whose center of mass is at $\vec{x}_0$ read

$$\frac{d\vec{J}}{dt} + \vec{\omega} \times \vec{J} = Mg\, \vec{x}_0 \times \vec{v}$$

$$\frac{d\vec{v}}{dt} = -\vec{\omega} \times \vec{v}$$

Here $\vec{J}$ is the angular momentum and $\vec{\omega}$ is the angular velocity with respect to axes fixed in the body. $\vec{J}$ and $\vec{\omega}$ are related by $\vec{J} = I\vec{\omega}$ where I is the moment of inertia tensor. Choosing coordinate axes along the principal axes of I, we have

$$I = \begin{Bmatrix} A & 0 & 0 \\ 0 & B & 0 \\ 0 & 0 & C \end{Bmatrix}$$

$\vec{v}$ is the unit vector which describes the direction of the gravitational force as seen in the coordinate system fixed in the body. The equation for $d\vec{v}/dt$ expresses the fact that $\vec{v}$ rotates without changing its length.

If we write $\vec{\omega} = \begin{Bmatrix} p \\ q \\ r \end{Bmatrix}$, $\vec{x}_0 = \begin{Bmatrix} x_0 \\ y_0 \\ z_0 \end{Bmatrix}$, and $\vec{v} = \begin{Bmatrix} \gamma_1 \\ \gamma_2 \\ \gamma_3 \end{Bmatrix}$

we have for our basic equations

$$A\dot{p} + (C-B)qr = (Mg)(y_0\gamma_3 - z_0\gamma_2)$$

$$B\dot{q} + (A-C)pr = (Mg)(z_0\gamma_1 - x_0\gamma_3)$$

$$C\dot{r} + (B-A)qP = (Mg)(x_0\gamma_2 - y_0\gamma_1)$$

$$\dot{\gamma}_1 = r\gamma_2 - qr_3$$

$$\gamma_2 = p\gamma_3 - r\gamma_1$$

$$\gamma_3 = q\gamma_1 - p\gamma_2$$

These are six equations in the six unknowns p,q,r $\gamma_1,\gamma_2,\gamma_3$. They have three standard integrals of motion, the energy, the magnitude of $\vec{v}$, and the component of the angular momentum along $\vec{v}$. The energy integral is most easily derived by taking the scalar product of the first Euler equation with $\vec{\omega}$. That yields

$$\frac{d}{dt}\left[\frac{1}{2}\vec{\omega}\cdot I\cdot\vec{\omega} + Mg\,\vec{x}_0\cdot\vec{v}\right] = 0$$

because $\vec{\omega}\cdot(\vec{x}_0 \times \vec{v}) = \vec{x}_0\cdot(\vec{v} \times \omega) = -\frac{d}{dt}\vec{x}_0 \cdot \vec{v}$. In the present notation it is

$$\frac{1}{2}(p^2A + q^2B + r^2C) + Mg(x_0\gamma_1 + g_0\gamma_2 + z_0\gamma_3) = \text{const} \quad .$$

The component of the angular momentum along $\vec{v}$ is $\vec{v}\cdot I\cdot\omega = \gamma_1Ap + \gamma_2Bq + \gamma_3Cr$.

We consider the Laurent series of the unknown functions around a pole which we may as well take to be the origin:

$$p = t^{-n_1}[p_0+p_1t+\ldots] \qquad \gamma_1 = t^{-m_1}[f_0+f_1t+\ldots]$$

$$q = t^{-n_2}[q_0+q_1t+\ldots] \qquad \gamma_2 = t^{-m_2}[g_0+g_1t+\ldots]$$

$$r = t^{-n_3}[r_0+r_1t+\ldots] \qquad \gamma_3 = t^{-m_3}[h_0+h_1t+\ldots]$$

Consistency with the equations of motion near $t = 0$ gives immediately

$$n_1 = n_2 = n_3 = 1 \qquad m_1 = m_2 = m_3 = 2$$

Equating the coefficients of the terms proportional to t^{-2} gives the equations

$$- Ap_0 = A_1 q_0 r_0 + y_0 h_0 - z_0 g_0$$

$$- 2f_0 = r_0 g_0 - q_0 h_0$$

$$- Bq_0 = B_1 r_0 p_0 + z_0 f_0 - x_0 h_0$$

$$- 2g_0 = p_0 h_0 - r_0 f_0$$

$$- Cr_0 = C_1 p_0 q_0 + x_0 g_0 - y_0 f_0$$

$$- 2h_0 = q_0 f_0 - p_0 q_0$$

where $A_1 = B-C$, $B_1 = C-A$, and $C_1 = A-B$, and for simplicity we have taken $Mg = 1$.

Similarly, by equating the powers t^{n-2} one gets a system of equations

$$\begin{bmatrix} (n-1)A & -A_1 r_2 & -A_1 q_2 & 0 & z_0 & -y_0 \\ -B_1 r_0 & (n-1)B & -B_1 p_0 & -z_0 & 0 & x_0 \\ -C_1 q_0 & -C_1 p_0 & (n-1)C & y_0 & -x_0 & 0 \\ 0 & h_0 & -g_0 & (n-2) & -r_0 & q_0 \\ -h_0 & 0 & f_0 & r_0 & (n-2) & -p_0 \\ g_0 & -f_0 & 0 & -q_0 & p_0 & (n-2) \end{bmatrix} \begin{bmatrix} p_n \\ q_n \\ r_n \\ f_n \\ g_n \\ h_n \end{bmatrix} = \begin{bmatrix} P_n \\ Q_n \\ R_n \\ F_n \\ G_n \\ H_n \end{bmatrix}$$

where on the right hand side are functions of $p_m, q_m, r_m f_m, g_m, h_m$ with $m < n$.

At the time of Kovalevskaya's work there were two special cases in which these equations were known to have solutions with the requisite number of parameters.

1) (Euler-Poinsot) $x_0 = y_0 = z_0 = 0$
   Then there is no external torque

2) (Lagrange) $A = B$, $x_0 = y_0 = 0$
   Spinning top

Kovalevskaya added a third case

3) $A = B = 2C$, $z_0 = 0$.

By a rotation of coordinates in the plane xy, which does not affect the diagonal form of the moment of inertia tensor, one can always arrange to have $y_0 = 0$ also. A choice of units will make $C = 1$. Then the differential equations become

$$2\frac{dp}{dt} = qr \qquad \frac{d\gamma_1}{dt} = r\gamma_2 - q\gamma_3$$

$$2\frac{dq}{dt} = -pr - C_0\gamma_3 \qquad \frac{d\gamma_2}{dt} = p\gamma_3 - r\gamma_1$$

$$\frac{dr}{dt} = C_0\gamma_2 \qquad \frac{d\gamma_3}{dt} = q\gamma_1 - p\gamma_2$$

where $Mg\, x_0 = C_0$. The three integrals of motion reduce to

$$(p^2+q^2) + \frac{1}{2}r^2 - C_0\gamma_1 = 3\ell_1$$

$$2(p\gamma_1+q\gamma_2) + r\gamma_3 = 2\ell$$

$$r_1{}^2 + \gamma_2{}^2 + \gamma_3{}^2 = 1$$

Kovalevskaya found another integral of motion

$$|(p+iq)^2 + C_0(\gamma_1+i\gamma_2)|^2 = k^2$$

where k is a real constant. She recognized that using the four integrals of motion she could express $p,q,r,\gamma_1,\gamma_2,\gamma_3$ in terms of generalized elliptic functions, and in forty-eight pages of manipulations showed explicitly how to do it. Nowadays, using the theory of abelian varieties one can see somewhat more geometrically the meaning of those manipulations.

Lecture IV The Inverse Sturm Liouville Problem

In 1836 in the first volume of Journal de Mathématiques (Liouville's journal), C. Sturm and J. Liouville studied the boundary value problem defined by

$$\cos\alpha\; y(0) + \sin\alpha\; y'(0) = 0$$

$$\cos\beta\; y(1) + \sin\beta\; y'(1) = 0$$

for the differential equation

$$-y'' + q(x)y = \lambda y$$

on the interval $0 \le x \le 1$. Here, for simplicity, it is assumed that $q \in L^2_{\mathbb{R}}([0,1])$. It is known that in general the set of all self-adjoint boundary conditions for the differential equation form a four-dimensional manifold .

Problem I Fix $b \in \mathcal{B}$. Characterize all sequences of λ's which are the spectrum for some $q \in L^2_{\mathbb{R}}([0,1])$.

$$\nu_0(q,b) < \nu_1(q,b) < \ldots .$$

Problem II Fix $b \in \mathcal{B}$ and $p \in L^2_{\mathbb{R}}([0,1])$. Define

$$M_b(p) = \{q \in L^2_{\mathbb{R}}([0,1]) \; \nu_n(q,b) = \nu_n(p,b) \quad n = 0,1,2,\ldots\}$$

Describe $M_b(p)$.

Let us begin with Dirichlet boundary conditions, for which $\alpha = \beta = 0$. For this special case, use the notation

$\mu_n(q) \quad n \geq 1$ is the Dirichlet spectrum of q

$$M(p) = \{q | \mu_n(q) = \mu_n(p) \quad \text{for} \quad n \geq 1\}$$

If one takes a geometrical point of view, these are to be thought of as defining surfaces in $L^2_{\mathbb{R}}([0,1])$. The normal to the surface $\mu_n(q) = \text{const}$ is the gradient $\partial\mu_n/\partial q(x)$ which we evaluated in Lecture II as

$$\frac{\partial \mu_n}{\partial q(x)} = \psi_n^2(x,q) \quad .$$

We claim that $\psi_n^2(x,q)$ is orthogonal to $\frac{d}{dx}(\psi_n^2)$ for all m and n .

$$\int_0^1 \psi_n^2(\psi_m^2)' \, dx = -\int_0^1 (\psi_n^2)'\psi_m^2 \, dx$$

$$= \frac{1}{2}\int_0^1 [\psi_n^2(\psi_m^2)' - (\psi_n^2)' \psi_m^2] \, dx$$

$$= \int_0^1 \psi_n\psi_m [\psi_n\psi_m' - \psi_n'\psi_m] \, dx$$

Now

$$(\psi_n\psi_m' - \psi_n'\psi_m)' = (\mu_n-\mu_m)\psi_n\psi_m$$

so the preceding expression is

$$\left.\frac{(\psi_n\psi_m' - \psi_n'\psi_m)^2}{2(\mu_n - \mu_m)}\right|_0^1 = 0$$

Thus, $(\psi_n^2)'$ is orthogonal to all the normals, $\partial\mu_n/\partial q(x)$, and therefore defines a tangent vector. $U_n(q) = 2(\psi_n^2)'$. Solve the ordinary differential equation

$$\frac{dq}{dt} = U_n(q) \quad \text{with the initial condition} \quad q(x,0) = r(x).$$

We get the solution

$$q_n^t(r) = r - 2\frac{d^2}{dx^2}\log\theta_n(x,t,r)$$

where

$$\theta_n(x,t,r) = 1 + (e^t - 1)\int_x^1 \psi_n^2(s,r)\,ds$$

and the theorem

<u>Theorem</u>

Let $k \in \ell_1^2(\mathbb{Z}^+)$ i.e. $\sum_{j=1}^{\infty} j^2 k_j^2 < \infty$ and set

$$\Theta(x,k,p) = \det[\delta_{ij} + (e^{k_i} - 1)\int_x^1 \psi_i(s,p)\psi_j(s,p)ds]$$

Then the mapping $\ell_1^2(\mathbb{Z}^+) \to M(p)$ defined by

$$k \to \exp_p(k) = p - 2\frac{d}{dx^2}\log\Theta(x,k,p)$$

is an analytic homeomorphism.

The even subspace, E, of $L^2([0,1])$ i.e. the subspace consisting of those potentials satisfying $q(x) = q(1-x)$ has the property that $M(p) \cap E$ consists of a single potential. Thus, we have a fibration of $L^2([0,1])$ with E, the space of even potentials as the base space and as fiber over an even potential, e, the manifold M(e), which we may regard as $\ell_1^2(\mathbb{Z}^+)$:

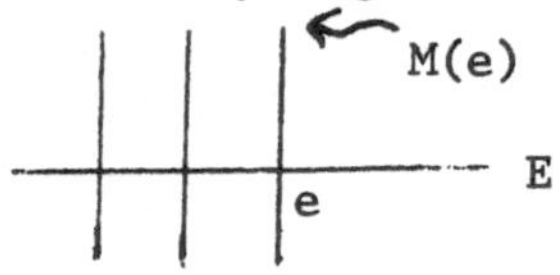

We have still to answer the question: which spectra can arise for Dirichlet boundary conditions? To answer it start with the potential $q = 0$. Then

$$-y'' = \lambda y \ , \ y(0) = y(1) = 0 \ , \ \lambda_n = \pi^2, 4\pi^2, 9\pi^2 \ldots$$

Generally, the μ_n depart from these values less and less as n increases.

<u>Theorem</u>

Let μ_n $n \geq 1$ be any strictly increasing sequence of real numbers such that

$$\mu_n = n^2\pi^2 + C + \eta_n$$

where

$$\sum_{n=1}^{\infty} \eta_n^2 < \infty$$

Then $\{\mu_n | n \geq 1\}$ is the spectrum for some $q \in L^2([0,1])$ with Dirichlet boundary conditions and conversely every spectrum for $- d^2/dx^2 + q$ with Dirichlet boundary conditions is of this form.

As a step toward general boundary conditions, consider the mixed boundary conditions

$$y(0) = 0 \ , \quad by(1) + y'(1) = 0 \quad .$$

If we define

$$M_b(p) = \{q | \nu_n(q,b) = \nu_n(p,b) \ n \geq 0\}$$

we have

<u>Theorem</u>

$M_b(p) \cap M(p)$ is either empty or contains a unique potential. In the latter case

$$\nu_{n-1}(b,p) < \mu_n(e) < \nu_n(b,p) \quad n \geq 1$$

and conversely.

For general boundary conditions

$$ay(0) + y'(0) = 0 \qquad by(1) + y'(1) = 0$$

define

$$M_{a,b}(p) = \{q | \nu_n(a,b,q) = \nu_n(a,b,p) \quad n \geq 1\}$$

If we attempt to go through the preceding argument, we find

$$\int_0^1 \psi_n^2(\psi_m^2)'dx \neq 0$$

so $(\psi_m^2)'$ does not define a tangent vector. We have to consider a bigger space:

$$\mathbb{R}^2 \times L_{\mathbb{R}}^2([0,1])$$

with elements $\underline{p} = \{a,b,p\}$ and a corresponding

$$\mathbb{M}(\underline{p}) = \{\underline{q} | \nu_n(\underline{q}) = \nu_n(\underline{p}) \quad n \geq 1 \quad .$$

Notice that $\mathbb{M}(p)$ collects operators with different boundary conditions as well as those with the same. The analogue of the vector field $(\psi_m^2)'$ is then

$$V_m(\underline{q}) = \begin{Bmatrix} -\psi_m(0,q)^2 \\ -\psi_m(1,q)^2 \\ 2\frac{d}{dx}(\psi_m(x,q)^2) \end{Bmatrix}$$

By integrating the ordinary differential equations belonging to these vector fields we get a flow carrying $\mathbb{M}(p)$ into itself. Furthermore, the flows for different m commute and using them altogether we get a global coordinate system on $\mathbb{M}(p)$.

Next we consider how to relate general boundary conditions to Dirichlet. Let $q \in M(a,b,p)$. Define

$$q^+ = q - 2\frac{d^2}{dx^2}\log \psi_0(x,a,b,q) \quad .$$

Then

$$\mu_n(q^+) = \nu_n(a,b,q) \qquad n \geq 1$$

and

$$\nu_0(-a,-b,q^+) = \nu_0(a,b,q) = \nu_0(a,b,p) \quad .$$

Define a mapping $M_{a,b,}(p) \to M(p^+)$

$$q \mapsto q^+ \; .$$

It maps onto

$$\{r \in M(p^+) \mid \nu_0(-a,-b,r) = \nu_0(a,b,p)\}$$

so we have to study the level sets of $\nu_0(a,b,\cdot)$. Where are the critical points?

Theorem

i) If $b = -a$, there is exactly one critical point $e \in M(e)$. It yields a global maximum

$$-\infty < \nu_0(a,b,\cdot) \leq \nu_0(a,b,e)$$

ii) If $b \neq -a$, there are no critical points.

In case ii) we try

$$\limsup_{\gamma \in M(p)} \nu_0(a,b,\cdot) = \nu_0\left(\frac{a-b}{2}, -\frac{(a-b)}{2}, e\right)$$

Theorem

Fix $a,b, \in \mathbb{R}$. The sequence $\nu_0, \nu_1, \ldots$ of real numbers is the spectrum for the boundary conditions a,b, if and only if

1) $\nu_n < \nu_{n+1} \qquad n \geq 0$

2) $\nu_n = n^2\pi^2 + c + r_n$ with $\sum_{n=0}^{\infty} r_n^2 < \infty$

3) Let e be an even function

$$e(x) = e(1-x) \qquad 0 \leq x \leq 1$$

with

$$\mu_n(e) = \nu_n \qquad n \geq 1$$

i) $b = -a, \quad -\infty < \nu_0 \leq \nu_0(a,b,e)$

ii) $b \neq -a, \quad -\infty < \nu_0 < \nu_0\left(\frac{a-b}{2}, -\frac{a-b}{2}, e\right)$

APPENDIX - (Summary of Seminars)

This brief report on the seminars given at the school is of mixed parentage; it was prepared by the organizers but is, in part, verbatim from brief summaries prepared by the speakers. It is hoped that these will be helpful to the reader who wants to get the full flavor of the subject from its most mathematical sources to its actual and hoped-for applications in solid state physics and hydrodynamics.

There were three talks on discrete dynamical systems originating in a problem of pure mathematics: what happens when one iterates a rational function of a complex variable? The resulting pictures contain extraordinary sets of the plane which are captivating both visually and conceptually.

ITERATION OF POLYNOMIALS OF DEGREE 2

ITERATIONS OF POLYNOMIAL-LIKE MAPPINGS

A. Douady - Ecole Normale Supérieure and Université de Paris-Sud
(Based on joint work with J.H. Hubbard)

If P is a polynomial, let K_P denote the set of values of z_0 such that $z_n = P^n(z_0)$ does not approach infinity. Here $P^n(z) = P(P^{n-1}(z))$ for $n = 1,2,...$ and $P^0(z) = z$. The <u>Julia set</u> is the boundary of K_P; it is also the closure of the set of repulsive (= unstable) periodic points. The set K_P is called the filled-in Julia set.

For $P = P_c : z \to z^2 + c$, K_P is connected iff $0 \in K_P$; otherwise, it is a Cantor set (Fatou, Julia 1919). The <u>Mandelbrot</u> set M is the set of values of c for which K_{P_c} is connected. If P_c admits an attracting cycle, c belongs to the interior $\mathring{M}$ of M. The converse is conjectured but not proved yet (problem of generic hyperbolicity). The set M is connected. It is believed to be locally connected (if we knew this, we could solve the generic hyperbolicity problem for polynomials of degree 2). A detailed combinatorial study of M (and the K_c's) has been made. It relies on the fact that we know the conformal representation of $\mathbb{C} \setminus M$ onto $\mathbb{C} \setminus \bar{D}$ ($\bar{D}$ is the closed unit disk).

The Julia sets corresponding to different values of c are very different from each other. So as Kadanoff pointed out in his lectures, individually Julia sets are not universal but the Mandelbrot set, and the family of Julia sets as a whole, does have some universality. Here is an example: Let $\{Q_\lambda\}$ be a family of polynomials of degree 3. (Hubbard studied $Q_\lambda(z) = (z-1)(z + 1/2 - \lambda)\ (z + 1/2 + \lambda)$). Call a point z_0 a bad point for Q_λ if the Newton method $z_n \to z_{n+1} = z_n - Q_\lambda(z_n)/Q'_\lambda(z_n)$ for solving Q_λ starting at z_0 does not lead to a root of Q_λ. Say Q_λ is a bad polynomial if the set of bad points has an interior point. In the set of bad values of λ appear detailed copies of M. This can be explained using the concept of polynomial-like mappings.

References

(Before 1977)

A.D. Brjuno, Analytical form of differential equations, Trans. Moscow Math. Soc., 25 (1971) 131-288

M. Brolin, Invariant sets under iteration of rational functions, Arkiv for Mathematik 6 (1966), 103-144

P. Fatou, Memoire sur les équations fonctionnelles, B.S.M.F. 47 (1919) 161-271, 47 (1920) 33-94 et 208 314

G. Julia, Itération des applications fonctionnelles, J. Math. Pures et Appl. (1918) 47-245

S. Lattes, Sur l'itération des substitutions rationnelles et les fonctions de Poincaré, C.R.A.S. 166 (1918) 26-28

H. Rüssmann, Kleine Nenner II: Bemerkungen zur Newton'schen Methode, Math. Phys. Kl. (1972) 1-20

C.L.C. Siegel, Iteration of analytic functions, Ann. Math. 43 (1942) 607-612

E. Zehnder, A simple proof of a generalization of a theorem by Siegel, Springer-Verlag, Lect. Notes in Maths. 597 (1977) 855-866

(Recent)

R. Bowen, Dimension of quasi-circles, Publ. I.H.E.S., Vol. 50 (1980)

A. Douady et J.H. Hubbard, Itération des polynômes quadratiques complexes, C.R.A.S., t. 294 (18 janvier 1982)

M.J. Feigenbaum, Quantitative universality for a class of non-linear transformations, J. Stat. Phys. 19-1 (1978) 25-52

M. Herman, Exemples de fractions rationnelles ayant une orbite dense sur la sphère de Riemann (manuscript)

O.E. Lanford, Smooth transformations of intervals, Sem. Bourbaki, exposé n° 563, Springer-Verlag, Lect. Notes in Maths. 901 (1981) 36-54

B. Mandelbrot, Fractal aspects of the iteration of $z \to \lambda z(1-z)$ for complex y and z, Annals NY Acad. of Sc. 357 (1980) 249-259

B. Mandelbrot, The Fractal Geometry of Nature, Freeman edit., San Francisco

J. Martinet, Normalisation des champs de vecteurs holomorphes (d'après A.D. Brjuno), Sém. Bourbaki, exposé n° 564, Springer-Verlag, Lect. Notes in Maths. 901 (1981) 55-70

R. Mañe, P. Sad and D. Sullivan, On the dynamics of rational maps, à paraître aux Annales de l'E.N.S.

D. Ruelle, Analytic repellers, J. of Ergodic Theory and Dynamical Systems, 1982

D. Sullivan, Density at infinity, Publ. I.H.E.S., Vol 50 (1980)

D. Sullivan, Conformal Dynamical Systems, preprint I.H.E.S.

D. Sullivan, Quasi-conformal homeomorphisms and dynamics I, II et III, preprints I.H.E.S., soumis aux Annals of Maths. en 1982

D. Sullivan and W. Thurston, Note on structural stability of rational maps, preprint I.H.E.S., 1982

A. Douady, Systèmes dynamiques holomorphes, Séminaire Bourbaki n° 599, Nov. 1982, to appear in "Asterisques" published by SMF, Paris

A. Douady and J.H. Hubbard, On the dynamics of polynomial-like mappings, to appear (A preprint of a first version of this can be found in Rencontres Mathématiciens Physiciens, IRMA Strasbourg, Nov. 1982

BOUNDARY OF THE STABILITY DOMAIN AROUND THE ORIGIN FOR CHIRIKOV'S STANDARD MAPPING

G. Dôme

SPS Division
CERN, Geneva

Chirikov's standard mapping is

$$y_{n+1} = y_n - K \sin x_n$$

$$x_{n+1} = x_n + y_{n+1} \qquad K > 0$$

The work reported on is an attempt to determine, by computation, the boundary of the region of stability around the origin, for which $-\pi \leq x_n \leq \pi$ and $-\pi \leq y_n \leq \pi$, for all n. For $0 < K < 4$, the origin is an elliptic point and the KAM theorem assures us that there is such a boundary. Of course, the stochastic sea beyond the boundary may contain other islands of stability; they are not discussed here. For $4 < K < 2\pi$ the elliptic point at the origin bifurcates into two whose x coordinates are $\pm x_s$ where $K = 4x_s/\sin x_s$.

The mapping is studied by computing its periodic orbits of

period N where N is some conveniently chosen number. The technique is to introduce the error function $D_N(x_0)$ which yields the difference between the initial coordinate, x_0 , and its value after N iterations, and to plot it as a function of x_0 . The orbits of period N are associated with zeros of $D_N(x_0)$.

The linear approximation, M, to the Nth iterate of the mapping classifies the behavior of the orbits in its neighborhood. When it is computed it is seen that tr M is of order of magnitude 1 for hyperbolic orbits in the interior of the region of stability around the origin, but becomes enormous outside.

This numerical technique is effective in locating the boundary of the region of stability.

An unsuccessful attempt was made to get a finite limit for the parameter R = 1/2 [1 - 1/2 trace M] as the periodic orbits approached that with a winding number the golden mean

$$R = \frac{\sqrt{5}-1}{2} .$$

References

G. Dôme Frontière du domaine de stabilité autour de l'origine pour la transformation "standard" de Chirikov CERN-SPS 182-20

J. Greene A Method for Determining A Stochastic Transition Jour. Math. Phys. 20 (1979) 1183-1201.

INCOMMENSURATE STRUCTURES IN SOLID STATE PHYSICS AND THEIR CONNECTION WITH TWIST MAPPINGS

S. Aubry

Laboratoire Leon Brillouin
Centre d'Études Nucléaires de Saclay
B.P. 2, 9110 Gif-sur-Yvette, France

In 1938 Frenkel and Kantorova made a one-dimensional model of atoms moving on the surface of a crystal. If the displacements of the atoms from the origin are u_i with i an integer, then the potential energy was assumed to be

$$\phi(\{u_i\}) = \sum_i \left[\frac{1}{2} (u_{i+1} - u_i)^2 - \mu(u_{i+1} - u_i) - \lambda \cos \frac{\pi u_i}{a}\right]$$

When $\lambda = 0$, the potential representing the crystal is absent and the atoms have a configuration of minimum potential energy for $u_{i+1} - u_i = \mu$. If $\lambda \neq 0$ and the separation μ is incommensurate with the period 2a of the lattice potential, the atoms are frustrated; they do not know which period to accommodate themselves to.

The condition that ϕ be stationary under variation of the $i^{\underline{th}}$ displacement is

$$\frac{\partial\phi}{\partial u_i} = 2u_i - u_{i-1} - u_{i+1} + \frac{\lambda\pi}{a}\sin\frac{\pi u_i}{a} = 0$$

(Notice that μ drops out.) If one writes $p_i = u_i - u_{i-1}$, these conditions become

$$\begin{Bmatrix} p_{i+1} \\ u_{i+1} \end{Bmatrix} = T_\lambda \begin{Bmatrix} p_i \\ u_i \end{Bmatrix} = \begin{Bmatrix} p_i + \frac{\lambda\pi}{a}\sin\frac{\pi u_i}{a} \\ u_i + p_i + \frac{\lambda\pi}{a}\sin\frac{\pi u_i}{a} \end{Bmatrix}$$

T_λ is just the standard area-preserving mapping of the plane studied at length by Chirikov, Greene and others. The main idea is to interpret orbits of the discrete dynamical system, $\{T_\lambda{}^n;\ n \in \mathbb{Z}\}$ as stationary configurations of atoms.

The procedure generalizes to potential energies

$$\phi = \sum_i L(u_{i+1}, u_i)$$

where the real-valued function L satisfies

$$L(x + 2\pi, y + 2\pi) = L(x,y)$$

$$-\frac{\partial^2 L}{\partial x \partial y} > c > 0$$

Variation of u_i yields

$$\frac{\partial L}{\partial u_i}(u_{i+1}, u_i) + \frac{\partial L}{\partial u_i}(u_i, u_{i-1}) = 0$$

or

$$\begin{Bmatrix} p_{i+1} \\ u_{i+1} \end{Bmatrix} \begin{matrix} = \\ = \end{matrix} \begin{Bmatrix} f_1(p_i, u_i) \\ f_2(p_i, u_i) \end{Bmatrix} = T \begin{Bmatrix} p_i \\ u_i \end{Bmatrix}$$

with $p_i = \frac{\partial L}{\partial u_i}(u_i, u_{i-1})$.

Aubry distinguishes two classes of orbits Q , the set of configurations $\{u_i\}$ which as orbits are recurrent and $\mathcal{G}$ those in Q which are ground states in the sense that

$$\sum_{i=N_1}^{N_2} [L(u_{i+1}+\delta u_{i+1}, u_i+\delta u_i) - L(u_{i+1}, u_i)] \geq 0$$

for all sufficiently small δu_j with $j = N_1, \ldots N_2+1$ and N_1 N_2 any integers.

Q and $\mathcal{G}$ decompose into a disjoint union of subsets Q_ℓ and $\mathcal{G}_\ell$ where $\ell/2\pi$ is the rotation number of the mapping.

The rich phenomena of hyperbolic and elliptic fixed points, homoclinic and heteroclinic points have interesting interpretations in terms of the ground and low lying states of this model. These are explored in a series of papers.

A New Concept of Transitions by Breaking of Analyticity in a Crystallographic Model
pp 264-277 in Solitons and Condensed Matter Physics
Springer Series in Solid-State Sciences 8

Many Defect Structures, Stochasticity and Incommensurability
pp 433-449 in Physique des Defauts, Les Houches Session XXXV
1980

The Twist Map, The Extended Frenkel Kontorova Model and The Devil's Staircase
Physica 7D (1983) 240-258

(with P. Le Daeron) The Discrete Frenkel Kontorova Model and Its Extensions Physica 8D (1983) 381-422

Devil's Staircase and Order without Periodicity in Classical Condensed Matter Journal de Physique 44 (1983) 147-162

A possible application to the theory of conductivity is

(with P. Le Daeron) The Metal Insulator Transition in the Peierls Chain Journal of Physics C 16 (1983) 4827-4838.

There are several papers in preparation which complete the proofs of some of the unproved assertions in the above.

These developments, in part, run parallel to work on the

existence of quasi-periodic orbits in Hamiltonian systems initiated by Percival

I.C. Percival A Variational Principle for Invariant Tori of Fixed Frequency, Jour. Phys. A12 (1979) LS7-L60

Variational Principles for Invariant Tori and Cantori pp 302-310 in Symposium on Nonlinear Dynamics and Beam-Beam Interaction Amer. Inst. Phys. Conf. Proc. No. 57.

and brought to full mathematical fruition by Mather and Katok.

J.N. Mather Existence of Quasi-Periodic Orbits for Twist Homeomorphisms of the Annulus Topolgy 21 (1982) 457-467

Concavity of the Lagrangian for Quasi-Periodic Orbits Comment. Math. Helv. 57 (1982) 356-376

A. Katok Some Remarks on Birkhoff and Mather Twist Map Theorems Ergodic Theory and Dynamical Systems 2 (1982) 185-194

Periodic and Quasi-Periodic Orbits for Twist Maps to appear

These results give some precise insight into the process by which a KAM torus disappears. It gets rougher and rougher and then changes into what Percival has dubbed a cantorus - a Cantor set in a neighborhood of the region where the torus was before. This is an important step beyond the KAM theory.

JULIA SETS - ORTHOGONAL POLYNOMIALS

PHYSICAL INTERPRETATIONS AND APPLICATIONS

D. Bessis

Service de Physique Theorique
CEN-SACLAY
9119 Gif sur Yvette, Cedex, France

Given a rational transformation, one can associate to it a Julia set [1] on which exists a unique invariant balanced probability measure [2]: $\mu(E)$ with support on the Julia set J.

When this transformation reduces to a polynomial one, there exists an electrostatic description [3] in which the previous measure is the electrical equilibrium measure for a two-dimensional electrical system. The Julia set, in this case, is the cross section of the corresponding grounded conductor. The various electrical functions of interest can be visualized on Table I. For the quadratic map, different shapes of the Julia sets which are in general fractals [4] can be found in [5].

A quantum mechanical description, still for polynomial transformations, is derived [6] from the fact that the iterates of any polynomial transformation, form subsets of orthogonal polynomials with respect to the invariant measure defined on the Julia set. Because orthogonal polynomials fulfill a three term recursive relation [7], it then follows that it is possible to linearize (in a larger space) the nonlinear iterative process. To this three term recursive relation one associates a discrete Schrödinger operator [8]. The corresponding physical functions are described in Table I. Also systems of coupled oscillators on fractal structures are described by these orthogonal polynomials which appear to be the eigenmodes of such systems [9]. An important consequence is that, in those models, the spectral dimensionality [10] has to be a complex number. Physically this means that the behaviour of the density of states near the end point of the spectrum, undergoes infinitely many oscillations, and simple scaling formula must be made more sophisticated. The Mellin transform technique [11] allows a clear understanding of this behaviour.

Finally a thermodynamical description, for the case of rational transformations, is provided by reinterpreting the invariant measure as the integrated density of zeroes of the partition function of a system which admits this transformation as a renormalization group [12]. The Julia set appears as the limiting distribution of those zeroes. An example is provided by the Diamond Hierarchical model [12]. The corresponding physical functions are to be found in Table I.

References

[1] Julia, G. : Mémoire sur l'iteration des fonctions rationnelles J. Math. Ser. 7 (Paris) 4, 47-245 (1918)
Fatou, P. : Sur les équations fonctionnelles, Bull. Soc. Math. France 47, 161-271 (1919) ; 48, 33-94 (1920) ; 48, 208-314(1920)
Brolin, H. : Invariant sets under iterations of rational functions : Ark. Mat. 6, 103-144 (1965)

[2] Freire, A., Lopez, A., Maue, R. : An invariant measure for rational map, Maryland University preprint 1982

[3] Barnsley, M.F., Geronimo, J.S., Harrington, A.N. : Geometrical and Electrical properties of some Julia sets, Georgia Inst. of Tchnology preprint, 1982, Submitted to Jour. of Stat. Phys.

[4] Mandelbrot, B. : The Fractal Geometry of Nature, San Francisco W.H. Freeman

[5] Barnsley, M.F., Geronimo, J.S., Harrington, A.N., Drager, L.D.: "Approximation theory on a Snowflake" from Multivariate Approximation theory II, edited by Walter Schempp and Karl Zeller Birkhäuser Verlag (Basel, Boston) 1982

[6] Bessis, D. and Moussa, P. : Orthogonal Properties of Iterated Polynomial Mappings. Comm. Math. Phys. 88, 503-529 (1983)
Bessis, D., Mehta, M.L., Moussa, P. : Orthogonal Polynomials on a Family of Cantor Sets and the Problems of Iterations of Quadratic Mappings. Lett. Math. Phys. 6, 123-140 (1982)

[7] Szegö, G. : Orthogonal polynomials, American Math. Soc. Colloquium Publications 23, 1939

[8] Bellissard, J., Bessis, D., Moussa, P. : Chaotic states of almost periodic Schrödinger operators, Phys. Rev.Lett. 49, 701-704 (1982)

[9] Rammal, R. : Spectrum of harmonic excitations on fractals Ecole Normale Supérieure, Paris, preprint 1983

[10] Bessis, D., Geronimo, J.S., Moussa, P., Complex spectral dimensionality on fractal structures, submitted for publication to Phys. Rev. Letters (1983)

[11] Bessis, D., Geronimo, J.S. and Moussa, P., Mellin transforms associated with Julia sets and Physical Application, to appear in Journal of Statistical Physics (Jan. 1984)

[12] Derrida, B., De Seze, L., Itzykson, C., Fractal structure of zeroes in Hierarchical models, Saclay preprint 1983 SPh.T/83/74

Table I

ELECTROSTATIC LANGUAGE	QUANTUM MECHANICAL LANGUAGE	THERMODYNAMICAL LANGUAGE
$\mu(E)$: electric charge supported by the set E	$\mu(E)$: integrated density of states	$\mu(E)$ integrated density of zeros of the partition function
$G_{ES}(z) = \int_J \ell n(z-x)\, d\mu$ = Electrostatic Green function	$L(z) = \int_J \ell n(z-x)\, d\mu$ = "Lyapounov" function	$F(z) = \int_J \ell n(z-x)\, d\mu$ = Free energy per site
Böttcher function = exponential of the Green function	Fredholm determinant = exponential of the "Lyapounov" function	Partition function per site = exponential of the free energy
$g(z) = \int_J \frac{d\mu}{1-zx}$ Generating function of the moments	$R(z) = \int_J \frac{d\mu}{1-zx}$ Resolvant, or quantum mechanical Green function	$U(z) = \int_J \frac{d\mu}{1-zx}$ Internal energy
$M_z(s) = \int_J (z-x)^s\, d\mu$ Mellin transform	$\zeta_z(s) = \int_J (z-s)^s\, d\mu$ zeta function	$M_z(s) = \int_J (z-x)^s\, d\mu$ Mellin transform

The remaining seminar dealt with a subject which has been one the principal inspirations for work on chaos in dynamical systems, and still provides many challenges and open problems: turbulence.

SCALING LAWS IN TURBULENCE

J.-D. Fournier, CNRS, Observatoire de Nice

Kolmogorov's scaling law for the energy spectrum of developed turbulence says that the energy density depends on the wave number, k, as a power $k^{-5/3}$. It is compatible with the experimental data, but has never been precisely tested nor disproved. Moreover, some of the phenomenology which predicts the scaling law also predicts other effects which are in contradiction with experiment. The phenomenon of intermittency is generally throught to be responsible for the discrepancies.

Four different phenomenologies used to "justify" Kolmogorov's law are i) dimensional analysis, ii) phenomenological non-linear dynamics, iii) eddy viscosity, iv) invariance properties of the equation of motion; each must contain some wrong assumption.

A recent test by F. Anselmet, Y. Gagne, E. Hopfinger, and R.A. Antonia, preprint Institut de Mecanique de Grenoble (1983) gives values shown in Table 1.

Table 1 Exponents of longitudinal structure functions

$S_p(\ell)$

$$S_p(\ell) \equiv \langle |v_i(\vec{x}+\vec{\ell}) - v_i(\vec{x})|^p \rangle < |\ell|^{\zeta_p}$$

The direction if i is parallel to $\vec{\ell}$. Kolmogorov scaling corresponds to $\zeta_p = p/3$.

n^p	2	3	4	5	6	7	8	9	10	12	14	16	18
R_λ=515 (duct)	0.71	1	1.33	----	1.8	----	2.27	----	2.64	2.94	3.32	----	----
R_λ=536 (jet)	0.71	1	1.33	1.54	1.8	2.06	2.28	2.41	2.60	2.74	----	----	----
R_λ=852 (jet)	0.71	1	1.33	1.65	1.8	2.12	2.22	2.52	2.59	2.84	3.28	3.49	3.71

References to recent introductory lectures on turbulence.

J.-D. Fournier in *Instabilités Hydrodynamiques et Applications Astrophysique* Spring School, Gontilas A. Baglin, ed., published by Societé Francaise des Spécialistes d'Astronomie (1983)

U. Frisch in *Chaotic Behavior in Deterministic Systems* Summer School, Les Houches, eds. G. Ioss, R. Helleman, R. Stora, North-Holland (1983)

R. Graham in *Nichtlineare Dynamik* Spring School, Jülich on *Festkörper Physik*, Jülich 1983

A famous theoretical model which exhibits simplified analogues of hydrodynamic phenomena is Burgers' equation

$$\frac{\partial v}{\partial t} + v\,\frac{\partial v}{\partial x} = \nu\,\frac{\partial^2 v}{\partial x^2}$$

One can study the complex and real singularities in the inviscid limit, $\nu \to 0$, and derive the resulting asymptotic behavior of the Fournier transform. Burgers' equation is interesting in its own right; moreover, it provides an example of scaling laws which follow from deterministic mechanisms. For real turbulent flows, we still do not know whether scaling laws hold, and, if they do, whether they arise from deterministic or statistical mechanisms.

Figure 1 *Motion and nature of complex and real singularities of Burgers' equation in the inviscid limit*

The first real singularity is called the preshock and has a cube root shape; it appears at t_* in the collision of two complex conjugate square root singularities. By continuing the scenario in the real domain, one gets the "formal solution" which is not a solution. In fact, the exact solution is not continuous for $t > t_*$, and a shock appears.

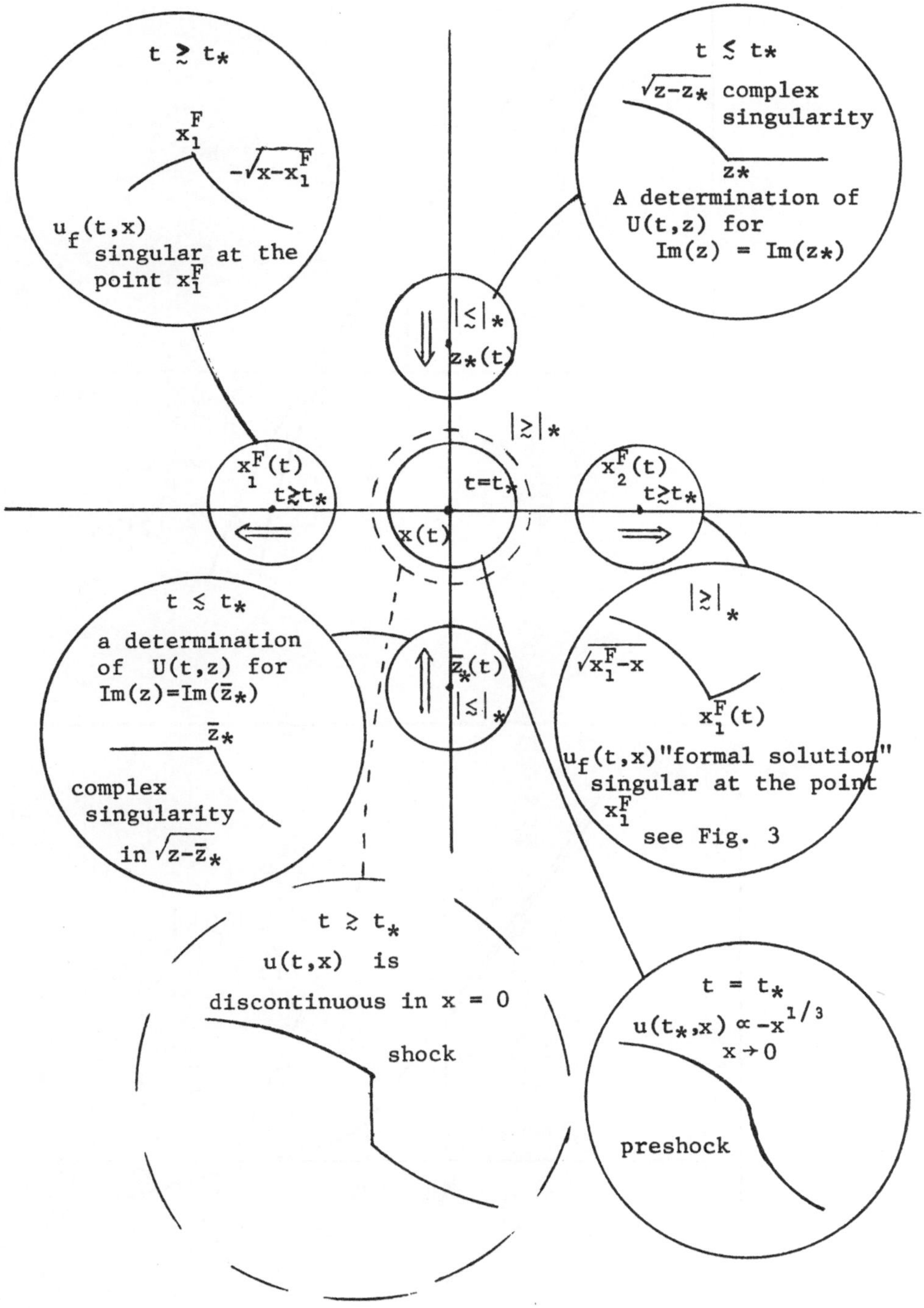
t ≳ t*
x_1^F
$-\sqrt{x-x_1^F}$
$u_f(t,x)$ singular at the point x_1^F
t ≲ t*
$\sqrt{z-z_*}$ complex singularity
z*
A determination of U(t,z) for Im(z) = Im(z*)
|≲|*
z*(t)
|≳|*
$x_1^F(t)$
t≳t*
t=t*
x(t)
$x_2^F(t)$
t≳t*
t ≲ t*
a determination of U(t,z) for Im(z)=Im($\bar{z}_*$)
$\bar{z}_*$
complex singularity in $\sqrt{z-\bar{z}_*}$
$\bar{z}_*(t)$
|≲|*
|≳|*
$\sqrt{x_1^F-x}$
$x_1^F(t)$
$u_f(t,x)$"formal solution" singular at the point x_1^F
see Fig. 3
t ≳ t*
u(t,x) is discontinuous in x = 0
shock
t = t*
$u(t_*,x) \propto -x^{1/3}$
x → 0
preshock

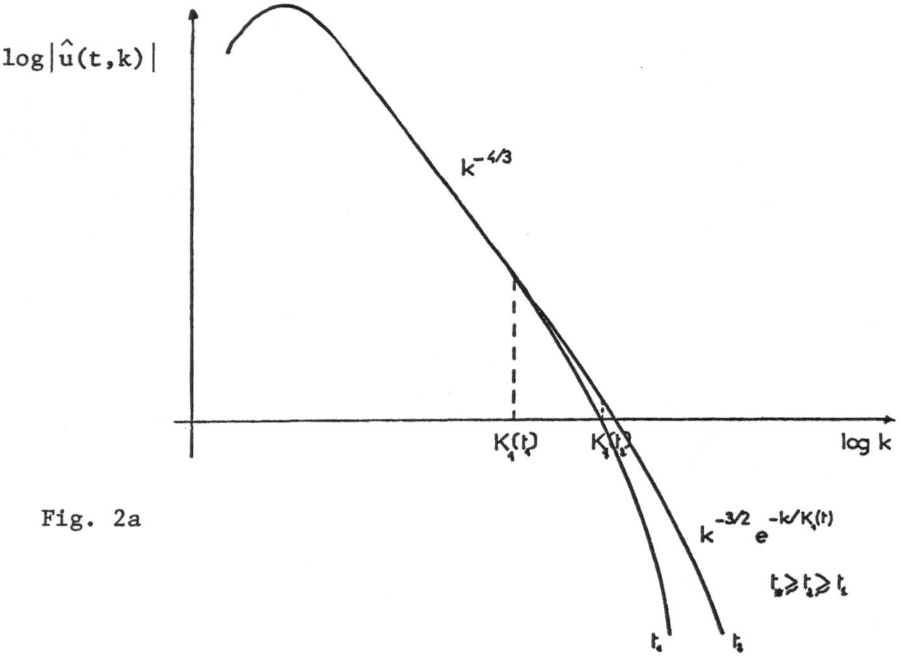

Fig. 2a

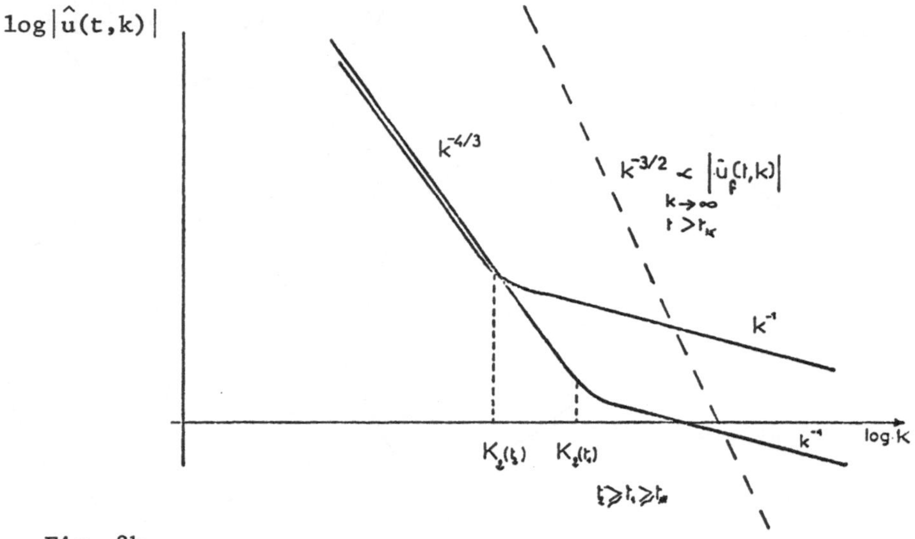

Fig. 2b

Figure 2 (a)&(b) ← Log-log plot of the ultra violet behavior of the Fourier amplitude $|\hat{u}(k,t)|$ for $t \simeq t_*$ (deterministic case)

A self-similar $k^{-4/3}$ behavior holds at this preshock time, t_*. For $t \lesssim t_*$ (Figure 2a), this range extends only up to a cutoff $K_1(t)$; beyond the cutoff, there is fast decrease with a shape governed by m pairs of complex square root singularities; for m = 1, this function is the product of a universal $k^{-3/2}$ prefactor by an exponentially decreasing function, with logarithmic decrement $K_1(t) \propto (t-t_*)^{-3/2}$. After t_* (Figure 2b) a k^{-1} inertial range due to shocks appears in the extreme ultraviolet, its amplitude is governed by the lagrangian length of the shocks and determines the transition wave number $K_2(t) \propto (t-t_*)^{-3/2}$; the $k^{-4/3}$ range, which is a reminiscence of the preshock, gradually disappears. The "formal solution" - which is not a solution - gives rise to an illusory $k^{-3/2}$scaling law, claimed to be universal by some authors, but actually related to fictitious square root singularities.

References on Burgers' equation

The above figures are taken from

J.-D. Fournier and U. Frisch, Jour. Méc. Th. Appl. 2 (1983) N° 5

Other relevant references are

J.M. Burgers The Non-linear Diffusion Equation, D. Reidel 1974

J.D. Cole, Quart. Appl. Math. 9 (1951) 225

D.& G. Chudnovsky, Il Nuovo Cim. 40B (1977) 339

A. Degasperis and J.J.P. Leon, Matrix spectral transform, reductions and the Burgers equation, Nuovo Cim. B (1983) to appear

U. Frisch and R. Morf, Phys. Rev. 23A (1981) 2673

E. Hopf, Commun. Pure and Appl. Math. 3 (1950) 201

E. Taflin, Phys. Rev. Lett. 47 (1981) 1425, erratum 48 (1982) 201

J. Weiss, M. Tabor, G. Carnevale, J. Math. Phys. 24 (1983) 522

INDEX

Abelian variety, 282
Algebraic variety, 280
Anosov diffeomorphism, 79, 135
Arnold-Liouville theorem, 224

Bernouilli systems, 18
Bowen-Ruelle measure, 234
Burgers' equation, 300

Cascade, 1
Chain recurrent point, 100
Chaos, 35
Chirikov-Taylor model, 46

Decay of correlations, 246
Dynamical system, 1
Dulac, 8

Elliptic fixed point, 8
Elliptic periodic orbit, 151
Ergodic hypothesis, 4
Ergodic theorems, 16
Escape rate, 63
Expanding maps, 236, 252

Feigenbaum number, δ, 38
Fibonacci numbers, 55
Flow, 1
Fixed point, 33, 61, 62

Geometric theorem
  Poincaré, 12
Birkhoff, 13
Golden mean, 53

Hamiltonian form, 108
Heteroclinic point, 10
Homoclinic orbit, 82
Homoclinic point, 10, 104, 138
Hyperbolic fixed point, 10
Hyperbolic set, 77, 138
Hyperregular Lagrangian, 109

Integrable system, 185
Intermittency, 41
Inverse Sturm-Liouville problem, 286
Isochronous system, 215
Iteration of polynomials, 293

Julia set, 68, 293, 296

KAM curve, 49
KAM theory, 20, 195
Kolmogorov's Law, 303
Korteweg-deVries equation, 272
K-S entropy, 18
K-system, 17

Lagrangian form, 107
Law of evolution, 1
Liouville operator, 15
Logistic equation, 30
Lyapounov exponents, 77, 141

Macroscopic observables, 20
Mandelbrot set, 293
Mixing, 17

Ornstein's theorem, 19

Pendulum, 45
Periodic orbit, 113, 116, 131, 168
Phase space, 1
Poincaré, 2, 8, 121
  Geometric theorem, 12
  Section, 8
  Mapping, 9
Pseudo-orbit, 82
Power spectrum, 28

Rectification theorem, 5, 269
Renormalization group, 60
Resonance, 8, 188, 189
Rigid body motion, 279, 283
Ruelle-Perron-Frobenius operator, 234

Shadowing theorem, 82
Siegel circle theorem, 21
Singular point of a vector field, 7
Small divisors, 8
Smale horseshoe, 80
Stabilité à la Poisson, 3
Stable manifold, 10
  Theorem, 96
Strange attractor, 14, 23
Structural stability, 90

Thermodynamic limit, 20
Turbulence, 27, 303

Unimodal mappings, 254
Universality, 61
Unstable manifold, 10

W-limit set, 98
Wandering point, 98